FERM'S FAST FINDER
INDEX

15th Edition,

Revised and Expanded Guide
to the 2020 *National Electrical Code*®

International Association of Electrical Inspectors
Richardson, Texas

Copyright © 1981, 1986 by Olaf G. Ferm

Copyright © 1990 by FFI Limited Partnership

Copyright © 1993, 1996, 1999, 2002, 2005, 2008, 2014, 2017, 2021 by
International Association of Electrical Inspectors
901 Waterfall Way, Suite 602
Richardson, TX 75080-7702

All rights reserved. First edition published 1966
Printed in the United States of America
25 24 23 22 21   5 4 3 2 1

ISBN-13: 978-1-890659-95-0

Notice to the Reader

This book has not been processed in accordance with NFPA Regulations Governing Committee Projects. Therefore, the text and commentary in it shall not be considered the official position of the NFPA or any of its committees and shall not be considered to be, nor relied upon as a formal interpretation of the meaning or intent of any specific provision or provisions of the 2020 edition of NFPA 70, *National Electrical Code*.[1]

Publishers do not warrant or guarantee any of the products described herein or perform any independent analysis in connection with any of the product information contained herein. Publisher does not assume, and expressly disclaims, any obligation to obtain and include information referenced in this work.

The reader is expressly warned to consider carefully and adopt all safety precautions that might be indicated by the activities described herein and to avoid all potential hazards. By following the instructions contained herein, the reader willingly assumes all risks in connection with such instructions.

THE PUBLISHERS MAKE NO REPRESENTATIONS OR WARRANTIES OF ANY KIND, INCLUDING, BUT NOT LIMITED TO, THE IMPLIED WARRANTIES OF FITNESS FOR PARTICULAR PURPOSE, MERCHANTABILITY OR NON-INFRINGEMENT, NOR ARE ANY SUCH REPRESENTATIONS IMPLIED WITH RESPECT TO SUCH MATERIAL. THE PUBLISHERS SHALL NOT BE LIABLE FOR ANY SPECIAL, INCIDENTAL, CONSEQUENTIAL OR EXEMPLARY DAMAGES RESULTING, IN WHOLE OR IN PART, FROM THE READER'S USES OF OR RELIANCE UPON THIS MATERIAL.

[1]*National Electrical Code* and *NEC* are registered trademarks of the National Fire Protection Association, Inc., Quincy, MA 02169.

Table of Contents

Resources 6

How to Use the Index 7

A 8
B 18
C 32
D 59
E 70
F 79
G 91
H 103
I 111
J 118
K 121
L 122
M 133

N 144
O 150
P 160
Q 169
R 170
S 182
T 201
U 214
V 217
W 219
X 226
Y 226
Z 226

Resources

[1]The following are registered trademarks of the National Fire Protection Association, Inc., Quincy, MA 02269.

NFPA 20, *Standard for the Installation of Stationary Fire Pumps for Fire Protection*

NFPA 30, *Flammable and Combustible Liquids Code*

NFPA 30A, *Code for Motor Fuel Dispensing Facilities and Repair Garages*

NFPA 33, *Standard for Spray Application Using Flammable and Combustible Materials*

NFPA 34, *Standard for Dipping and Coating Processes Using Flammable or Combustible Liquids*

NFPA 54, *National Fuel Gas Code*

NFPA 70, *National Electrical Code* 2002, 2005, 2008, 2011, 2014, 2017, 2020.

NFPA 70-E, *Standard for Electrical Safety in the Workplace*

NFPA 72, *National Fire Alarm Code*

NFPA 79, *Electrical Standard for Industrial Machinery*

NFPA 88A, *Standard for Parking Structures*

NFPA 99, *Standard for Health Care Facilities*

NFPA 101, *Life Safety Code*

NFPA 220, *Standard on Types of Building Construction*

NFPA 303, *Fire Protection Standard for Marinas and Boatyards*

NFPA 325, *Guide to Fire Protection Hazard Properties of Flammable Liquids, Gases, and Volatile Solids contained in the NFPA "Fire Protection Guide to Hazardous Materials–2001 edition"*

NFPA 497, *Recommended Practice for the Classification of Flammable Liquids, Gases, or Vapors and of Hazardous (Classified) Locations for Electrical Installations in Chemical Process Areas*

NFPA 499, *Recommended Practice for the Classification of Combustible Dusts and of Hazardous (Classified) Locations for Electrical Installations in Chemical Process Areas*

NFPA 780, *Standard for the Installation of Lightning Protection Systems*

The following are registered trademarks of the Institute of Electrical and Electronics Engineers New York, NY 10016-5997.

ANSI C2, *National Electrical Safety Code*

ANSI 18, *Shunt Power Capacitors*

ANSI Z535.4, *Product Safety Signs and Labels*

IEEE 835-1994 IEEE Standard Power Cable Ampacity Tables*

The *NEC–IRC* Cross References, used in Section 3, are registered trademarks of the International Code Council, and are used with permission.

The following are registered trademarks of Underwriters Laboratories Inc. Northbrook, IL 60062-2096.

Fire Resistance Directory*

UL Product iQ®

How to Use the Index

Ferm's Fast Finder Index is an alphabetized index of important words and phrases within the 2020 *National Electrical Code*. It also contains cross-references to other codes that are referenced within the *NEC*.

Look for the tab with the letter associated with the main section of what you are looking for. For example, Luminaires are located under "L." The main subheads are in bold and are flush against the left margin of the page. Any indented text under the main subheads are subsections of that group.

Example:

LUMINAIRES . . . . Art. 410	← Main Category
As a Raceway 410.64	← First subsection of main category
Ballast Type (Electric Discharge and LED)	
Calculations, Inductive and LED Loads . . . . . . 220.18(B)	← Subsection of the first subsection.

Under each main category heading are:

- references in the 2020 *National Electrical Code*, NFPA 70
- references to other NFPA standards and documents that are included in fine print notes throughout the *NEC* 2020. These include their numbers and editions.
- UL Guide Card letters for all references to the UL ProductiQ, available online at https://iq2.ulprospector.com/session/new?redirect=http%3a%2f%2fiq.ulprospector.com%2fen%2f.
- references to additional codes and standards, such as those by ANSI and IEEE.

Readers may contact the National Fire Protection Association at (617) 770-3000 to verify current dates or proposed revisions.

For further information on ordering UL publications, write or call:
Underwriters Laboratories, Inc.
Attn: Publications Stock
333 Pfingsten Road
Northbrook, IL 60062-2096
(847) 272-8800, ext.42899

A

ABANDONED CABLES & WIRING

Abandoned Audio Distribution Cable . . . . 640.6(C)

    Definitions. 640.2

Abandoned Class 2, Class 3, and PLTC Cables 725.25

    Definition 725.2

Abandoned Coaxial Cable 800.25

    Definition . . . . Article 100

Abandoned Communication Cable. 800.25

    Definition 800.2

Abandoned (Data Processing) Information Technology Cables . . 645.5(G)

    Definition 645.2

Abandoned Fire Alarm Cable 760.25

    Definition 760.2

Abandoned Network-Powered Broadband Communication Systems 800.25

    Definition 830.2

Abandoned Optical Fiber Cable 770.25

    Definition 770.2

Abandoned Premises-Powered Broadband Communications Systems 840.25

Circuit Conductors

    Cellular Concrete Floor Raceways 372.18

    Cellular Metal Floor Raceways. 374.18

    Underfloor Raceways 390.57

Temporary Wiring . . . . .590.3(D)

ABOVEGRADE, RACEWAYS, WET LOCATIONS

1000 Volts or Less 300.9

Over 1000 Volts 300.38

AC RESISTANCE & REACTANCE . . . . . Chapter 9 Table 9

AC SYSTEMS

Conductor to be Grounded. 250.26

Conductors of the Same Circuit . . . . 300.3(B)

    Class 1 Circuits . . . . 725.48(A)

    Underground . . . . 300.5(I)

Grounding Electrode Conductor, Size 250.66

Grounding of 250.20

Grounding Separately Derived Systems 250.30

Grounding Service-Supplied Systems 250.24

Induced Currents in Ferrous Metal Enclosures or Ferrous Metal Raceways. 300.20

ACCESS & WORKING SPACE

Boxes Behind Electric-Discharge Luminaires (Lighting Fixtures) 410.24(B)

Behind Panels Designed to Allow Access to

Electrical Equipment 300.23

    Audio Signal Processing, Amplification and Rep.. . . . .640.5

    Cables . . . . 300.4(C)

    CATV and Radio Distribution Systems 800.21

    Class 1, 2, and 3 Remote-Control, Signaling, and Power-Limited Circuits 725.21

    Communications Circuits 800.21

    Fire Alarm Systems. 760.21

    Optical Fiber Cables and Raceways. 770.21

    Network-Powered Broadband Comm. Systems 800.21

    Premises-Powered Broadband Comm. Systems. 800.21

Cable Trays . . . . 392.18(E) & (F)

Cranes and Hoists. 610.57

Dedicated Space . . . . 110.26(E)

Duct Heaters 424.66

    Limited Access Working Space . . . . 110.26(A)(4)

Elevators, Dumbwaiters, etc. 620.5

Energized Parts, Service Equipment 230.62

Hydromassage Bathtub Electrical Equipment. 680.73

Manholes. .Art. 110 Part V

    Access to 110.75

    Cabling Work Space 110.72

    Equipment Work Space 110.73

Metal Poles, to Supply or Cable Terminations. . . . . 410.30(B)

Panels Designed to Allow Access 300.23

Over 1000 Volts

    Entrance and Access to Work Space 110.33

    Controlled by Various Means. 110.31

    Guarding to Prevent Access, Temporary Installations . .590.7

    Work Space and Guarding 110.34

    Work Space About Equipment. 110.32

Overcurrent Devices 230.93

    Access to Occupants . . . . 230.72(C)

    Occupant to Have Ready Access . . . . 240.24(B)

Solar Photovoltaic System Boxes 690.34

Splices and Taps

    Cellular Concrete Floor Raceways 372.56

    Cellular Metal Floor Raceways. 374.56

    Metal Wireways 376.56

    Surface Metal Raceways 386.56

Surface Nonmetallic Raceways 388.56
　　Surface Metal Raceways. . . . . 386.10(4)
Surface Nonmetallic Raceways . . . . 388.10(2)
Vaults and Tunnels 110.76
Wireways, Metal . . . . 376.10(4)
　　Nonmetallic . . . . 378.10(4)
Working Space 600 Volts, Nominal, or Less . . . . . 110.26(A)

ACCESSIBLE
Attics (Wiring in)
　　Armored Cable. 320.23
　　Concealed Knob-and-Tube Wiring. 394.23
　　Metal Clad Cable. 330.23
　　Nonmetallic-Sheathed Cable 334.23
　　Open Wiring on Insulators. 398.23
　　Service Entrance Cable . . . . 338.10(B)(4)(a)
　　Underground Feeder and Branch-Circuit Cable . . 340.10(4)
Boxes Behind Electric-Discharge and LED Luminaries 410.24(B)
Cable Trays . . . . 392.18(E) & (F)
Conduit Bodies, Junction, Pull and Outlet Boxes 314.29
　　Over 1000 Volts . . . . 314.72(D)
Definition of . . . . Art. 100 Part I
Disconnect for Cord- and Plug-Connected Appliances . . 422.33
Grounding Electrode Connection. . . . . 250.68(A)
HACR Receptacle Outlet 210.63
Handhole Enclosures 314.29
Handhole for Metal Poles Supporting Lighting Luminaires (Fixtures)
. . . . 410.30(B)
Intersystem Bonding Connection 250.94
Limited Access Working Space . . . . 110.26(A)(4)
Metallic Water Pipe Bonding Jumper . . . . 250.104(A)
Other Metal Piping Bonding Jumper . . . . 250.104(B)
Overcurrent Devices, Integrated Electrical Systems 685.10
Raised Floors, Wiring Under . . . . .645.5(E)(1)
Sealoffs, Hazardous (Classified) Locations. . . . . . .501.15(C)(1)
　　Class II, Divisions 1 and 2 . . . . 502.15
　　Class I, Zones 0, 1, and 2 . . . . .505.16(D)(1)
Sign Ballasts, Transformers and Power Supplies . . . . 600.21(A)
Sign Outlet . . . . 600.5(A)
Solar Photovoltaic System Overcurrent Devices. . . . . 690.9(B)
Splices & Taps
　　Auxiliary Gutters . . . . 366.56(A)
　　Cable Trays — Accessible and above side rails 392.56
　　Strut-Type Channel Raceway. 384.56
　　Surface Metal Raceways 386.56

Surface Nonmetallic Raceways 388.56
Wireways
　　Metal Wireways . . . . 376.56(A)
　　Nonmetallic Wireways 378.56
Structural Steel Bonding Jumper Connection . . . .250.104(C)
Supplementary Overcurrent Protection . . . . . . 422.11(F)(1)
　　Overcurrent Protective Devices. . . . . 424.22(C)
System Grounding Connection . . . . .250.24(A)(1)

ACCESSIBLE (READILY ACCESSIBLE)
Arc-Fault Circuit-Interrupter Protection 210.12
Definition of . . . . Art. 100 Part I
Disconnecting Means for
　　Agricultural Buildings . . . . 547.9(A)(8)
　　Air Conditioning or Refrigeration Equipment 440.14
　　　Information Technology Equipment Rooms 645.10
　　Carnival Ride and Concession . . . . 525.21(A)
　　Crane or Hoist Runway Conductor . . . . 610.31(1)
　　Electric Vehicle Charging System. 625.42
　　Electrically Driven Irrigation Machines . . . . 675.8(B)
　　Electrified Truck Parking Space Supply Equipment 626.22(D)
　　Elevator, Dumbwaiters, Escalators, etc. . . . . . . 620.51(C)
　　Fuel Cell Systems. 692.17
　　More Than One Building or Other Structure 225.32
　　Motor and Controller Disconnecting Means 430.107
　　Outside Feeder Taps . . . . 240.21(B)(5)(4)
　　Outside Secondary Conductors . . . . 240.21(C)(4)(4)
　　Over 1000 Volt Service Conductors 230.205
　　　Permitted to be Not Readily Accessible . . . . 230.205(A)
　　Phase Converters. . . . . 455.8(A)
　　Portable Switchboards 520.51
　　Room Air Conditioners 440.63
　　Service Disconnecting Means . . . . .230.70(A)(1)
　　Shore Power Connections . . . . 555.33(A)
　　Solar Photovoltaic Systems . . . . 690.13(A)
　　Storage Batteries . . . . 480.7(A)
　　Swimming Pool, Spa, Hot Tub, etc. 680.12
　　　Emergency Switch for Spas and Hot Tubs 680.41
　　Wind Electric Systems . . . . .694.22(A)(1)
　　X-Ray Equipment, Medical . . . . 517.72(B)
　　　Non-Medical. 660.5
Dispensing Equipment, Motor Fuel 514.11
Ground-Fault Circuit-Interrupter Protection for Personnel 210.8
　　GFCI for Appliances 422.5

Locked Service Overcurrent Devices 230.92
Overcurrent Devices . . . . 240.24(A)
    Fuel Cell Systems. . . . . 692.9(B)
Receptacle Outlets in Guest Rooms of Hotels and Motels 210.60(B)
Switches and Circuit Breakers. . . . . 404.8(A)
Transformers and Transformer Vaults 450.13

AC-DC GENERAL USE SNAP SWITCHES
Marking. 404.20
Motor Controllers . . . . 430.83(C)(1)
Motor Disconnecting Means . . . . 430.109(C)(1)
Ratings, Type Loads. 404.14

ADA STANDARDS FOR ACCESSIBLE DESIGN
ADA Accessibility . . . . 110.1 Info. Note
ADA Standards for Accessible Design . . . Informative Annex J

ADJUSTABLE SPEED DRIVES
Adjustable Speed Drive (Definition of) . . . . Article 100
Adjustable Speed Drive System (Definition of) . . . . Article 100
Adjustable Speed Drive Systems . . . . Art. 430 Part X
Branch-Circuit Short-Circuit and Ground-Fault Protections
    Several Motors or Loads 430.131
    Single Motor Circuits 430.130
Bypass Circuit/Device . . . . 430.130(B)
Conductors. 430.122
Disconnecting Means. 430.128
Motor Overtemperature Protection 430.126
Output Conductors. . . . . 430.122(B)
Overload Protection 430.124
Several Motors (Conductor Ampacity) . . . . 430.122(D)

ADMINISTRATION AND ENFORCEMENT
. . . . **Informative Annex H**

AERIAL CABLE
CATV and Radio Distribution Systems. 820.44
Coaxial Cables 800.44
Communication Wires and Cables 800.44
Identification . . . . 200.6(A)(8)
Instrumentation Tray Cable . . . . 727.4(6)
Messenger Supported Wiring . . . . Art. 396
Network-Powered Broadband Communications Systems 800.44
Nonmetallic Extensions . . . . 382.12(2)
Optical Fiber Cables 770.44
Type MC Cable . . . . 330.10(A)(8)

AFCI
See ARC-FAULT CIRCUIT INTERRUPTERS . Ferm's Finder

AGRICULTURAL BUILDINGS . . . . Art. 547
Bonding . . . . 547.10(B)
    Site-Isolating Device . . . . 547.9(A)(4)
Distribution Point, Definition 547.2
Electrical Supply
    Feeders to Two or More Buildings or Structures Art. 225 Part II
    From Distribution Point 547.9
    Service Disconnecting Means & Overcurrent Protection at the
    Building(s) or Structure(s) . . . . 547.9(B)
    Service Disconnecting Means & Overcurrent Protection at the
    Distribution Point . . . . 547.9(C)
Equipment Enclosures, Boxes, Conduit Bodies, and Fittings . . . 547.5(C)
Equipment Grounding Conductor . . . . 547.5(F)
Equipotential Plane 547.10
    Definition 547.2
    Where Required Indoors/Outdoors . . . . 547.10(A)
Ground-Fault Protection for Receptacles . . . . 547.5(G)
Horticultural Lighting Equipment . . . . . . Art. 410, Part XVI
Identification — Where More Than One Distribution Point . . . 547.9(D)
Luminaires (Lighting Fixtures) 547.8
    See LUMINAIRES (LIGHTING FIXTURES) . Ferm's Finder
Marking of Site-Isolation Device . . . . 547.9(A)(10)
Motors 547.7
    See MOTORS . . . . Ferm's Finder
Physical Protection of Wiring and Equipment . . . . 547.5(E)
Site Isolating Device . . . . 547.9(A)
    Definition 547.2
Wiring Methods 547.5

AIR-CONDITIONING & REFRIGERATING EQUIPMENT
. . . . .Art. 440
Branch-Circuit Conductors. . . . . Art. 440 Part IV
Branch-Circuit, Short-Circuit and Ground-Fault Protection . . .
. . . . Art. 440 Part III
    Note: If nameplate calls for fuses as the overcurrent protection then fuses must be used.
Conductor Ampacity & Rating. 440.6
Controllers for Motor-Compressors. . . . . . . .Art. 440 Part V
Disconnecting Means . . . . .Art. 440 Part II
    Cord Connected Room Air Conditioners 440.63
    Information Technology Equipment Rooms 645.10

Grounding and Bonding 440.9
   Cord- and Plug-Connected. 250.114
   Fixed 250.110
   Fixed (Specific) 250.112
   Non-threaded fittings (EGC Required for Rooftops). . .440.9
   Room Air Conditioners 440.61
   See GROUNDING, Fixed Equipment . . . . . Ferm's Finder
Hermetic Refrigerant Motor-Compressor, Definition. . Art. 100
Highest Rated (Largest) Motor 440.7
Lighting Outlet Required . . . . 210.70(C)
Nameplate and Marking Requirements 440.4
   For Controllers 440.5
   Not Permitted to be Obscured by Disconnect Switch . 440.14
Overload Protection . . . . Art. 440 Part VI
Protection Devices 440.65
   Arc-Fault Circuit Interrupter . . . . 440.65(2)
   Heat Detecting Circuit Interrupter . . . . 440.65(3)
   Leakage-Current Detector-Interrupter . . . . . . 440.65(1)
Rated Load Current. 440.2
Receptacle Required 210.63
Recreational Vehicles, Pre-Wired . . . . 551.47(Q)
Room Air Conditioners. . . . . Art. 440 Part VII
   As a Single Motor Unit. . . . . 440.62(A)
   Maximum Loads . . . . 220.18(A)
   Total Marked Rating . . . . 440.62(B)&(C)
Short-Circuit Current Rating 440.10

AIRCRAFT HANGARS . . . . **Art. 513**
Aircraft Electrical Systems . . . . 513.10(A)
Battery and Charging . . . . 513.10(B)
Classification of Locations 513.3
Equipment in Class I Locations 513.4
Equipment Not within Class I Locations 513.7
External Power Sources for Energizing Aircraft . . . . 513.10(C)
Grounding and Bonding 513.16
   Bonding in Hazardous (Classified) Locations. 250.100
   Class I, Division 1 and 2 . . . . 501.30
   Class I, Zone 1 and 2 . . . . 505.25
   Main and System Bonding Jumpers 250.28
   Method of Bonding at the Service . . . . 250.92(B)
Mobile Servicing Equipment . . . . 513.10(D)
Pendants . . . . 513.7(B)
   Class I, Division 1 Locations . . . . 501.130(A)(3)
   Class I, Division 2 Locations . . . . 501.130(B)(3)

Sealing 513.9
Sealing and Drainage 501.15
Wiring Methods in Class I Locations 513.4
Wiring Not within Class I Locations 513.7
Wiring Under Hazardous Areas 513.8
   Raceways Embedded in or under a Concrete Floor. . 513.8(A)
   Uninterrupted Raceways Embedded in or under a Concrete Floor 513.8(B)
See HAZARDOUS (CLASSIFIED) LOCATIONS Ferm's Finder

AIR-HANDLING SPACES, PLENUM **300.22**
Definition of Plenum . . . . Art. 100 Part I

AIRPORT RUNWAYS
Airfield Lighting Cables. . . . . 300.37 Ex.
Minimum Cover Requirements . . . . Table 300.5
Minimum Cover Requirements Over 1000 Volts . . Table 300.50

ALARM INDICATION OR SYSTEMS
Burglar Alarm Systems . . . . Art. 725
Class 1, 2 and 3 Circuits . . . . Art. 725
   See CLASS 1, 2 & 3 REMOTE CONTROL CIRCUITS . . .
. . . . Ferm's Finder
Common Area Branch Circuits . . . . 210.25(B)
Conductors, Separation from Electric Light and Power, Class l, etc.
. . . . 725.136
   With Optical Fiber Cables . . . . .770.133(A)
Emergency Systems .700.6
   See EMERGENCY SYSTEMS . . . . .Ferm's Finder
Fire Alarm Systems . . . . Art. 760
   See FIRE ALARM SYSTEMS . . . . Ferm's Finder
Fire Pump Circuits . . . . 695.4(B)(3)(e)
Fixed Electric Heating Equip. for Pipelines and Vessels . . 427.22
Generator Overload Indication . . . . 445.12 Ex.
Hazardous (Classified) Locations
   Class I Locations 501.150
   Class II Locations 502.150
   Class III Locations 503.150
Health Care Facilities . . . . Art. 517 Part VI
   Connection to Life Safety Branch . . . . 517.43(C)
Legally Required Standby Systems 701.6
Motor Circuit Over 1000 Volts . . . . 430.225(A), Ex.
Motor Orderly Shutdown. 430.44
Motors and Generators, Class I, Division 1 . . . . . 501.125(A)
Optional Standby Systems 702.6
Outside Wiring of Alarm Systems. 800.1

See Life Safety Code, NFPA 101 . . . . NFPA 101
See National Fire Alarm Code, NFPA 72. . . . . NFPA 72

ALTERNATE POWER SOURCE **517.2**
   Critical Operations Power Systems (COPS). . . Art. 708 Part III
   Direct Current Microgrids . . . . Art. 712
   Emergency Systems . . . . Art. 700
   Energy Storage Systems . . . . Art. 706
   Legally Required Standby Systems . . . . Art. 701
   Optional Standby Systems . . . . Art. 702
   Stand-Alone Systems . . . . Art. 710

ALUMINUM CONDUCTORS **310.3**
   Aluminum to Copper Connections 110.14
   Ampacity of 310.14
      Bars in Sheet Metal Auxiliary Gutters . . . . . . 366.23(A)
   CO/ALR Marking on
      Receptacles . . . . 406.3(C)
      Snap Switches . . . . 404.14(C)
   Dimensions of
      Conductors . . . . Chapter 9 Table 8
      Compact Conductors . . . . Chapter 9 Table 5A
         See Aluminum Conductor Dimension, Compact . . . . . .
         Ferm's Charts and Formulas
      Insulated Conductors . . . . Chapter 9 Table 5
   Dwelling Service and Feeder 310.12
      See Dwelling Unit Services . . . . . *Ferm's Charts and Formulas*
   Minimum Size and Rating
      Outside Branch Circuits and Feeders, Overhead Spans 225.6(A)
      Overhead Service Conductors . . . . 230.23(B)
      Underground Service Conductors . . . . 230.31(B)
   Oxide Inhibitor on Aluminum Terminations
      Where Required by Listing . . . . 110.3(B)
      Where Used 110.14
      See (ZMVV) . . . . *UL Product iQ*
      See (DVYW) . . . . *UL Product iQ*
   Properties of . . . . Chapter 9 Table 8
   Restrictions Where Used as Grounding Electrode Conductor . . 250.64(A)
   Terminals, Identified for . . . . 110.14(A)

ALUMINUM CONDUIT AND FITTINGS
   Fittings and Enclosures with Corrosive Environments 358.10(B)
      Dissimilar metals 358.14
      Intermediate Metal Conduit 342.14
      Steel Rigid Metal Conduit 344.14
   Installation in Concrete or in Contact with Earth
      See (DYWV) . . . . *UL Product iQ*
   Protection against Corrosion . . . . 300.6(B)
   Suitable for Installation Condition . . . . 344.10(B)

ALUMINUM DUST
   Class II, Group E . . . . 500.6(B)(1)
   Class II Locations . . . . Art. 502

ALUMINUM SIDING, GROUNDING OF 250.116 Info. Note

AMBIENT TEMPERATURE . 310.14(A)(3) Info. Note No. 1
   Correction Factors for Conductor Ampacities 0-2000 volts
      Calculation . . . . 310.14(B)
      Table . . . . Tables 310.15(B)(2)
   Correction Factors for Conductor Ampacities 2001 to 35,000 volts
      Calculation . . . . 311.60(B)
      Table . . . . Table 311.60(C) to (86)
   Correction Factors for Raceways and Cables Exposed to
      Sunlight on or Above Rooftops . . . . 310.15(B)(2)
   Electrical Nonmetallic Tubing, Uses Not Permitted . 362.12(3)
   High Density Polyethylene Conduit, Type HDPE, Uses Not Permitted . . . . 353.12(4)
   Liquidtight Flexible Metal Conduit, Uses Not Permitted . 350.12
   Liquidtight Flexible Nonmetallic Conduit, Uses Not Permitted . 356.12(2)
   Nonmetallic Auxiliary Gutters, Indoor and Outdoor . 366.10(B)
   Other Space Used for Environmental Air . . . . . . 300.22(C)(3)
   Rigid Polyvinyl Chloride Conduit: Type PVC,
   Uses Not Permitted . . . . 352.12(D)
   Reinforced Thermosetting Resin Conduit: Type RTRC
      Uses Not Permitted . . . . 355.12(D)
   Solar Photovoltaic Circuit Requirements . . . . . . 690.7 & 8
   Surface Nonmetallic Raceways, Uses Not Permitted . 388.12(6)

AMBULATORY HEALTH CARE OCCUPANCY **517.2**
   Essential Electrical System 517.45

AMPACITY, DEFINED . . . . **Art. 100**

AMPACITY OF
   Aluminum Conductors, Allowable . . Tables 310.16 through 21)
   Aluminum Conductors, Allowable, 2001 to 35000 Volts
   . . . . . Tables 311.60(C)(68) through (C)(86) (even numbered)
   Armored Cable 320.80
   Bare or Covered Conductors . . . . 310.15(D)

Bus Bars and Conductors in Auxiliary Gutters 366.23

Cable (in Cable Tray)

    2000 Volts or Less . . . . 392.80(A)

    2001 Volts or Over – Type MC and Type MV . . . 392.80(B)

    Power and Control Tray Cable 336.80

    Type MC Cable 330.80

Cablebus 370.20

Conductors for Individual Dwelling Unit or Mobile Home 310.12

Continuous Load at 125 Percent

    Branch Circuit . . . . 210.19(A)(1)(a)

    Feeders . . . . 215.2(A)(1)(a)

    Services . . . . .230.42(A)(1)

Copper Conductors, Allowable . . . . Tables 310.16 through 21

Copper Conductors, Allowable, 2001 to 35000 Volts . . . . Tables 311.60(C)(67) through (C)(85) (odd numbered)

Crane & Hoist Conductors . . . . Table 610.14(A)

Derating Allowable Ampacity

    Ambient Temperature Correction Factors Tables 310.15(C)(1)

    Flexible Cords and Cables . . . . 400.5(A)

    Grounding or Bonding Conductor . . . . 310.15(F)

    Neutral Conductor . . . . 310.15(E)

    Nonmetallic Sheathed Cable 334.80

    Number of Conductors . . . . 310.15(C)

    Solar Photovoltaic Systems . . . . 690.31(E)

    Wiring above Heated Ceilings

        Electric Space-Heating Cables 424.36

        Electric Radiant Heating Panels and Heating Panel Sets 424.94

        Wiring in Walls . . . . 424.95(B)

Determination of 310.14

    2001 to 35,000 Volts 311.60

Feeder to Dwelling Unit 310.12

    See Allowable Ampacity Tables of 0 to 2000 Volts *Ferm's Charts and Formulas*

Feeder to Mobile Home 310.12

    Allowable Demand Factors 550.31

    Capacity and Rating . . . . 550.33(B)

    See DEMAND FACTORS . . . . Ferm's Finder

Fire Alarm (NPLFA) Circuits 760.43

Fire Alarm (NPLFA) Circuit Conductors . . . . 760.49(A)

    NPLFA Conductors (Allowable) 18 AWG & 16 AWG . . Table 402.5

    Larger than 16 AWG 310.14

    NPLFA Circuits and Class 1 Circuits in Same Raceway 760.51(A)

    NPLFA and Power-Supply Conductors in Same Raceway . . . . 760.51(B)

(Fixture) Luminaire Wire . . . . Table 402.5

Flexible Cords and Cables (Adjustment Factors) 400.5

    Flexible Cords and Cables (Allowable) . . . Table 400.5(A)(1)

    Types SC, SCE, SCT, PPE, G, G-GC, and W Table 400.5(A)(2)

Fuel Cell Systems

    Conductors . . . . 692.8(B)

    Grounded or Neutral Conductor . . . . 692.8(C)

Generator Conductors to First Overcurrent Device 445.13

High Voltage Conductors, 2001 to 35, 000 Volts 311.60

Integrated Gas Spacer Cable 326.80

Machine Tool Wire . . . . 310.15(A) IN 2

    See Electrical Standard for Industrial Machinery, NFPA 79 . . .

Metal-Clad Cable 330.80

    Allowable, 14 AWG and Larger, 0 through 2000 Volts 310.14

    Allowable, 14 AWG and Larger, 2001 to 35,000 Volts 311.60

    Allowable, 18 AWG and 16 AWG 402.5

    Installed in Cable Tray . . . . 330.80(A)

    Single Conductors Grouped Together . . . . 330.80(B)

Mineral-Insulated Cable, Metal-Sheathed Cable 332.80

    Installed in Cable Tray . . . . 332.80(A)

    Single Conductors Grouped Together . . . . 332.80(B)

Mobile Home (Service or Feeder) . . . . . . 215.2(A)(4), 310.12

Modifications to Tables 310.16 through 310.21 . . . 310.15(A)

Modifications to Tables 311.60(C)(67) through 311.60(C)(86) . . 311.60

Nonmetallic Sheathed Cable 334.80

Overhead Service Conductors . . . . 230.23(A)

    Single Insulated Cond. in Free Air 60 to 90°C . . Table 310.17

    Single Insulated Conductors in Free Air 150 to 250°C . . . . Table 310.19

Service-Entrance Cable: Types SE . . . . .338.10(B)(4)

Solar Photovoltaic System . . . . 690.9(B)

    Flexible Cords and Cables . . . . 690.31(E)

Tray Cable . . . . 392.80(A)

    Power and Control Tray Cable 336.80

    Type MV and Type MC Cables (2001 Volts or Over) 392.80(B)

Underground Feeder and Branch-Circuit Cable (Type UF) 340.80

AMERICANS WITH DISABILITY ACT

    ADA Accessibility . . . . 110.1 Info. Note

    ADA Standards for Accessible Design . . . Informative Annex J

AMMONIA

    Hazardous (Classified) Locations

    Class and Division

    Class I, Group D . . . . 500.6(A)(4) Info Note 2
    Refrigerant Machinery Rooms . . . . 500.5(A)
  Zones
    Group IIA . . . . 505.6(C)
    Refrigerant Machinery Rooms . . . . 505.5(A)

AMPACITY CALCULATION APPLICATION INFORMATION
. . . . INFORMATIVE ANNEX B

ANESTHETIZING LOCATIONS, INHALATION . . . .
. . . . **Art. 517 Part IV**
  Battery-Powered Lighting Units . . . . 517.63(A)

ANNEXES *See* **Informative Annexes** . . . . Ferm's Finder

ANTENNA DISCHARGE UNITS
  Receiving Stations. 810.20
  Transmitting Stations 810.57
  ANTENNA SYSTEMS, RADIO & TELEVISION . . Art. 810
  Grounding Devices to be Listed, Where Required 810.7
  Lead-In Protectors 810.6
  Used in Audio Systems . . . . 640.3(K)

APPLICATOR, DEFINED 665.2

APPLIANCES . . . . Art. 422
  Attachment Fitting 422.33
  Branch Circuit Installation . . . . Art. 422 Part II
    See CALCULATIONS . . . . Ferm's Finder
  Branch Circuit Ratings 422.10
  Branch Circuits Required 210.11
  Ceiling Fans. 422.18
  Central Vacuum Outlet Assemblies 422.15
  Cord- and Plug-connected Appliances Subject to Immersion 422.41
  Cord Connected 422.16
  Disconnecting Means . . . . Art. 422 Part III
    Cord- and Plug-Connected Appliances. 422.33
    Motor-Driven Appliances . . . . 422.31(C)
    Permanently Connected Appliances 422.31
  Electric Drinking Fountains (water coolers) . . . . . 440.3(C)
  Ground-Fault Circuit-Requirement
    Accessibility (Readily Accessible) 422.5
    Drinking water coolers . . . . 422.5(A)(2)
    High-Pressure Spray Washers . . . . 422.5(A)(3)
    Tire Inflation (For Public Use) . . . . 422.5(A)(4)
    Type . . . . 422.5(B)
    Vacuum Machines (Automotive for Public Use) . 422.5(A)(1)
    Vending Machines . . . . 422.5(A)(5)
  Grounding of . . . . Art. 250 Part VI
    Cord Connected Equipment 250.114
    Fixed Equipment – General 250.110
    Fixed Equipment – Specific 250.112
    Mobile Homes 550.16
    Park Trailers . . . . 552.56(F)
    Ranges & Dryers 250.140
    Recreational Vehicles . . . . 551.55(F)
  Household Cooking Used in Instructional Programs . 220.14(B)
  Installation of . . . . Art. 422 Part II
  Listing 422.6
  Load Calculations . . . . Art. 220
    See CALCULATIONS . . . . Ferm's Finder
    Commercial Cooking Equipment 220.56
    Demand Factors (Four or More) 220.53
    Dryers (Clothes) 220.54
    Electric Heat 220.51
    Laundry Load . . . . 220.52(B)
    New Restaurants 220.88
    Ranges 220.55
    Small Appliance (Dwellings) . . . . 220.52(A)
  Marking . . . . Art. 422 Part V
  Mobile Homes 550.14
  Overcurrent Protection of 422.11
    48 Ampere Maximum Load . . . . 422.11(F)
  Park Trailer 552.58
  Polarity in Cord- and Plug-Connected Appliances . . . 422.40
  Recreational Vehicles 551.57
  Vending Machines . . . . 440.3(C)

APPROVED 90.4
  Approval 110.2
  Definition of . . . . Art. 100 Part I
  Examination of Equipment for Safety 90.7
  Installation and Use . . . . 110.3(B)
  Labeled, Definition of . . . . Art. 100 Part I
  Listed, Definition of . . . . Art. 100 Part I

ARC ENERGY REDUCTION
  Documentation . . . . 240.87(A)
  Fuses 240.67
  How it Operates . . . . 240.87(B) Info Note 2
  Method to Reduce Clearing Time . . . . 240.87(B)

ARC WELDERS . . . . .Art. 630 Part II
   See WELDERS, Electric . . . . Ferm's Finder

ARCS & ARCING PARTS 110.18
   Aircraft Hangers . . . . 513.7(C)
   Arcing or Suddenly Moving Parts. 240.41
   Arc Lamps 520.61
      Portable Arc Lamps. 530.17
   Bulk Storage Plants . . . . 515.7(B)
   Commercial Garages . . . . 511.7(B)(1)(a)
   Conduit Seals, Class I Divisions 1 & 2 . . . . . . 501.15(C)(5)
   Health Care Facilities . . . . 517.61(B)(2)
   Motion Picture Projectors . . . . 540.11(A)
   Motors . . . . 430.14(B)
   Spray Application, Dipping and Coating . . . . . . 516.7(B)

ARC-FAULT CIRCUIT INTERRUPTERS 210.12
   Branch Circuit Extensions and Modifications, Dwelling Units . . 210.12(D)
   Combination Type . . . . 210.12(A)
   Definition of . . . . Art. 100
   Direct Current 690.11
   Dormitory Units . . . . 210.12(B)
   Dwelling Units . . . . 210.12(A)
      Branch Circuit Extensions or Modifications . . . . 210.12(D)
      Outlets and Devices (supplying) 210.12
   Extensions and Modifications . . . . 210.12(D)
   Guest Rooms and Guest Suites . . . . 210.12(C)
   Mobile Homes and Manufactured Homes . . . . . 550.25(B)
   NPLFA Circuit Power Source Requirements . . . . 760.41(B)
   Outlet Type Device . . . . 210.12(A)
   PLFA Circuits, Power Sources for . . . . 760.121(B)
   Readily Accessible Location 210.12
   Receptacles
      Branch Circuit Protections . . . . 406.4(D)(4)
      Dormitory Units . . . . 210.12(B)
      Dwelling Units . . . . 210.12(A)
      Extensions and Modifications . . . . 210.12(B)
      Replacement . . . . 406.4(D)
      Readily Accessible Location . . . . 406.4(D)
      Replacement Impracticable . . . . 406.4(D)(3)Ex.
      Room Air Conditioners 440.65
      Solar Photovoltaic Systems (Direct Current) 690.11

ARC FLASH HAZARD WARNING 110.16

ARC LAMPS, PORTABLE 530.17

AREA IN CIRCULAR MILLS OF CONDUCTORS . . . . . .
. . . . Chapter 9 Table 8

AREA IN SQUARE INCHES
   Auxiliary Gutters 366.22
   Cable Trays 392.22
   Compact Copper and Aluminum Conductors Chapter 9 Table 5A
   Concentric Conductors . . . . Chapter 9 Table 5
   Conduit Fill Tables
      See Compact Copper and Aluminum . . . .
. . . . Ferm's Charts and Formulas
      See Conduit and Tubing . . . . . . Ferm's Charts and Formulas
      See Dwelling Services . . . . Ferm's Charts and Formulas
      See Equipment Ground, with . . . . Ferm's Charts and Formulas
      For Conductors of the Same Size . . . . Informative Annex C
      Percent of Cross Section . . . . Chapter 9 Table 1
   Conduit or Tubing . . . . Chapter 9 Table 4
   Strut-Type Channel Raceway 384.22
   Wireways, Metal . . . . 376.22(A)
   Wireways, Nonmetallic 378.22

ARENAS
   Emergency Systems . . . . 700.2 IN
   Places of Assembly . . . . Art. 518
   Theaters, Audience Areas, Performance Areas, etc. . . . Art. 520

ARMORED CABLE, TYPE AC
Art. 320
   Adjustment Factors 320.80
   Bending Radius 320.24
   Boxes and Fittings 320.40
   Equipment Grounding Conductor 320.108
   Exposed Work 320.15
   Grounding 320.108
   In Accessible Attics 320.23
   Listing Requirements 320.6
   Protection against Physical Damage 300.4
      Exposed Work 320.15
      Through or Parallel to Framing Members 320.17
   Securing and Support of 320.30
   Thermal Insulation (in contact) . . . . 320.80(A)
   Uses Not Permitted 320.12
   Uses Permitted 320.10
   See (AWEZ) . . . . UL Product iQ

ARMORIES . . . . 518.2(A)
   General Lighting Load . . . . Table 220.12
   Wiring Methods 518.4

ARRAY, SOLAR PHOTOVOLTAIC SYSTEMS 690.2
   Installation and Service . . . . Art. 690

ARRESTERS, SURGE — OVER 1000 VOLTS . . . . Art. 242
   Bonding and Grounding at Mobile Homes 800.106
   Communications Circuits. . . . . Art. 800 Part III
   Network-Powered Broadband Communications Systems . 830.90
   Premises-Powered Broadband Communications Systems . 840.90
   Surge Protection, Class I Locations 501.35
   Surge Protection, Class II Locations. 502.35
   See (VZQK). . . . . UL Product iQ

ARTICLES AND PARTS IN *NEC*
   Arrangement of 90.3
   Listing of (Front of *National Electrical Code*) . . . . . . Contents

ASKAREL
   Definition of . . . . Art. 100 Part I
   Transformers 450.25

ASSEMBLY HALLS . . . . 518.2(A)
   General Lighting Load . . . . Table 220.12
   Temporary Wiring .518.3
   Wiring Methods 518.4

ASSEMBLY OCCUPANCIES . . . . ARTICLE 518
   Outdoor Working Space 518.6

ASSEMBLY TYPE CORD, LISTED . . . . 400.10(A)(11)

ASSURED EQUIPMENT GROUNDING CONDUCTOR PROGRAM . . . . .590.6(B)(2)

ATTACHMENT PLUGS (CAPS)
   Class I Locations 501.145
   Class II Locations 502.145
   Class III Locations 503.145
   Definition of . . . . Art. 100
   Grounding-Pole Identification . . . . 406.10(B)
   Grounding Requirements on Flexible Cord 400.24
   Grounding Type 406.9
   Listing and Marking 406.7
   Polarized Terminal Identification . . . . 200.10(B)
   Where Required on Flexible Cords . . . . 400.10(B)

ATTICS
   Access to Working Space . . . . 110.33(B)
   Armored Cable in. 320.23
   Knob-and-Tube Wiring in 394.23
   Lighting Outlet
      All Occupancies . . . . 210.70(C)
      Dwelling Units . . . . .210.70(A)(2)
   Metal Clad Cable in 330.23
   Nonmetallic Sheathed Cable in 334.23
   Open Wiring in 398.23
   Service Entrance Cable . . . . 338.10(B)(4)(a)
   Sign Transformers in . . . . 600.21(E)

AUDIO SIGNAL PROCESSING, AMPLIFICATION, AND REPRODUCING EQUIPMENT . . . . Art. 640
   Access to Equipment Behind Access Points 640.5
   Abandoned Audio Distribution Cables . . . . 640.6(C)
   Audio Systems Near Bodies of Water 640.10
   Combination Systems (with Fire Alarm Systems) . . . 640.3(J)
   Covered by Article . . . . 640.1(A)
   Definitions 640.2
   Flexible Cords and Cables
      Permanent Installations 640.21
      Temporary Installations 640.42
   Grounding 640.7
      Isolated Ground Receptacles . . . . 640.7(C)
      Separately Derived Systems, 60 Volts to Ground . . . 640.7(B)
   Grouping of Conductors 640.8
   Mechanical Execution of Work 640.6
   Near Bodies of Water 640.10
   Neat and Workmanlike Manner . . . . 640.6(A)
      Distribution Cables . . . . 640.6(B)
      Identified for Future Use . . . . 640.6(C)
   Not Covered by Article . . . . 640.1(B)
   Other Articles 640.3
   Output Wiring and Listing of Amplifiers . . . . 640.9(C)
   Permanent Installations . . . . .Art. 640 Part II
      Conduit or Tubing 640.23
      Loudspeakers, in Fire Resistance-Rated Walls, etc. . . . 640.25
      Use of Flexible Cords and Cables. 640.21
      Wireways, Gutters, and Auxiliary Gutters 640.24
      Wiring of Equipment Racks 640.22
   Portable and Temporary Installations . . . . Art. 640 Part III
      Environmental Protection of Equipment 640.44
      Equipment Access 640.46

| Multipole Branch-Circuit Cables Connectors 640.41
| Protection of Wiring 640.45
| Use of Flexible Cords and Cables 640.42
| Wiring of Equipment Racks 640.43
Protection of Equipment 640.4
Use of Audio Transformers and Autotransformers . . . 640.9(D)
Wiring Methods 640.9

AUDITORIUMS 518.2
General Lighting Load . . . . Table 220.12
Wiring Methods 518.4

AUTOMATIC, DEFINED . . . . Art. 100

AUTOMOTIVE
Tire Inflation Machines (For Public Use) . . . . . . 422.5(A)(4)
Vacuum Machines (For Public Use) . . . . 422.5(A)(1)

AUTOTRANSFORMERS 450.4
Audio Systems . . . . 640.9(D)
Ballast for Lighting Units 410.138
Branch Circuits 210.9
Feeders 215.11
Grounding 3-Phase, 3-Wire Systems 450.5
Motor Starting . . . . 430.82(B)
Motor Disconnecting Means . . . . 430.109(D)
Not Permitted
| Park Trailers . . . . 552.20(E)
| Recreational Vehicles . . . . 551.20(E)
Overcurrent, Protection of . . . . 450.4(A)
See BALLASTS . . . . Ferm's Finder

AUXILIARY GROUNDING ELECTRODES 250.54

AUXILIARY GUTTERS . . . . Art. 366
Audio Systems 640.24
Clearance, Bare Live Parts . . . . 366.100(E)
Conductors of the Same Circuit . . . . 300.3(B)(4)
Corrosion Protection .300.6
Definition
| Metallic Auxiliary Gutter 366.2
| Non-Metallic Auxiliary Gutter 366.2
Deflection of Conductors . . . . 366.58(A)
Grounding 366.60
Nonmetallic
| Ampacity of Conductors . . . . 366.23(B)
| Indoor Use . . . . 366.10(B)(2)
| Number of Conductors . . . . 366.22(B)
| Outdoor Use . . . . 366.10(B)(1)
| Securing and Supporting . . . . 366.30(B)
Sheet Metal
| Ampacity of Conductors . . . . 366.23(A)
| Indoor and Outdoor Use . . . . 366.10(A)(1)
| Number of Conductors . . . . 366.22(A)
| Securing and Supporting . . . . 366.30(A)
| Wet Locations . . . . 366.10(A)(2)
Splices and Taps 366.56
Used as Pull Boxes . . . . 366.58(B)
Wire Bending Space . . . . 366.58(A)
| Available Fault Current 110.24
| Emergency Systems 700.4
| Industrial Control Panel 409.22
| Irrigation Machines 670.5
| Legally Required Standby Systems 701.4
| Modular Data Centers 646.7
| Series Ratings, Breakers 240.86
| Surge Protective Devices . . . . Art. 242
| Transfer Equipment 702.5

AVAILABLE FAULT CURRENT 110.24
Definition . . . . Art. 100
Emergency Systems .700.4
Field Marking . . . . 110.24(A)
Industrial Control Panel 409.22
Interconnected Electric Power Prod. Sources . . 705.12(B)(2)(e)
Irrigation Machines 670.5
Label
| Arc-Flash Hazard Warning 110.16
| Service Equipment 110.24
Legally Required Standby Systems 701.4
Modifications . . . . 110.24(B)
Modular Data Centers 646.7
Motor Control Centers 430.99
Series Ratings, Breakers 240.86
Surge Protective Devices . . . . Art. 242
Transfer Equipment 702.5

B

BACK-FED OVERCURRENT DEVICES
Equipment Over 1000 Volts, Nominal 490.25
In Panelboards . . . . 408.36(D)
Interconnected Electric Power Production Sources . . . .
. . . . 705.12(B)(2)(3)(b)
Stand Alone Systems . . . . 710.15(E)

BACK-FED DUE TO LOSS OF PRIMARY SOURCE OF POWER
Electric Vehicle Charging Systems 625.46
Electrified Truck Parking Places. 626.26

BACKFILL . . . . 300.5(F)
Underground Conductors, Over 1000 Volts . . . . . 300.50(E)

BALCONIES, DECKS, AND PORCHES RECEPTACLE OUTLETS REQUIRED . . . . 210.52(E)(3)

BALLASTS
Calculations of Load . . . . 220.18(B)
Conductors within 75 mm (3 in.) of a Ballast Must Be
Rated Not Lower Than 90°C 410.68
Electric-Discharge Lighting Systems 1000 Volts or less Art. 410 Part XII
    Disconnecting Means for . . . . 410.130(G)
Electric-Discharge Lighting Systems
    More than 1000 Volts . . . . Art. 410 Part XIII
Luminaire (Fixture) Mounted on Fiberboard . . . . .410.136(B)
High-Intensity Discharge Luminaires (Fixtures). . . . 410.130(F)
Sign Ballasts Listed and Thermal Protection. 600.22
Sign Ballasts, Locations 600.21
Thermal Prot. of Fluorescent Luminaires (Fixtures) . .410.130(E)
Wired Luminaire (Fixture) Sections. . . . . 410.137(C)
See LUMINAIRES, Ballast Type . . . . Ferm's Finder

BARE CONDUCTOR
Ampacity. . . . . 310.15(D)
Auxiliary Gutters . . . . 366.56(B)
Basic Requirement . . . . 310.3(D)
Contact Conductors, Permitted . . . . 610.13(B)
    Installation, Cranes and Hoists. . . . . Art.610 Part III
Definition of . . . . Art. 100
Dimensions. . . . . Chapter 9 Table 8
Lighting Systems Operating at 30 Volts or Less . . . . 411.6(C)
Service Entrance, Types SE, USE 338.100
Service-Entrance Conductors . . . . 230.41 Ex.
Underground Service Conductors, Grounded Conductor . . . .
. . . . .230.30(A)Ex.

BARE LIVE PARTS
Auxiliary Gutters .366.100(E)
Cabinets, Cutout Boxes, etc. . . . . .312.11(A)(3)
Clearances . . . . .110.26(A)(1)
Guarding 110.27
    Over 1000 Volts 110.34
Guarding, within Compartments 490.32
Guarding, within Compartments, Low Voltage 490.33
Minimum Space Separation 490.24
Switchboards or Panelboards 408.56

BARE NEUTRAL (Permitted) . . . . 310.3(D)
Over 1000 Volts . . . . 250.184(A) Ex. 1 & 2
    Grounded Neutral Conductor . . . . . . .250.184(A) Ex. 3
Overhead Service Conductors. . . . . 230.22 Ex.
Service-Entrance Conductors . . . . 230.41 Ex.
Underground Service Conductors. . . . . 230.30(A) Ex.

BARRIERS
Box Volume Calculations . . . . 314.16(A)
Busways, Over 1000 Volts . . . . .368.234(B)
Cabinets, Cutout Boxes and Meter Sockets . . . . . . 312.11(D)
Cable Tray, Cables Rated Over 1000 Volts . . . . . 392.20(B)(2)
Emergency System Feeder Wiring. . . . . 700.10(D)(1)(3)
ENT and RNC, Places of Assembly. . . . . 518.4(C)
ENT, Thermal Barrier. 362.10
Fire Barriers 300.21
Nonmetallic-Sheathed Cable, Thermal Barrier, Types III, IV, V
Buildings . . . . 334.10(3)
Motor Control Center . . . . 430.97(A) Ex.
    Service-Entrance Motor Control Centers . . . . . 430.97(E)
Permanent Barriers in Boxes . . . . 314.28(D)
Service Conductors in Cable Trays . . . . 230.44 Ex.
Service Equipment . . . . 230.62(C)
Service Switchboards . . . . 408.3(A)(3) Ex.
Spas, Hot Tubs, Bonding, Indoor . . . . 680.43(D)
Spas, Hot Tubs, Receptacles, Indoor . . . . .680.43(A)(4)
Swimming Pool Covers, Electrically Operated . . . 680.27(B)(1)
Swimming Pool Junction Boxes. . . . . 680.24(A)(2)(b)
    Other Enclosures. . . . . 680.24(B)(2)(b)

| Swimming Pool Switching Devices . . . . 680.22(C)
| Swimming Pool Receptacles. . . . . 680.22(A)(53)
| Transformer, Dry-Type, Not Over 112½ kVA . . . . 450.21(A)
| Transformer, Oil-Insulated Installed Outdoors 450.27
| Voltage Between Adjacent Devices — Receptacles . . . 406.5(J)
| Voltage Between Adjacent Devices — Switches . . . . . 404.8(B)

BASEMENT
| Armored Cable in. 320.15
| Nonmetallic Extensions Not Permitted in . . . . . . 382.12(1)
| Nonmetallic Sheathed Cable in . . . . 34.15(C)
| Receptacles in
| GFCI Protection Required, Unfinished . . . . . 210.8(A)(5)
| Required . . . . 210.52(G)(3)

BATHROOM(S)
| Branch Circuit Required . . . . 210.11(C)(3)
| Definition of . . . . Art. 100 Part I
| GFCI Protection of Receptacles
| In Dwellings . . . . 210.8(A)(1)
| Other Than Dwellings . . . . 210.8(B)(1)
| Lighting Outlet, Wall Switched in Dwelling . . . . 210.70(A)(1)
| Luminaires (Fixtures) in, ... 410.10(D)
| Mobile and Manufactured Homes . . . . 550.14(D)
| Overcurrent Devices Not Permitted in . . . . 240.24(E)
| Receptacles Prohibited in Bathtub and Shower Space . . 406.9(C)
| Mobile and Manufactured Homes . . . . 550.13(F)(1)
| Park Trailers . . . . 552.41(F)(1)
| Recreational Vehicles . . . . 551.41(C)
| Receptacles Within 6 feet of Bathtub or Shower Stall, GFCI Protection . . . . 210.8(A)(9)
| Requirement for Receptacle Outlets in Dwellings. . . 210.52(D)
| Required in Mobile and Manufactured Homes. .550.13(D)(9)
| Service Equipment Not Permitted in . . . . 230.70(A)(2)

BATHTUBS
| Hydromassage . . . . Art.680 Part VII
| Definition 680.2
| Luminaires (Fixtures), Ceiling Fans . . . . 410.10(D)
| Mobile and Manufactured Homes . . . . 550.14(D)
| Mobile and Manufactured Homes, Receptacle Not Permitted
| In Tub or Shower Space . . . . 550.13(F)(1)
| Recreational Vehicles Tub or Shower Space. . . . . 551.41(C)
| Receptacle Not Permitted in Tub or Shower Space . . . 406.9(C)
| Receptacles Within 6 feet of Bathtub or Shower Stall, GFCI Protection . . . . 210.8(A)(9)

BATTERIES (Storage) . . . . **Art. 480**
| Aircraft Batteries . . . . 513.10(A)(2)
| Battery and Cell Terminations 480.4
| Battery System (Definition). . . . . Art. 100
| Commercial Garages . . . . 511.10(A)
| Critical Operations Power Systems (COPS). . . . . . 708.20(E)
| For Generator Sets . . . . 708.20(F)(4)
| Diesel Engine Drives for Fire Pumps . . . . 695.12(C)
| Disconnecting Means. 480.6
| Emergency System Power Source . . . . 700.12(A)
| For Generator Sets . . . . .700.12(B)(4)
| Maintenance . . . . 700.3(C)
| Emergency System Unit Equipment . . . . 700.12(F)
| Energy Storage Systems . . . . Art. 706
| Battery Interconnections 706.32
| Charge Control 706.23
| Disconnect and Overcurrent Protection 706.15
| Dwellings . . . . 706.30(A)
| General. 706.50
| Installation 706.30
| IEEE Standards . . . . 480.1 Info. Note
| Legally Required Standby System Power Source . . . 701.12(A)
| For Generator Sets . . . . 701.12(B)(5)
| Tests and Maintenance . . . . 701.3(C)
| Legally Required Standby System Unit Equipment . . 701.12(G)
| Less Than 50 Volts 720.9
| Listed 480.3
| Overcurrent Protection . . . . 240.21(H)
| Recreational Vehicles . . . . 551.30(C)
| Solar Photovoltaic Systems . . . . Figure 690.1(b)
| Ventilation
| Locations. . . . . 480.10(A)
| Electric Vehicles 625.52
| Vents (Cells) 480.10
| Wind Electric Systems . . . . Figure 694.1(b)

BELL CIRCUITS . . . . **Art. 720**
| Class 1, 2 & 3 Circuits . . . . Art. 725
| *See* Articles 501, 502, 503, for Installation within Hazardous (Classified) Locations . . . . *Ferm's Charts and Formulas*
| *See* CLASS 1, 2 & 3 REMOTE CONTROL CIRCUITS . . . . *Ferm's Finder*

BENDING CONDUIT AND TUBING
Electrical Metallic Tubing 358.24
Electrical Nonmetallic Tubing 362.24
Flexible Metal Conduit 348.24
Flexible Metallic Tubing, Fixed . . . . 360.24(B)
   Infrequent Flexing Use . . . . 360.24(A)
      High Density Polyethylene Conduit 353.24
Intermediate Metal Conduit 342.24
Liquidtight Flexible Metal Conduit 350.24
Liquidtight Flexible Nonmetallic Conduit 356.24
Nonmetallic Underground Conduit with Conductors . . 354.24
Reinforced Thermosetting Resin Conduit (Fiberglass) Type RTRC
. . . . 355.24
Rigid Metal Conduit 344.24
Rigid PVC Conduit 352.24

BENDING SPACE, CONDUCTORS
Auxiliary Gutters . . . . 366.58(A)
Bus Enclosures, Entering 408.5
Cabinets, Cutout Boxes and Meter Sockets 312.6
Manholes 110.74
Motor Control Centers . . . . 430.97(C)
Motor Controllers . . . . 430.10(B)
Motor Terminal Housings . . . . 430.12(B)
Over 1000 Volts 300.34
Panelboard Construction 408.55
Panelboards and Switchboards . . . . 408.3(G)
Pull and Junction Boxes 314.28
   Power Distribution Blocks . . . . 314.28(E)(3)
Switches . . . . 404.3(A)
Switches, Construction . . . . Art.404 Part II
Transformers 450.12
Under 1000 Volts . . . . 110.3(A)(3)
Wireways
   Metal . . . . 376.23(A)
      Power Distribution Blocks . . . . 376.56(B)(3)
   Nonmetallic . . . . 378.23(A)

BENDS
Cables, in
   Armored Cable, Type AC 320.24
   High Voltage Cable 300.34
   Integrated Gas Spacer Cable 326.24
   Metal-Clad Cable, Type MC 330.24
   Mineral-Insulated Cable, Type MI 332.24
   Nonmetallic-Sheathed Cable (Type NM) 334.24
   P Cable 337.24
   Power and Control Tray Cable 336.24
   Service-Entrance Cable 338.24
   Underground Feeder and Branch-Circuit Cable (Type UF) 340.24
Number of Bends Permitted in
   Electrical Metallic Tubing 358.26
   Electrical Nonmetallic Tubing 362.26
   Flexible Metal Conduit 348.26
   High Density Polyethylene Conduit 353.26
   Integrated Gas Spacer Cable 326.26
   Intermediate Metal Conduit 342.26
   Liquidtight Flexible Metal Conduit 350.26
   Liquidtight Flexible Nonmetallic Conduit 356.26
   Nonmetallic Underground Conduit with Conductors 354.26
   Reinforced Thermosetting Resin Conduit 355.26
   Rigid Metal Conduit 344.26
   Rigid PVC Conduit 352.26

BLOCK (City, Town or Village)
Definition of for Communication Circuits 800.2
Definition of for Network-Powered Broadband Circuit . . . 830.2
Primary Protector Requirements Broadband Systems . 830.90(A)
Primary Protector Requirements Comm. Circuits . . . 800.90(A)
Underground Block Distribution . . . . 800.47(B)

BOAT HOISTS
GFCI Protection Required, Dwelling Unit Locations . . . 555.9

BOATYARDS AND MARINAS . . . . Art. 555
Bonding of Non-Current-Carryig Metal Parts 555.13
Disconnecting Means for Shore Power Connections . . . 555.36
Electrical Connections 555.30
Electrical Datum Plane 555.3
Enclosures 555.31
Equipment Grounding Conductor, Type of Conductor . . 555.37
Floating Buildings, Covered by Code . . . . 90.2(A)(1)
Floating Buildings 555.50
GFCI Protection 555.35
   Other Than Shore Power . . . . 555.35(A)
Grounding 555.54
Hoists
   Marine Hoists, Railways, Cranes, and Monorails 555.8
   Boast Hoists 555.9
Load Calculations for Service and Feeder Conductors . . . 555.6

Maximum Voltage 555.5

Motor Fuel Dispensing Stations. . . . .

    Hazardous (Classified) Locations. 555.11

Portable Power Cables . . . . .555.34(B)(3)

Repair Facilities — Hazardous (Classified) Locations . . . 555.12

Service Equipment Location 555.4

Shore Power Receptacles . . . . 555.33(A)

Signage 555.10

Transformers 555.7

Wiring Methods 555.34

    Installation . . . . 555.34(B)

See Fire Protection Standard for Marinas and Boatyards, NFPA 303 . . . . NFPA 303

BOILERS

Branch Circuits

    Electrode Type 425.82

    Resistance Type. . . . . 425.72(D)

Electrode-Type, 600 Volts or Less . . . . . . Art. 424 Part VIII

Electrode-Type, Over 1000 Volts . . . . .Art. 490 Part V

Fixed Industrial Process Resistance-Type Boilers Art. 425 Part VI

Fixed Industrial Process Electrode-Type Boilers Art. 425 Part VII

Resistance-Type . . . . Art. 424 Part VII

Resistance-Type in ASME Stamped and Rated Vessel 422.11(F)(3)

BONDING . . . . Art. 250 Part V

Agricultural Buildings . . . . 547.10(B)

Bonding Conductors and Jumpers 250.102

    Conductor Sizing Load Side 250.122

    Conductor Sizing Supply Side 250.66

    Definition . . . . Art. 100

    Main Bonding Jumper. 250.28

    Other Than Service Enclosures . . . . 250.96(A)

    Over 250 Volts. 250.97

    Service Equipment Enclosures 250.92

    System Bonding Jumper 250.28

    System Bonding Jumper . . . . .250.30(A)(1)

    Supply-Side Bonding Jumper. . . . . .250.30(A)(2)

        Definition . . . . Art. 100

    Size of on Load Side of Service . . . . 250.102(D)

    Size of on Supply Side of Service . . . . 250.102(C)

    Sizing of . . . . Table 250.102(C)(1)

    To Other Systems 250.94

    Note: Installing supplementary equipment grounding conductors in metal raceways is no substitute for effectively bonding those raceways . . . . .250.96(B) IN

Cable Trays . . . . 392.60(A)

    Where Permitted as Equipment Grounding Conductor . . . . .392.60(B)(4)

CATV Systems . . . . 800.100(D)

    Bonding at Mobile Homes . . . . .800.106(B)

Communications Circuits 250.94

    Bonding of Electrodes . . . . 800.100(D)

    Bonding at Mobile Homes . . . . .800.106(B)

Concentric and Eccentric Knockouts . . . . 250.92(B)

    Over 250 Volts. 250.97

Conductor Sizing

    Line Side of Service Equipment . . . . .250.102(C)

    Installation . . . . .250.102(E)

        Minimum Size . . . . Table 250.66

    Load Side of Service Equipment . . . . 250.102(D)

    Installation . . . . .250.102(E)

    Minimum Size . . . . Table 250.122

Connections 250.8

Definition of . . . . Art. 100

Electrically Conductive Materials and Other Equipment

    Grounded Systems . . . . 250.4(A)(4)

    Ungrounded Systems. . . . . 250.4(B)(3)

Electrical Equipment

    Grounded Systems . . . . 250.4(A)(3)

    Ungrounded Systems . . . . 250.4(B)(2)

Enclosures for Grounding Electrode Conductors . . . 250.64(E)

Expansion Fittings 250.98

Exposed Structural Steel . . . . .250.104(C)

Flexible Metal Conduit for Service . . . . .230.43(15)

Floating Buildings 555.54

Fountains 680.53

    Signs in, ... 680.57(E)

Grounding Electrode System 250.50

Grounding Electrodes Together 250.58

    CATV Systems. . . . . 800.100(D)

    Communications Circuits . . . . 800.100(D)

    Intersystem 250.94

    Network-Powered Broadband Communications Systems. . . . . 800.100(D)

    Radio and Television Equipment. . . . . 810.21(J)

    Use of Strike Termination Devices 250.60

Hazardous (Classified) Locations 250.100

    Class I Locations . . . . 501.30(A)

Class I, Zone 0, 1, and 2 Locations . . . . 505.25(A)
Class II Locations . . . . 502.30(A)
Class III Locations . . . . 503.30(A)
Intrinsically Safe Systems 504.60
Health Care Facilities (Panelboard) 517.14
Critical Care (Category 1) Spaces, Equipment Grounding and Bonding . . . . 517.19(E)
Critical Care (Category 1) Spaces, Patient Vicinity in (Optional) . . . . 517.19(D)
Hydromassage Bathtubs. 680.74
Installed With Circuit Conductors, General . . . . . . 300.3(B)
Installed With Circuit Conductors, Underground . . . 300.5(I)
Intersystem . . . . 250.94(A)
Communication . . . . 800.100(B)(1),(2) and (3)
Community Antenna Television and Radio Distribution Systems 800.110
Network-Powered Broadband Communications Systems. . . . 800.100(B)(1),(2) and (3)
Optical Fiber Cables . . . . 770.100(B)(1),(2) and (3)
Radio and Television Equipment . . . 810.21(F)(1),(2) and (3)
Jumper, Supply-Side . . . . Art. 100
Liquidtight Flexible Metal Conduit 350.60
Used for Service Entrance . . . . 230.43(15)
Loosely Jointed Metal Raceways 250.98
Main Bonding Jumper
Definition . . . . Art. 100
Installation 250.28
Serviced-Supplied AC Systems 250.24
Size. 250.102
Manufactured Buildings 545.11
Metal Raceways
Enclosing Grounding Electrode Conductors . . . . 250.64(E)
Used for Other Than Service Entrance 250.96
Used for Service Entrance 250.92
Metallic Piping Systems
Other Piping Systems . . . . 250.104(B)
Water Piping Systems . . . . 250.104(A)
In Multiple Occupancy Buildings . . . . . . 250.104(A)(2)
In Multiple Buildings Supplied from a Feeder or Branch Circuit. . . . . 250.104(A)(3)
Separately Derived Systems . . . . 250.104(D)(1)
Mobile Homes . . . . 550.16(C)
Multiple Raceways
Load Side. . . . . 250.102(D)
Supply Side . . . . 250.102(C)

Other Systems 250.94
CATV and Radio Distribution Systems . . . . . 800.100(D)
Communications Circuits . . . . 800.100(D)
Network-Powered Broadband Communication Systems . . . . 800.100(D)
Radio and Television Equipment. . . . . 810.21(J)
Over 250 Volts to Ground 250.97
Park Trailers. 552.57
Portable and Vehicle-Mounted Generators 250.34
Premises-Powered Broadband Communications Systems
Mobile Homes . . . . 840.106(B)
Recreational Vehicles 551.56
Separate Building 250.32
Separately Derived Systems 250.30
Service Equipment Enclosures . . . . 250.92(A)
Common Grounding Electrode 250.58
Concentric or Eccentric Knockouts . . . . 250.92(B)
Equipment Bonding Jumpers . . . . 250.102(C)
Main Bonding Jumper 250.28
Metal Enclosures Protecting Grounding Electrode Conductor . . . . 250.64(E)
Installation . . . . 250.64(E)
Switchboards and Panelboards Used for Service . . . 408.3(C)
Service Equipment
Main Bonding Jumper 250.28
Neutral Must Be Bonded to Service Equipment 250.24
Size on Line Side, Calculation . . . . 250.102(C)
Size on Line Side, Table . . . . Table 250.66
Use of Grounded Circuit Conductor on Line Side of Mains . . . . 250.92(B)(1)
Sizing Table . . . . 250.102(C)(1)
Solder Not Permitted in Bonding Connection(s) . . . . 250.8(B)
Spas & Hot Tubs Outdoor Locations . . . . 680.42(B)
General Requirements 680.26
Spas and Hot Tubs Indoor Locations . . . . 680.43(D)
Methods . . . . 680.43(E)
Swimming Pools 680.26
Cord-Connected Equipment. . . . . 680.8(B)
Lifts, Electrically Powered 680.83
Underwater Lighting. . . . . 680.26(B)(4)
Underwater Audio Equipment . . . . 680.27(A)
See SWIMMING POOLS, Grounding . . . . . Ferm's Finder
System Bonding Jumper
Definition . . . . Art. 100
Installation 250.28

Therapeutic Pools & Tubs
   Permanently Installed Therapeutic Pools 680.61
   Therapeutic Tubs (Hydrotherapeutic Tanks) . 680.62(B)&(C)
Unions in Water Line . . . . 250.52(A)(1)
   Effective Bonding Path . . . . 250.68(B)
Water Meter or Filter . . . . .250.53(D)(1)
   Effective Bonding Path . . . . 250.68(B)
Water Pipe (Metal) . . . . 250.104(A)
   Permitted Connection Locations . . . . 250.68(C)
See GROUNDING, Separately Derived Systems . . Ferm's Finder

BONDING JUMPER
Conductor or Jumper . . . . Art. 100
Connections 250.8
Electrode System . . . . 250.53(C)
Equipment . . . . Art. 100
Fences 250.194
Main . . . . Art. 100
Material . . . . .250.102(A)
Piping Systems 250.104
Receptacle to Box 250.146
Service . . . . 250.92(B)
Sizing Table . . . . 250.102(C)(1)
Supply Side (Definition) . . . . Art. 100
System . . . . Art. 100

BORED HOLES THROUGH STUDS, JOISTS . 300.4(A)(1)

BOWLING LANES . . . . 518.2(A)
Wiring Methods .518.4

BOXES & FITTINGS . . . . Art. 314
6 mm (¼ in.) Setback or Flush Mounting Walls and Ceilings . 314.20
Accessible 314.29
Accessible After Electric-Discharge and LED Luminaires
   (Fixtures) Installed . . . . 410.24(B)
Barrier . . . . 314.16(A)
Boxes at Ceiling-Suspended (Paddle) Fan Outlets . . . 314.27(C)
Cables Entering Boxes 314.17
Ceiling Fan Outlet . . . . 314.27(C)
Ceiling Outlet . . . . .314.27(A)(2)
Conductors Entering Boxes 314.17
   Bushings on Intermediate Metal Conduit 342.46
   Bushings on Rigid Metal Conduit 344.46
   Bushings on Rigid Nonmetallic Conduit 352.46

Insulated Fittings on Raceways . . . . 300.4(G)
Conductors Entering Cabinets, Cutout Boxes, and Meter Socket
Enclosures 312.5
   Conductors 4 AWG and Larger . . . . 300.4(G)
Conductors, Number in 314.16
Conduit Bodies, Where Required 300.15
   Splices and Taps, Electrical Metallic Tubing 358.56
   Splices and Taps, Electrical Nonmetallic Tubing 362.56
   Splices and Taps, Flexible Metal Conduit 348.56
   Splices and Taps, Flexible Metallic Tubing 360.56
   Splices and Taps, High Density Polyethylene Conduit . 353.56
   Splices and Taps, Intermediate Metal Conduit 342.56
   Splices and Taps, Liquidtight Flexible Metal Conduit . 350.56
   Splices and Taps, Liquidtight Flexible Nonmetallic Conduit . .
   . . . . 356.56
   Spices and Taps, Nonmetallic Underground Conduit
     With Conductors 354.56
   Splices and Taps, Type RTRC Conduit 355.56
   Splices and Taps, Rigid Metal Conduit 344.56
   Splices and Taps, Rigid PVC Conduit 352.56
Conduit Bodies 314.1
   Accessible 314.29
   Conductors Entering 314.17
   Definition . . . . Art. 100
   Number of Conductors . . . . .314.16(C)(1)
   Over 1000 Volts . . . . Art. 314 Part IV
   Size of . . . . 314.28(A)
   Unused Openings . . . . 110.12(A)
Continuity of Metal Enclosures 250.90
   Bonding Other Enclosures 250.96
   Class I, Divisions 1 & 2 Locations . . . . 501.30(A)
   Class I, Zone 0, 1, and 2 Locations . . . . 505.25(A)
   Class II, Divisions 1 and 2 Locations . . . . 502.30(A)
   Class III, Divisions 1 and 2 Locations . . . . 503.30(A)
   Couplings and Connectors, EMT 358.42
   Couplings and Connectors, IMC 342.42
   Couplings and Connectors, RMC 344.42
   Electrical Continuity 300.10
   Hazardous Locations 250.100
   Loosely Jointed Raceways 250.98
   Mechanical Continuity 300.12
   Method of Bonding at the Service . . . . 250.92(B)
   Over 250 Volts 250.97
   Threaded Conduit Hazardous Locations . . . . . .500.8(E)(3)

Covers and Canopies 314.25
   Construction of 314.41
   Extensions from . . . . 314.22 Ex.
   For Pull and Junction Boxes . . . . 314.28(C)
   For Systems Over 1000 Volts. . . . . 314.72(E)
   Grounding of . . . . 314.25(A)
   Manholes. . . . . 110.75(D)
   Requirement For, Completed Installations 410.22
Cutout . . . . Art. 312
Drainage Opening, Field Installed 314.15
Emergency Systems, Identification . . . . 700.10(A)
Extension Rings. . . . . 314.16(A)
Surface Extensions 314.22
Fill Calculations. . . . . 314.16(B)
Floor Boxes . . . . 314.27(B)
Free Length of Conductor in 300.14
Free Length of Conductor in, Electric Heat Cables . . 424.43(B)
Free Length of Conductor in, Deicing & Snow-Melting . 426.23(A)
Free Length of Conductor in, Pipeline Heating . . . 427.18(A)
Grounding . . . . Art. 250 Part IV
   Continuity and Attachment to Boxes. 250.148
   Electrical Continuity 300.10
   Methods of Equipment Grounding . . . . . Art. 250 Part VII
   Other Than Service Enclosures 250.86
   Requirement for Grounding Metal Boxes 314.4
   Service Enclosures 250.80
Hazardous (Classified) Locations . *See* Articles 500 through 517
High Voltage Systems . . . . Art. 314 Part IV
   See OVER 1000 VOLTS, NOMINAL . . . . . Ferm's Finder
Luminaire (Lighting Fixture) or Lampholder Outlets . 314.27(A)
   Ceiling Outlets . . . . 314.27(A)(2)
   Maximum Luminaire (Fixture) Weight. . . . . . 314.27(A)
   Means of Support . . . . 410.36(A)
Metallic Boxes
   Conductors Entering . . . . 314.17(B)
   Conductor Openings. . . . . 314.17(A)
   Corrosion Resistant . . . . 314.40(A)
   Equipment Grounding Conductor Conn. . . 250.148(A)&(C)
   Grounded 314.4
   Grounding Provisions . . . . 314.40(D)
   Mounting . . . . 110.13(A)
   Securing Raceways or Cables to . . . . 314.17(B)
   Supports 314.23
   Thickness of Metal . . . . 314.40(B)
   Over 1650 cm^3 (100 $in.^3$) . . . . 314.40(C)
   Thickness of Metal, Cabinets and Cutout Boxes . . 312.10(B)
   Unused Openings . . . . 110.12(A)
Minimum Depth of Boxes 314.24
Nonmetallic Boxes
   Arrangement of Grounding Conductors . . . . . 250.148(D)
   Conductor Openings. . . . . 314.17(A)
   Mounting . . . . 110.13(A)
   Permitted. 314.3
   Securing Permitted Wiring Methods to . . . . . . 314.17(C)
   Support 314.23
      Provisions for Support 314.43
   Unused Openings . . . . 110.12(A)
Number of Conductors in 314.16
Required 300.15
Round Boxes (Not Always Permitted) 314.2
Separable Attachment Fittings . . . . 314.27(E)
Sizing of Outlet, Device and Pull & Junction Boxes and Conduit Bodies. 314.16
Sizing of Pull and Junction Boxes 314.28
   Over 1000 Volt Systems 314.71
   Conductor Bending Radius 300.34
Sizing of Cabinets and Cutout Boxes 312.6
Snap Switches and Adjacent Devices, Over 300 Volts Between . . 404.8(B)
Support of Ceiling-Suspended (Paddle) Fans . . . . . 314.27(C)
   Weight Limits 314.27
Support of, General . . . . 110.13(A)
Support of Luminaires (Lighting Fixtures) . . . . . . 410.36(A)
   Designed for . . . . 314.27(A)
   Weight Limits . . . . 314.27(A)(1)&(2)
Support of, Means 314.23
Under Roof Decking . . . . 300.4(E)
Unused Openings Boxes and Conduit Bodies . . . . 314.17(A)
Unused Openings Closed, General . . . . 110.12(A)
Utilization Equipment . . . . 314.27(D)
Vertical Surface . . . . 314.27(A)(1)
Volume Required Per Conductor . . . . Table 314.16(B)
Wall or Ceiling Use 314.20
Wet Locations 314.15
   Cabinets and Cutout Boxes 312.2
   Corrosion Protection 300.6
      Cabinets and Cutout Boxes . . . . 312.10(A)
   Enclosures for Switches or Circuit Breakers 404.4

Nonmetallic Conduit Systems and Boxes 352.100
See SWIMMING POOLS, Junction Boxes Feeding . .Ferm's Finder
See SWIMMING POOL, Underwater Lighting . . Ferm's Finder

BOXLESS DEVICES . . . . 300.15(E)
Fitting Only . . . . 300.15(F)
Flat Conductor Cable. 324.42
Manufactured Buildings 545.10
Mobile Homes .550.15(I) Ex.
Nonmetallic-Sheathed Cable . . . . 334.30(C)
    Devices of Insulating Material . . . . 334.40(B)
    Devices with Integral Enclosures . . . . 334.40(C)
Nonmetallic-Sheathed Cable Interconnector . . . . . 334.40(B)
Park Trailers. . . . . .552.48(I)
    Component Interconnections . . . . 552.48(N)
Receptacles, Self-Contained . . . . 334.40(B)
Recreational Vehicles . . . . .551.47(J)
    Component Interconnections . . . . 551.47(O)
Switches, Self-Contained . . . . 334.40(B)

BRANCH CIRCUITS . . . . Art. 210
Air Conditioners . . . . Art. 440 Part IV
    Room Air Conditioners 440.62
Arc-Fault Circuit-Interrupter Protection 210.12
Appliances
    Installation of for Appliances. . . . . .Art. 422 Part II
    Number Required, Small Appliances. . . . . . .210.11(C)(1)
    Outlets Served, Small Appliances. . . . . 210.52(B)
    Small Appliance Loads Dwellings . . . . 220.52(A)
    Specific Appliances. . . . . 220.14(A)
Balanced Load . . . . 210.11(B)
Bathroom Circuit Required, Dwelling Units . . . .210.11(C)(3)
Busways as . . . . 368.17(C)
    Overcurrent Protection. . . . . 368.17(D)
Calculations of 220.10
    See Also CALCULATIONS . . . . .Ferm's Finder
Classification & Rating of 210.3
    Ratings of . . . . .Art. 210 Part II
Color Coding
    Conductor Identification. 310.6
    Equipment Grounding Conductor 250.119
    Grounded Conductor 200.6
    High-Leg. 110.15
    Identification. .210.5
    Intrinsically Safe Systems . . . . 504.80(C)

    Isolated Power Systems . . . . 517.160(A)(5)
    Sensitive Electronic Equipment . . . . 647.4(C)
Common Area (Branch and Dwelling Unit Branch) Circuits . 210.25
Conductors (Ampacity & Size) . . . . 210.19(A)
    Over 600 Volts. . . . . 210.19(B)
Continuous Loads . . . . .210.19(A)(1)
    Overcurrent Protection. . . . . 210.20(A)
        Individual Appliance Branch Circuit . . . . . . 422.10(A)
Critical Care (Category 1) Spaces, Health Care Facilities . 517.19(A)
Direct-Current Ungrounded Conductor Identification . 210.5(C)(2)
Dryers, Demand Loads 220.54
Dwellings, Required in
    Bathroom Receptacle Outlets . . . . .210.11(C)(3)
    Central Heating Equipment 422.12
    General Lighting . . . . .220.14(J)
    Laundry Equipment . . . . .210.11(C)(2)
    Minimum Number. . . . . 210.11(A)
    Small Appliance . . . . 210.11(C)(1)
Electric Vehicle, For Charging Purposes 625.40
Electrified Truck Parking Space Wiring Systems. 626.10
Emergency Systems
    Emergency Lighting 700.17
    Emergency Power 700.18
    Limitations on Loads Supplied. 700.15
    Wiring of. 700.10
Extensions (Grounding of) . . . . .250.130(C)
Fixed Electric Space Heating 424.3
    Electric Pool Water Heaters 680.10
Garage Branch Circuits . . . . .210.11(C)(4)
General Care (Category 2) Spaces, Health Care Facilities . .517.18(A)
Ground-Fault Circuit-Interrupter Protection (GFCI)
Bathtubs, Hydromassage 680.71
Boat Hoist . . . . 210.8(C)
Commercial Garages 511.12
Construction Sites 590.6
Definition of . . . . Art. 100
    Dwellings . . . . 210.8(A)
    Elevator Pits, Cartops, etc. 620.6
    Feeders, Protection in Lieu of Branch Circuit Protection 215.9
    Fountains . . . . 680.51(A)
        Cord- and Plug-Connected Equipment . . . . . 680.56(A)
        Signs within Fountains . . . . 680.57(B)
    HACR Equipment Receptacles installed Outdoors. . . 210.63
    HACR Equipment Receptacles installed Outdoors

(Dwellings). . . . . 210.8(A)(3)
Health Care Facilities Critical Care (Category 1) . . . 517.21
Health Care Facilities Therapeutic Pools and Tubs . 680.62(A)
    Receptacles within 1.83 m (6 ft) . . . . 680.62(E)
    Therapeutic Tubs. . . . . 680.62(A)
Health Care Facilities Wet Locations . . . . 517.20(A)
High Pressure Spray Washers . . . . 422.5(A)(3)
Hotels 210.18
Hot Tubs Outdoors 680.44
    See SWIMMING POOLS . . . . Ferm's Finder
Hot Tubs Indoors
    Luminaires (Lighting Fixtures), Outlets, and Ceiling Fans .
    . . . . 680.43(B)(1)
    Receptacles for Hot Tubs . . . . 680.43(A)(3)
    Receptacles within 3.0 m (10 ft). . . . . . . 680.43(A)(2)
    Required Protection 680.44
Hydromassage Bathtub 680.71
Marinas & Boat Yards 555.35
Mobile Homes. . . . . 550.13(B)
Mobile Home Service Equipment . . . . 550.32(E)
Other Than Dwelling Units . . . . 210.8(B)
Overcurrent Protective Device, definition . . . . . . Art. 100
Park Trailers . . . . 552.41(C)
    Pipe Heating Cable Outlet . . . . 552.41(D)(3)
Receptacle Replacement . . . . 406.4(D)(2) and (3)
Recreational Vehicles
    Shower Luminaires (Fixtures) . . . . 551.53(B)
    Specific Outlets . . . . 551.41(C)
    With One Circuit . . . . 551.40(C)
Recreational Vehicle Park 551.71
Residential Occupancies . . . . 210.8(A)
    See (KCXS) . . . . UL Product iQ
    Signs (Outdoor Portable) . . . . 600.10(C)(2)
    at Fountains . . . . 680.57(B)
Spas Indoors
    Luminaires (Lighting Fixtures), Outlets, and Ceiling Fans .
    . . . . 680.43(B)(1)
    Receptacles for Spas . . . . 680.43(A)(3)
    Receptacles within 3.0 m (10 ft). . . . . . . 680.43(A)(2)
    Required Protection 680.44
Spas Outdoors 680.44
    See SWIMMING POOLS . . . . Ferm's Finder
Swimming Pools
    Electrically Operated Pool Covers . . . . . 680.27(B)(2)

    Luminaires (Lighting Fixtures), Outlets, & Ceiling Fans . .
    . . . . 680.22(B)
    Motors. . . . . 680.21(C)
    Receptacles Protected . . . . 680.22(A)(1) through (A)(4)
    Separation of Conductors for Underwater Lighting from GFCI 680.23(F)(3)
    Storable Pool Luminaires (Lighting Fixtures)
        Over the Low Voltage Contact Limit to 150 Volts . . 680.33(B)
        Note: Open-neutral protection required by this section
    Storable Pools 680.32
    Types Permitted 680.5
    Underwater Luminaires (Fixtures) More than the Low Voltage Contact Limit . . . . 680.23(A)(8)
    Underwater Luminaires (Fixtures) Relamping. . .680.23(A)(3)
    Therapeutic Pools & Tubs . . . . 680.62(A)
Identification of Circuits 110.22
    In Panelboards 408.4
Identification of Conductors
    Equipment Grounding Conductors 250.119
    Equipment Grounding Conductors, Branch Circuits 210.5(B)
    General Rules 310.6
    Grounded Conductor of Branch Circuit . . . . . . 210.5(A)
    Grounded Conductors 200.6
    Ungrounded Conductors, All Branch Circuit . . . . 210.5(C)
Individual (Branch Circuit)
    Definition of . . . . Art. 100 Part I
    Overcurrent Protection. 210.20
    Permissible Loads. 210.23
    Rating or Setting Motor Circuit 430.52
    Receptacle Rating . . . . 210.21(B)(1)
    Required
        Central Heating Equipment 422.12
        Electric Signs . . . . 600.5(A)
        Electrode-Type Boiler Over 1000 Volts . . . . . 490.72(A)
        Elevator Car Air Conditioning and Heating . . 620.22(B)
        Elevator Car Lighting . . . . 620.22(A)
        Elevator Machine Room. . . . . 620.23(A)
        Hoistway Pit Lighting and Receptacle(s) . . . . 620.24(A)
        Hydromassage Bathtubs. 680.71
        Isolated Power System, Health Care Facility . . . .
        . . . . .517.31(C)(2)
        Marinas and Boatyards Shore Power Receptacles 555.33(A)
        Office Furnishings. . . . . 605.9(B)
        Transport Refrigerated Units (TRUs) 626.30
Isolated Power Systems, Health Care Facilities 517.160

Laundry Area (Dwellings) Required . . . . 210.11(C)(2)
   Load Calculation. . . . . 220.52(B)
   Load for Electric Dryers 220.54
   Location, Appliance Receptacle within 6 Feet . . . 210.50(C)
   Not Required . . . . 210.52(F) Ex. 1&2
Maximum Loads 220.18
   Summary. 210.24
Maximum Voltage 210.6
   120 Volts Between Conductors . . . . 210.6(B)
   277 Volts to Ground . . . . .210.6(C)
   600 Volts Between Conductors . . . . 210.6(D)
   Between Adjacent Switches and Devices . . . . . . 404.8(B)
   Elevators, etc. .620.3
   Lighting Equipment Outdoors 225.7
   Occupancy Limitations. . . . . 210.6(A)
   Over 600 Volts Between Conductors. . . . . 210.6(E)
Minimum Ampacity & Size
   Appliances 422.10
   Electric Radiant Heating Panels and Sets . . . . . 424.95(B)
   Electrically Driven or Controlled Irrigation Machines . .675.9
   Electrode-Type Boilers 424.82
   Elevators, etc. . . . . 620.12 & 13
   Fixed Electric Heating Equipment for Pipelines and Vessels . 427.4
   Fixed Electric Space-Heating Equipment . . . . . 424.4(B)
   Fixed Outdoor Electric Deicing and Snow-Melting
   Equipment. 426.4
   Household Ranges and Cooking Appliances . . . 210.19(A)(3)
   Information Technology Equipment . . . . 645.5(A)
   Multi-Receptacle Circuits . . . . .210.19(A)(2)
   Not Over 600 Volts . . . . .210.19(A)(1)
   Other Loads . . . . .210.19(A)(4)
   Over 600 Volts . . . . 210.19(B)
   X-Ray Equipment, Non-Medical . . . . 660.6(A)
Motor Circuit Conductors . . . . .Art. 430 Part II
Multioutlet Assembly, Calculations of . . . . 220.14(H)
Multioutlet Branch Circuits, Rating 210.3
   Conductors . . . . 210.19(A)(2)
   Outlet Devices . . . . 210.21(B)(2)&(3)
   Permissible Loads. 210.23
   Protection of Conductors 240.4
   Summary. 210.24
Multiple Circuits 210.4
   Equipment Grounding Connections 250.144
      Simultaneous Disconnecting Means 210.7
      Devices or Equipment. . . . . 210.4(B)
   Size of Equipment Grounding Conductors . . . .250.122(C)
Multiwire Branch Circuits 210.4
   Definition of . . . . Art. 100 Part I
   Neutral Continuity Must Not Be Dependent
      on Device Connections . . . . 300.13(B)
   Simultaneous Disconnecting Means Required
      (Line-to-Line Loads). . . . . 210.4(C)
   Circuit Breaker as Overcurrent Device. . . . . . 240.15(B)
   Devices or Equipment on Same Yoke 210.7
   Tapped from Grounded Systems. 210.10
Not More Than One Dwelling Unit 210.25
Number of, Required 210.11
   Central Heating Equipment 422.12
   Calculation of Loads to Determine Number . Art. 220, Part II
   Permissible Load 210.23
Number of, Supplied to More Than One Buildings. . . . 225.30
Outside . . . . Art. 225
   Overhead, Over 1000 Volts . . . . 399.10(2)
   See OUTSIDE BRANCH CIRCUIT & FEEDERS. . Ferm's Finder
Overcurrent Protection 210.20
   Outlet Devices 210.21
   Protection of Conductors 240.4
   Protection of Equipment. 240.3
Patient Bed Location
   Critical Care (Category 1) Spaces . . . . 517.19(A)
   General Care (Category 2) Spaces . . . . 517.18(A)
Permissible Loads 210.23
   Computation of Loads . . . . Art. 220, Part II
   Loads Evenly Proportioned. . . . . 210.11(B)
   Maximum Load 220.18
   Outlet Devices 210.21
   Specific Circuits 210.23
   Summary. 210.24
Ranges . . . . 210.19(A)(3)
   Branch Circuit Rating . . . . 422.10(A)
   Computations . . . . 220.14(B)
   Demand Loads, Dwelling Units 220.55
   Demand Loads, Other Than Dwelling Units 220.56
   Neutral Load 220.61
Rating & Classification of 210.3
   Branch Circuit Ratings . . . . Art. 210, Part II
Required 210.11
   See LIGHTING OUTLETS REQUIRED . . . Ferm's Finder

See RECEPTACLE, Outlets Required . . . . . . Ferm's Finder
Sign Circuit . . . . 600.5(A)
Requirements General 210.11
Requirements Summary 210.24
Selection Current, Hermetic Refrigerant Motor-Compressors. . .
. . . . 440.4(C)
   Definition . . . . Art. 100
Small Appliance (Dwelling Units). . . . . .210.11(C)(1)
   Load for Calculation . . . . 220.52(A)
   Outlets Served . . . . 210.52(B)
Specific Purpose 210.2
   Air Conditioning & Refrigeration 440.6
      Branch Circuit Conductors, General 440.31
      Room Air Conditioners 440.62
      Single Motor-Compressor 440.32
   Busways 368.17
   Class 1, 2 & 3 Remote Control . . . . Art. 725
   Cranes & Hoists . . . . .Art. 610 Part II
      Overcurrent Protection 610.42
   Deicing & Snow-Melting Equipment (Continuous Load) . . 426.4
   Dryers (Location), Receptacle Outlet Within 6 Feet . . 210.50(C)
      Loads 220.54
   Dumbwaiters. Art. 620
   Electric Heat (Space) 424.3
   Elevators . . . . Art. 620
   Escalators. . . . . Art. 620
   Fire Alarms . . . . Art. 760
   Heating Equipment for Pipelines and Vessels (Continuous Load)
. . . . 427.4
   Induction & Dielectric Heating . . . . Art. 665
   Information Technology Equipment 645.5
   Infrared Heating Equipment 422.48
      Branch Circuits 424.3
   Low Voltage Systems . . . . Art. 720
   Low Voltage Lighting Systems 411.7
   Marinas & Boatyards 555.53
   Mobile Homes . . . . Art. 550
   Motion Picture & TV Studios . . . . Art. 530
   Motors. .Art. 430 Part II
   Moving Walks . . . . Art. 620
   Office Furnishings, Fixed-Type. 605.7
   Office Furnishings, Freestanding-Type 605.8
   Office Furnishings, Freestanding-Type Cord & Plug . . . 605.9
   Pipe Organs . . . . Art. 650

Ranges . . . . .210.19(A)(3)
   Demand Load 220.55
   Overcurrent Protection Branch Circuit . . . . . 422.11(B)
   Rating of Receptacle. . . . . 210.21(B)(4)
Recreational Vehicles 551.42
Refrigeration (HACR) Equipment Receptacle 210.63
Signaling Systems . . . . Art. 725
   Fire Alarm Systems . . . . Art. 760
Signs 600.5
Sound Recording Equipment . . . . Art. 640
Space Heating Equipment 424.3
Switchboards & Panelboards Instrument Circuits . . . 408.52
Systems Over 1000 Volts . . . . Art. 110 Part III
   Conductors 311.60
      Outdoor Overhead Conductors . . . . Art. 399
   Conductor Ampacity . . . . 210.19(B)
   Equipment . . . . Art. 490
   General . . . . .Art. 300 Part II
   Motors. . . . . Art. 430 Part XI
   Outside Branch Circuits. . . . . Art. 225 Part III
Systems Under 50 Volts . . . . Art. 720
Theaters . . . . Art. 520
   Fixed Stage Equipment . . . . Art. 520 Part III
   Fixed Stage Switchboard . . . . .Art. 520 Part II
   General 520.9
   Portable Stage Equipment . . . . .Art. 520 Part V
   Portable Stage Switchboard . . . . Art. 520 Part IV
Water Heaters . . . . 422.10(A)
   Pool Heaters. 680.10
   Storage-Type. 422.13
Welders . . . . Art. 630
   Arc Welders (Motor and Nonmotor Generator) 630 Part II
   Resistance Welders. . . . . 630 Part III
X-Ray Equipment, Medical . . . . .Art. 517 Part V
X-Ray Equipment, Nonmedical . . . . Art. 660
   Fixed and Stationary Equipment. . . . . 660.4(A)
   Over 1000 Volts . . . . 660.4(C)
   Portable, Mobile, and Transportable . . . . 660.4(B)
   Rating of Supply Conductors . . . . 660.6(A)
Specific Requirements . . . . .Art. 210 Part II
Summary 210.24
Supplementary Overcurrent Protection 240.10
   Definition . . . . .Art. 100, Part I
   Motor Control Circuits 430.72

Protection of Flexible Cords and Fixture Wires. 240.5
Taps Permitted
   Conductor Protection 210.20
   Conductor Protection . . . . 240.4(E)
   Motor Circuits. 430.53(D)
     Other Loads . . . . .210.19(A)(4) Ex. 1
   Overcurrent Protection. . . . . 240.21(A)
   Protection of Flexible Cords and Fixture Wires. . . . 240.5(B)
   Ranges and Cooking Appliances . . . . . 210.19(A)(3) Ex. 1
   Temporary Installations . . . . .590.4(C)
Uses Permitted of Type SE Cable . . . . 338.10(B)
   Frames of Ranges and Clothes Dryers 250.140
Voltage Drop on . . . . 210.19(A) IN 4
   Fire Pumps. 695.7
   Not Considered for Ampacity Considerations . 310.14(A)(1) IN 1
   Sensitive Electronic Equipment . . . . .647.4(D)
     Cord-Connected Equipment . . . . 647.4(D)(2)
     Fixed Equipment . . . . 647.4(D)(1)
     Utilization Equipment. . . . . 110.3(B)
   Note: Manufacturers of electrical luminaries and other equipment will often specify the minimum acceptable operating voltage for their product to function correctly.
Voltage Limitations. .210.6
   Elevators, etc. 620.3
Water Heaters . . . . 422.10(A)
Pool Heaters 680.10
Storage-Type 422.13

BRANCH CIRCUITS & FEEDERS
Calculation of Loads . . . . Art. 220
   See CALCULATIONS . . . . Ferm's Finder
   Outside Wiring . . . . Art. 225
     Attached to Buildings or Structures 225.11
     Circuit Exits and Entrances 225.11
     Clearance Overhead Conductors and Cables 225.18
       Over 1000 Volts. 225.60
     Clearance from Windows . . . . .225.19(D)(1)
     Clearance above Roofs . . . . 225.19(A)
     Clearance over Swimming Pools 680.9
     Conductor Covering. 225.4
     Entering a Building or Structure 225.11
     Exiting a Building or Structure 225.11
     Minimum Size Conductor 225.5
       Festoon Lighting. . . . . 225.6(B)
       Overhead Spans . . . . 225.6(A)
       Over 1000 Volts. 225.50
     More Than One Building or Other Structure . . Art. 225 Part II
     Open-Conductor Spacing 225.14
     Outdoor Overhead Conductors over 1000 Volts . . . Art. 399
     Point of Attachment . . . . 225.16(A)
     Vegetation Not Permitted as Means of Support 225.26
       Temporary Wiring . . . . 590.4(J)
     Wiring on Outside of Building 225.10
       In Raceways 225.22
       Installation of Types SE and USE Service Cable 338.10(B)(4)(b)
       Mounting Supports 230.51
       Multiconductor Cables 225.21
       Protecting Open Conductors and Cables 230.50
       Use Permitted for Service Entrance Cable . . . . 338.10(B)
   See BRANCH CIRCUITS . . . . Ferm's Finder
   See FEEDERS . . . . Ferm's Finder

BREAKERS & FUSES . . . . **Art. 240**
Accessibility 240.24
   Access to Service Disconnecting Means . . . . . . 230.72(C)
   Circuit Breakers Used as Switches . . . . 404.8(A)
   Location . . . . 230.70(A)
Backfed Overcurrent Devices . . . . 408.36(D)
   Solar Photovoltaic Systems . . . . 710.15(E)
Breakers Used as Switches, Lighting Loads
   HID . . . . 240.83(D)
   SWD . . . . 240.83(D)
Continuous Loads
   Branch Circuits . . . . 210.20(A)
   Feeders. . . . . 215.2(A)(1)
   Fixed Electric Heating Equipment for Pipelines and Vessels . . 427.4
   Fixed Electric Space-Heating Equipment . . . . . . 424.4(B)
   Fixed Outdoor Electric Deicing and Snow-Melting Equipment 426.4
   Non-Motor Operated Appliance . . . . 422.10(A)
   Pool Water Heaters. 680.10
   Solar Photovoltaic Systems . . . . 690.8(B)
   Water Heaters 422.13
Definitions of Different Types . . . . Art. 100 Part I
Delta Breakers (Not Permitted) . . . . 408.36(C)
Grounded Conductor, Overcurrent Device Not in, General . 240.22
   Service Equipment. . . . . 230.90(B)
Identification of 110.22
   Circuits in Switchboards and Panelboards 408.4

In Parallel (Not Permitted) 240.8
Interrupting Capacity 110.9
    Cartridge Fuses & Fuse Holders . . . . Art. 240 Part VI
    Circuit Breakers . . . . .Art. 240 Part VII
    Impedance and Other Characteristics 110.10
    See Short-Circuit Calculations Formula (Point to Point) . . .
    . . . . *Ferm's Charts and Formulas*
    See Short-Circuit Current in Amperes Tables . . . .
    . . . . *Ferm's Charts and Formulas*
    Series Ratings. 240.86
        Selected Under Engineering Supervision, Existing Installations 240.86(A)
Locked or Sealed 230.92
    Specific Circuit Permitted to Be Locked 230.93
Mounting Height . . . . 240.24(A)
Next Higher Size Permitted . . . . 240.4(B)
    Autotransformers. . . . . 450.4(A)
    Motor Circuits . . . . .430.52(C)(1)
    Transformers 450.3
Plug Fuses, Fuseholders, and Adapters . . . . . .Art. 240 Part V
Standard Ratings . . . . 240.6(A)
    Adjustable-Trip Circuit Breakers . . . . 240.6(B)
    Restricted Access . . . . 240.6(C)
Tie Bars (Handle Ties) Required
    Multiwire Branch Circuits Temporary Installations. . 590.4(E)
    Multiwire Circuits, Dwelling Units. . . . . 210.4(B)
    Multiwire Circuits, General 210.7
    Multiwire Circuits Other Than Line-to-Neutral Loads . . . . .
    . . . . 210.4(C) Ex. 2
    Permitted for Specific Line-to-Line Loads . . . . . 240.15(B)
    Service Disconnects, Not Over 1000 Volts 230.74
    Service Disconnects, Over 1000 Volts . . . . . .230.205(B)
    Single-Pole Units for Service Disconnects . . . . . 230.71(B)
    Taps From Grounded Systems 210.10
Up Position — On 240.81
    Circuit Breakers Used as Switches 404.7
Wet Locations 240.32
    Installation of Cabinets or Cutout Boxes 312.2
    In Enclosure or Cabinet 404.4
Wire Bending Space
    At Terminals. . . . . 312.6(B)
    Enclosures for Motor Circuits . . . . 430.10(B)
    Enclosures for Switches and Circuit Breakers. . . . . 404.3(A)
    In an Enclosure Containing a Panelboard 408.55
    Industrial Control Panels . . . . .409.104(B)

See FUSES . . . . *Ferm's Finder*
See Circuit Breakers (DHJR) . . . . *UL Product iQ*
See Fuses (JCQR) . . . . *UL Product iQ*
See OVERCURRENT PROTECTION . . . . . . *Ferm's Finder*

BTU/HR TO TONS FORMULA *Ferm's Charts & Information*

BTU/HR TO WATTS FORMULA *Ferm's Charts & Information*

BUCK AND BOOST TRANSFORMERS
Audio Systems . . . . 640.9(D)
Autotransformers. .450.4
Ballast for Lighting Units 410.138
Branch Circuits 210.9
Feeders 215.11
Grounding Autotransformers 450.5
Motor Starting . . . . 430.82(B)
Overcurrent, Protection of 450.4
See AUTOTRANSFORMERS . . . . *Ferm's Finder*

BULK STORAGE PLANTS . . . . **Art. 515**
Class I Locations, General . . . . 500.5(B)
Class I Locations, Specific 515.3
Conductors (Gas & Oil Resistant) 501.20
    See (ZLGR) . . . . *UL Product iQ*
Gasoline Dispensing at Bulk Stations 515.10
    See Code for Motor Fuel Dispensing Facilities and Repair Garages, NFPA 30A . . . . NFPA 30A
Gasoline Dispensing General . . . . Art. 514
Grounding & Bonding . . . . Art. 250
    Grounding Regardless of Voltage. 515.16
    In Class I, Division 1 & 2 Locations 501.30
    In Class I, Zone 0, 1, & 2 Locations 505.25
    In Hazardous Locations 250.100
    Using Nonmetallic Wiring Methods . . . . 515.8(C)
Lightning Protection (Surge Arresters over 1 kV) . . . . Art. 242
    See Standard for the Installation of Lightning Protection Systems, NFPA 780 . . . . NFPA 780
    Surge Protection 501.35
Sealing, General. 501.15
Sealing, Specific. .515.9
Static Protection
    See Flammable and Combustible Liquids Code, NFPA 30
Wiring Methods
    Above Class I Locations 515.7
    Class l, Division 1 & 2 Locations . . . . Art. 501

Class I, Zone 0, 1, & 2 Locations . . . . Art. 505
Farms, *See Flammable and Combustible Liquids Code, NFPA 30* . . . . NFPA 30
Underground Wiring .515.8
    Seals. 515.9
Within Class I Locations. 515.4
See HAZARDOUS (CLASSIFIED) LOCATIONS Ferm's Finder

BULL SWITCH (Definition) **530.2**
Current-Carrying Parts . . . . 530.15(D)

BURGLAR ALARMS . . . . **Art. 725**
Dwelling Units- Not GFCI Protected. . . 210.8(A)(5) Ex. to (5)
Not Covered by Article 640. . . . . 640.1(B)
Outside Wiring. . . . . Art. 805
See ALARM INDICATION OR SYSTEMS . . . . Ferm's Finder

BURIED CONDUCTORS
Ampacity Calculations, 0 to 2000 Volts Under
    Engineering Supervision . . . . 310.15(B)(1)
        Application Information. . . . . Informative Annex B
Ampacities, 2001 to 35,000 Volts. . . . . 311.60(B)
Backfill for 1000 Volts and Above. . . . . 300.50(E)
Backfill for Under 1000 Volts . . . . .300.5(F)
Bushing on Open End of Conduit . . . . .300.5(H)
Conductors of Same Circuit . . . . 300.5(I)
Cover
    Less Than 1000 Volts . . . . Table 300.5
    1000 Volts and Above . . . . Table 300.50
Directional Boring . . . . 300.5(K)
Earth Movement . . . . 300.5(J)
Conductors 1000 Volts and Above . . . . 300.50(C)
Identified for Such Use . . . . 310.10(E)
    1000 Volts and Above . . . . 300.50(A)
    1000 Volts and Above (Shielding) 311.44
    See OVER 1000 VOLTS, NOMINAL . . . . . Ferm's Finder
Protection of Conductors
    Less Than 1000 Volts. . . . . .300.5(D)
    1000 Volts and Above . . . . 300.50(C)
Sealing Underground Raceways, Feeders 225.27
Sealing Underground Raceway, Services 230.8
    Sealing Underground Raceways, 1000 Volts and Above 300.50(F)
    Sealing Underground Raceways, Under 1000 Volts . 300.5(G)
Splicing of, 1000 Volts and Above . . . . 300.50(D)
Splicing of, Under 1000 Volts. . . . . 300.5(E)

Swimming Pools, Fountains and Similar Installations . . . . . .
. . . . 680.11 and Table 300.5
UF Cable. . . . . Art. 340
    See (YDUX) . . . . *UL Product iQ*
Under Buildings . . . . 300.5(C)
    Service Conductors. 230.6
Under Bulk Storage Plants 515.8
    Sealing 515.9
Under Swimming Pools 680.11
USE & UF . . . . Table 310.4(A)
USE, Type 338.2
    See (TXKT) . . . . *UL Product iQ*

BUSBARS
Aboveground Wiring Method 1000 Volts and Above . . . 300.37
Ampacity for Service Entrance . . . . 230.42(A)
Ampacity of Conductors, Auxiliary Gutters. . 366.23(A) and (B)
Busways . . . . Art. 368
Clearance in Switchboards 408.5
Expansion and Contraction . . . . .366.100(E)
Grounding Electrode Conductor Installations . . . 250.64(C)(2)
Mounting in Switchboards 408.3
    Rigidly Mounted. 408.51
Phase Arrangement, Switchboards and Panelboards. . . 408.3(E)
Support and Arrangement 408.3
Support and Arrangement, Motor Control Centers. . 430.97(A)
    Phase Arrangement. . . . . 430.97(B)

BUSHINGS
At Boxes
    Cabinets, Cutout Boxes, and Meter Socket Enclosures 312.6(C)
    Metal Boxes or Conduit Bodies . . . . 314.17(B)
    On Armored Cable. 320.40
    On Electrical Nonmetallic Tubing 362.46
    On High Density Polyethylene Conduit 353.46
    On Intermediate Metal Conduit 342.46
    On Nonmetallic Underground Conduit with Conductors . 354.46
    On Reinforced Thermosetting Resin Conduit 355.46
    On Rigid Metal Conduit. 344.46
    On Rigid Polyvinyl Chloride Conduit 352.46
At Box Covers & Fittings for Flexible Cord 314.42
    Protection from Damage, Flexible Cords and Cables. . 400.14
Fixed Outdoor Electric Deicing and Snow-Melting Equipment. .
. . . . 426.22(C)
Insulating (4 AWG Conductors & Larger) . . . . . . 300.4(G)

Open End of Conduit Entering Equipment . . . . . 300.16(B)
Open End of Conduit for Support . . . . 300.15(C)
Open End of Conduit Underground . . . . 300.5(H)

BUSWAYS . . . . **Art. 368**
Ampacity for Service Entrance . . . . 230.42(A)
Branches from 368.56
Definition 368.2
Disconnects . . . . 368.17(C)
Expansion Joints
    Over 1000 Volts 368.244
    Provisions for in Auxiliary Gutters 366.44
        Grounding 368.60
Overcurrent Protection (General) 368.17
    Rating for Branch Circuits . . . . 368.17(D)
    Rating for Feeders . . . . 368.17(A)
    Where Required for Reduction in Size . . . . . . 368.17(B)
    Where Used as Feeder or Branch Circuit . . . . . 368.17(C)
Over 1000 Volts, Nominal . . . . 368 Part IV
    Aboveground Wiring Methods 300.37
    Switches 368.239
    See OVER 1000 VOLTS, NOMINAL . . . . . *Ferm's Finder*
    Reduction in Ampacity . . . . 368.17(B)
Sealing of, Over 1000 Volts, Nominal . . . . . . 368.234(A)
    Prevent Accumulation of Flammable Gases 368.238
Service-Entrance Conductors . . . . 230.43(9)
    Permitted as Wiring Method, Over 1000 Volts . . 230.202(B)
Support of 368.30
Terminations and Connections Over 1000 Volts 368.238
Use Permitted 368.10
Uses Not Permitted 368.12
See (CWFT) . . . . *UL Product iQ*

"BX CABLE," *See* **Armored Cable, Type AC** . . . . . . **Art. 320**
See ARMORED CABLE, TYPE AC . . . . *Ferm's Finder*

BYPASS ISOLATION SWITCH
Critical Operations Power Systems . . . . 708.24(B)
    Definition . . . . Art. 100 Part I
    Emergency Systems . . . . 700.5(B)
    Legally Required Standby Systems . . . . 701.5(B)

C

CABINETS, CUTOUT BOXES & METER SOCKET ENCLOSURES . . . . **Art. 312**
Conductors Entering 312.5
Conductors Entering, 4 AWG and Larger . . . . 300.4(G)
Insulating Bushings . . . . 312.6(C)
Mounting of 110.13
Wet Locations 312.2
    Enclosures for Overcurrent Devices 240.32
    Enclosures for Switches and Circuit Breakers 404.4
        Flush-Mounted . . . . 404.4(B)
        Surface-Mounted . . . . 404.4(A)
        Tub and Shower Enclosures . . . . 404.4(C)
Wire Bending Space . . . . 312.6(B)
Conductor Bending Radius Over 1000 Volts 300.34

CABLE END FITTINGS **300.16**

CABLE, ENTERING BOXES **314.17**
Entering Cabinets and Cutout Boxes . . . . 312.5(C)
    Cables 4 AWG and Larger . . . . 300.4(G)
Insulated Fittings . . . . 300.4(G)

CABLE LIMITERS . . . . **230.82(1)**
Permitted without Disconnecting Means 240.40
See (CYMT) . . . . *UL Product iQ*

CABLE MANAGEMENT SYSTEM
Electric Vehicle Supply Equipment 625.2
Electrified Truck Parking Spaces 626.2

CABLE, PHYSICAL PROTECTION
Agricultural Buildings . . . . 547.5(E)
Carnival and Amusement Rides . . . . 525.20(A)&(G)
CATV Systems 800.24
    Grounding Conductor . . . . 820.100(A)(6)
Class 1, 2, & 3 Circuits 725.24
    Safety-Control Circuits . . . . 725.31(B)
Communication Circuits 805.90
    Grounding Electrode Conductors . . . . 800.100(A)(6)
Elevators, Not Required to be in Raceway . . . . . 620.21 Ex.
Equipment Grounding Conductors Smaller Than 6 AWG . . . .
. . . . 250.120(C)
Fire Alarm Systems, Mechanical Execution of Work . . . 760.24
Fire Pump Control Wiring . . . . 695.14(A)

Fire Pump Engine Controllers and Batteries 695.12
Fire Pump Power Wiring . . . . .695.6(D)
Fitting Required Cables Entering or Exiting Conduit . 300.15(C)
General 300.4
Grounding Electrode Conductors. . . . . 250.64(B)
Network-Powered Broadband Systems 830.24
Open Wiring . . . . 398.15(C)
Optical Fiber Cables and Raceways 770.24
Service Entrance Cable . . . . .338.10(B)(4)
Service Entrance Conductors and Cables Aboveground 230.50(B)
Service Entrance Conductors, Underground . . . . . 230.50(A)
Solar Photovoltaic Systems (PV) . . . . 690.31(E)
Temporary Locations . . . . 590.4(J)
Type AC Cable . . . . 320.12(1)
Type AC Cable Exposed Work 320.15
Type AC Cable in Accessible Attics 320.23
Type MC Cable . . . . 330.12(1)
Type MC Cable in Accessible Attics. 330.23
Type MC Cable Through or Parallel to Framing Members 330.17
Type MI Cable Through or Parallel to Framing Members 332.17
Type NM Cable Exposed Work. . . . . 334.15(B)
Type NM Cable in Accessible Attics 334.23
Type NM Cable in Crawl Space . . . . 334.15(C)
Type NM Cable Through or Parallel to Framing Members 334.17
Type NM Cable in Unfinished Basements and Crawl Spaces . . .
. . . . 334.15(C)
Under Roof Decking . . . . 300.4(E)
Underground Installations . . . . .300.5(D)
Underground Service Lateral 230.32
Wind Electric Systems . . . . 694.30(B)

CABLE ROUTING ASSEMBLY
Defined. . . . . Art. 100

CABLE, SECURING AND SUPPORTING OF
Armored Cable 320.30
Communication Circuits 800.24
Exiting From Cable Trays . . . . 392.18(G)
In Vertical Raceways 300.19
Messenger Supported Wiring . . . . Art. 396
Metal-Clad Cable 330.30
Mineral-Insulated, Metal-Sheathed Cable 332.30
Nonmetallic-Sheathed Cable 334.30
PV Cables. . . . . 690.4 & 690.31
Temporary Locations . . . . 590.4(J)

CABLE TIES AND CABLE ACCESSORIES
Armored Cable . . . . 320.30(A)
Audio Signal Processing Cables . . . . 640.6(A)
Class 1, Class 2, and Class 3 Circuits 725.24
Communication Circuits,
   Mechanical Execution 800.24
   Plenum Rated . . . . .800.170(C)
Community Antenna Television and Radio Distribution
Systems 800.24
Electrical Nonmetallic Tubing . . . . 362.30(A)
Fire Alarm Cables . . . . 760.24(A)
Flexible Metal Conduit . . . . 348.30(A)
Liquidtight Flexible Metal Conduit 350.30
Liquidtight Flexible Nonmetallic Conduit 356.30
Low-Voltage Suspended Ceiling Power Distribution Systems . . .
. . . . 393.14(A)
Medium Voltage Cable 311.40
Metal-Clad Cable. . . . . 330.30(A)
Motion Picture and TV Studios
   Portable Wiring . . . . 530.12(B)
Multiple Circuits (Neutral Identifications) . . . . 200.4(B)
Nonmetallic-Sheathed Cable 334.30
Network-Powered Broadband Communications Systems . 830.24
Nonmetallic, Ducts and Plenums . . . . .300.22(C)(1)
Optical Fiber Cables 770.24
Photovoltaic Systems (Conductor Grouping) . . . . 690.31(B)(2)
Signs
   Class 2 (Cables) . . . . 600.33(B)(1)
Temporary Installations (Support) . . . . 590.4(J)

CABLE TRAYS . . . . Art. 392
Above Side Rails Permitted, Splices 392.56
Accessible Splices 392.56
Airfield Lighting Cable Tray. . . . . 392.10(E)
Ampacity of Cables
   2000 Volts or Less . . . . 392.80(A)
   2001 Volts or Over . . . . 392.80(B)
Bushed Conduit and Tubing Transitions 392.46
Combination of Multi and Single Conductor Cables
   2000 Volts or Less . . . . .392.80(A)(3)
   2001 Volts or Over . . . . .392.80(B)(1)
Expansion Splice Plates 392.44
Fill Calculations. 392.22
   Dividers (applied to each side) . . . . 392.22(A)
Grounding and Bonding of 392.60

Installation of Cables . . . . 392.18(B)
Installation of Tray . . . . 392.18(A)
Labeling Where Service and Non-Service Conductors are Present 230.44
Marking-Over 600 Volts . . . . 392.18(H)
    Industrial Establishments . . . . 392.18(H) Ex.
    Requirements for Marking Labels . . . . 110.21(B)
Mixture of Multiconductor Cables
    2000 Volts or Less . . . . .392.80(A)(3)
    2001 Volts or Over . . . . .392.80(B)(1)
Number of Multiconductor Cables
    2000 Volts or Less . . . . .392.80(A)(3)
    2001 Volts or Over . . . . .392.80(B)(1)
Number of Single Conductors
    2000 Volts or Less . . . . .392.80(A)(2)
    2001 Volts or Over . . . . .392.80(B)(2)
Over 600 Volts in Same Cable Tray . . . . 230.44 Ex.
Power & Control Tray Cable . . . . Art. 336
Service Entrance Conductors 230.44
Single Conductors (1/0 AWG or Larger) . . . . . 392.10(B)(1)
Splices Permitted 392.56
Temporary Wiring in Assembly Occupancies . . . . 518.3(B) Ex.
Under 1000 Volts in Same Cable Tray . . . . 230.44 Ex.
Used as Supports . . . . 392.18(G)
Uses Not Permitted 392.12
Uses Permitted 392.10
Welding Cables 630.42

CABLEBUS . . . . **Art. 370**
Ampacity 370.80
Definition of 370.2
Fittings 370.42
Grounding 370.60
Marking 370.120
Overcurrent Protection 370.23
Support . . . . 370.30(A)
    Conductor . . . . 370.30(B)
Uses, Not Permitted 370.12
Uses, Permitted 370.10

CABLES
Abandoned
    Abandoned Audio Distribution Cable . . . . 640.6(C)
        Definition 640.2
    Abandoned Class 2, Class3, and PLTC Cable 725.25
        Applications 725.154
        Definition 725.2
    Abandoned Coaxial Cable 800.25
        Applications 820.154
        Definition 800.2
    Abandoned Communication Cable 800.25
        Applications 800.154
        Definition 800.2
    Abandoned Fire Alarm Cable 760.25
        Applications 760.154
        Definition 760.2
    Abandoned Optical Fiber Cable 770.25
        Applications 770.154
        Definition 770.2
Circuit Conductors
    Cellular Concrete Floor Raceways 372.58
    Cellular Metal Floor Raceways 374.58
Temporary Wiring . . . . 590.3(D)
Under Raised Floors . . . . 645.5(G)
Aerial Entrance . . . . 820.133(B) Ex.
    Lead-in Clearance . . . . 800.44(A)(4)
Aerial, Network-Powered Broadband Systems 800.44
Armored, Type AC . . . . Art. 320
Border Lights, Theater 520.44
Bundled
    Definition of 520.2
    Derating Required . . . . 310.15(C)(1)
Cable Trays 392.10
Carnivals, Circuses, Fairs and Similar 525.20
Circuit Integrity Cables (Communications Cables) 725.179(F)
Circuit Integrity Cables, (Communications, Definition of) 800.2
Circuit Integrity Cables (Fire Alarm Cables) . . . . . 760.176(F)
Circuit Integrity Cables, (Fire Alarm, Definition of) 760.2
Circuit Integrity Cables (Remote Control and Signal Cables) . . . . .725.179(F)
Circuit Integrity Cables, (Definition of) . . . . Art. 100
Class 1, 2 & 3 Remote Control and Signaling . . . . . Art. 725
Community Antenna TV & Radio Distribution Systems Art. 820
Continuity 300.12
Electric Vehicle Charging Systems 625.17
Elevators, Not Required in Raceways . . . . 620.21 Ex.
Fire Alarm System Cables . . . . Art. 760
Flat Cable Assemblies, Type FC . . . . Art. 322
Flat Conductor, Type FCC . . . . Art. 324

Grouped 520.2
Heating, *See* HEATING CABLES . . . . .Ferm's Finder
Installed Under Floor Coverings 424.45
Installed Under Roof Decking . . . . 300.4(E)
Instrumentation Tray Cable, Type ITC . . . . Art. 727
Medium Voltage Cable, Type MV . . . . Art. 328
Metal-Clad, Type MC . . . . Art. 330
Mineral-Insulated, Metal-Sheathed Type MI . . . . . . Art. 332
Nonmetallic Extensions . . . . Art. 382
Nonmetallic-Sheathed Cable, Types NM, NMC and NMS Art. 334
Optical Fiber . . . . Art. 770
Power and Control Tray, Type TC . . . . Art. 336
Protection Against Physical Damage 300.4
Roof Decking, Installed Under . . . . 300.4(E)
Rooftops, Sunlight and Ambient Temperature Correction . . . .
. . . . .310.15(B)(2)
Sealing, Hazardous (Classified) . . . . 501.15(D)&(E)
Secured . . . . 300.11(A)
Service Entrance Cable, Types SE and USE . . . . . . Art. 338
Shallow Grooves . . . . 300.4(F)
Splices in Boxes 300.15
Stage . . . . 530.18(A)
Supported from Cable Trays . . . . 392.10(A)
Through Studs, Joists, Rafters 300.4
Underground . . . . Art. 230 Part III
General. 300.5
Identified for Such Use . . . . 310.10(F)
Over 1000 Volts 300.50
Splicing Means Listed . . . . 110.14(B)
Underground Feeder and Branch-Circuit, Type UF . . Art. 340
Under Floor Coverings (Space Heating) 424.45

CALCULATIONS . . . . **Art. 220**
Air Conditioning Equipment . . . . Art. 440
Branch Circuit Loads. . . . . Art. 440 Part IV
Ampacity Calculation Application Information. . . . .
Informative Annex B
Appliance Loads 210.23
Branch Circuit Computations 220.10
Branch Circuit Ratings. 422.10
Dryers, Dwelling Units. 220.54
Laundry Circuit, Dwellings . . . . 220.52(B)
Maximum Loads 220.18
More Than Four, Dwelling Units. 220.53
Ranges, Dwelling Units 220.55
Small Appliance Loads, Dwelling Units . . . . . 220.52(A)
Storage-Type Water Heaters 422.13
Ballasts . . . . 220.18(B)
Commercial Kitchen Equipment 220.56
Optional Calculation New Restaurants. 220.88
Continuous Loads (Definition) . . . . Art. 100 Part I
Branch-Circuit Conductors . . . . 210.19(A)(1)
Branch-Circuit Overcurrent Device . . . . 210.20(A)
Electric Vehicle Supply Equipment. 625.17
Feeder Conductors . . . . 215.2(A)(1)
Feeder Overcurrent Device. 215.3
Fixed Electric Space-Heating . . . . 424.4(B)
Individual Branch Circuit, Appliances . . . . 422.10(A)
Pool Water Heaters. 680.10
Storage-Type Water Heaters 422.13
Wind Electric Systems 694.12
Cranes & Hoists . . . . 610.14(E)
Demand Factors
Appliance and Laundry Loads (Dwellings) 220.52
Four or More on Same Circuit 220.53
Permitted to Apply Table 220.42 . . . . 220.52
Commercial Cooking Equipment . . . . Table 220.56
Optional Calculation New Restaurant. 220.88
Cranes and Hoists . . . . Table 610.14(E)
Dryers, Clothes (Dwellings) 220.54
Elevators 620.13
Feeder Demand Factor 620.14
Motor Circuit Conductors. . . . . 430 Part II
Farm Loads
Dwelling Unit. . . . . 220.102(A)
Other Than Dwelling Unit . . . . 220.102(B)
Total. 220.103
Household Electric Range . . . . Table 220.55
LED . . . . 220.18(B)
Lighting Load Demand Factors . . . . Table 220.42
Marinas and Boatyards. 555.6
Mobile Home Parks . . . . Table 550.31
Mobile Homes. 550.18
Motors . . . . .Art. 430 Part II
Multifamily Dwellings, More Than Three Units
(Optional) . . . . Table 220.84
Park Trailers 552.47
Park Trailers DC Converter Rating. . . . . 552.20(B)
Receptacle Loads (Dwellings) . . . . Table 220.12

Dwelling Occupancies . . . . .220.14(J)
    Included with General Lighting Loads . . . . Table 220.42
    Small Appliance & Laundry Included in General Load 220.52
Receptacle Loads (Non-Dwelling) . . . . .220.14(I)
    Demand Factor Receptacles Only . . . . Table 220.44
    Permitted Demand Factor with General Lighting . . 220.44
    Permitted To Include with General Lighting. . Table 220.42
    Unit Loads. . . . . Table 220.12 Note
Recreational Vehicles DC Converter Rating . . . . 551.20(B)
Recreational Vehicle Parks . . . . Table 551.73(A)
Restaurants, New (Optional) 220.88
Schools, Service Conductors or Feeders (Optional). . . 220.86
Schools, Service Conductors or Feeders (Standard) . .Art. 220 Part II
See Additional Loads
    All Occupancies, Optional 220.87
    Dwellings, Optional 220.82
    Dwellings, Standard (Additions to Existing Installations) . . 220.16(A)
    Other Than Dwellings, Standard (Additions to Existing Installations) . . . . 220.16(B)
    See Optional Calculations . . . . Art. 220 Part IV
Studio or Stage Set Lighting 530.19
Welders
    Arc Welders 630.11
    Resistance Welders. 630.31
Dwelling Unit . . . . Art. 220
    Additional Loads to Existing Unit (Standard) . . . 220.16(A)
        Optional Method 220.83
        Permitted 220.87
    Optional Method . . . . Art. 220 Part IV
        Multifamily, Three or More Units 220.84
        Single Unit 220.82
        Two Units 220.85
    Standard Method. . . . . .Art. 220 Part II
Electric Heat 220.51
    Optional Dwelling Units. . . . . 220.82(C)
    Optional for Adding Loads in Existing. . . . . . . 220.83(B)
    Optional, Multifamily 220.84
    See ELECTRIC HEAT (SPACE). . . . . .Ferm's Finder
Electric Vehicles. . . . . Art. 625
    Other Applicable Articles . . . . Table 220.3
Elevators, Escalators, Dumbwaiters & Moving Walks. . . 620.13
    Feeders 620.14
    Motor-Circuit Conductors . . . . .Art. 430 Part II
    Motor Controller Rating 620.15
Examples . . . . Informative Annex D
Farm Loads . . . . .Art. 220 Part V
Feeders . . . . .Art. 220 Part II
    Examples. . . . . Informative Annex D
    Optional Methods . . . . Art. 220 Part IV
        Fractions of an Ampere . . . . 220.5(B)
Hotels 220.12
    General Lighting Demand Factors 220.42
Laundry Loads (Dwellings) . . . . 220.52(B)
    Electric Dryers 220.54
Lighting Loads 220.12
    Permissible Demand Factors 220.42
Lighting, Track . . . . 220.43(B)
Marinas and Boatyards 555.6
Mobile Homes 550.18
Mobile Home Parks. 550.31
Motels 220.12
    General Lighting Demand Permitted. 220.42
Motors . . . . .Art. 430 Part II
Multifamily Dwelling (Optional Method) 220.84
Multioutlet Assemblies . . . . 220.14(H)
Neutral Feeder Load 220.61
    Adjustment Factor . . . . 310.15(C)
        When Considered Current-Carrying . . . . . 310.15(E)
Noncoincident Loads 220.60
Ranges (Household) . . . . .210.19(A)(3)
    Branch Circuit Loads. . . . . 220.14(B)
    Commercial 220.56
    Individual Circuits . . . . 422.10(A)
    Optional Calculation New Restaurants. 220.88
    Permissible Demands (Dwelling Units) 220.55
    Receptacles . . . . .210.21(B)(4)
Receptacle Outlets
    Dwelling Units . . . . .220.14(J)
    Nondwelling Units. . . . . .220.14(I)
        Demand Factor Permitted 220.44
        See Note [b] Under Table 220.12 . . . . 220.14(K)
Recreational Vehicle Parks. 551.73
Recreational Vehicles 551.42
Schools (Optional Method) 220.86
Show Windows. . . . . 220.14(G)
    Load Per Linear Foot . . . . 220.43(A)
Signs and Outline Lighting . . . . 220.14(F)

Solar Photovoltaic Systems . . . . .Art. 690 Part II
Track Lighting . . . . 220.43(B)
Voltages. . . . . 220.5(A)
Water Heaters
    Individual Circuit for. . . . . 422.10(A)
        Pool Water Heaters 680.10
        Storage-Type. 422.13
Welders, Electric . . . . Art. 630
    Arc Welders 630.11
    Resistance Welders 630.31
Wind Electric Systems 694.12
See DEMAND FACTORS . . . . Ferm's Finder

CAMPING TRAILER (Definition of) 551.2

CANOPIES
Attachment Methods (Screws) 314.25
Boxes and Fittings 314.25
    Box Fill Calculation . . . . .314.16(B)(1) Ex.
Combustible Finishes
    Canopies at Boxes . . . . 314.25(B)
    Canopies at (Fixtures) Luminaires 410.23
Conductors, Space for 410.20
Cover, at Boxes 410.22
Electric Discharge Luminaires (Fixtures) . . .410.62(C)(1)(2)(c)
Live Parts Exposed 410.5
See Capacitance (Formulas) . . . . . . Ferm's Charts and Formulas

CAPACITORS . . . . Art. 460
1000 Volts and Under. . . . . Art. 460 Part I
    Ampacity of Conductors . . . . 460.8(A)
    See . . . . Ferm's Charts and Formulas
    Discharge of Stored Energy 460.6
    Disconnecting Means . . . . 460.8(C)
    See . . . . Ferm's Charts and Formulas
    Grounding 460.10
    Marking 460.12
    Motor Overload Device 460.9
    Overcurrent Protection . . . . 460.8(B)
    With Motors 430.27
Accidental Contact . . . . 460.2(B)
Containing More than 11 L (3 Gal) Flammable Liquid 460.2(A)
Hazardous (Classified) Locations
    Class I 501.100
    Class II. 502.100
    Class III 503.100
Over 1000 Volts . . . . .Art. 460 Part II
    Grounding 460.27
    Identification 460.26
    Means for Discharge 460.28
    Overcurrent Protection. 460.25
    Switching. 460.24
        Isolation . . . . 460.24(B)
        Load Current . . . . 460.24(A)
        Series Capacitors. . . . . 460.24(C)
See (CYWT) . . . . UL Product iQ

CARNIVALS, CIRCUSES, FAIRS, AND SIMILAR EVENTS . Art. 525
Attractions Using Pools, Fountains, etc. . . . . 525.3(D)
Disconnecting Means . . . . 525.21(A)
GFCI Protection 525.23
    General-Use 15- and 20-ampere, 125-volt . . . . 525.23(A)
    Not Permitted . . . . 525.23(C)
    Not Required . . . . 525.23(B)
    Receptacles Supplied by Portable Cords . . . . . 525.23(D)
Grounding and Bonding . . . . Art. 525 Part IV
    Equipment Bonding 525.30
    Equipment Grounding 525.31
    Equipment Grounding Conductor Continuity Assurance . 525.32
Overhead Conductor Clearances 525.5
    Portable Structure Clearances
        Over 600 Volts . . . . 525.5(B)(2)
        Under 600 Volts . . . . 525.5(B)(1)
    Vertical Clearances . . . . 525.5(A)
Permanent Amusement Attractions . . . . Art. 522
Portable Distribution or Termination Boxes 525.22
Portable Wiring Inside Tents and Concessions . . . . 525.21(B)
Power Sources 525.10
Protection of Electrical Equipment 525.6
Wiring Methods . . . . Art. 525 Part III
    Boxes and Fittings . . . . 525.20(H)
    Cord Connectors . . . . 525.20(E)
    Flexible Cord . . . . 525.20(A)
    Open Conductors . . . . 525.20(C)
    Protection . . . . 525.20(G)
    Single-Conductor . . . . 525.20(B)
    Splices . . . . 525.20(D)
    Support . . . . 525.20(F)

CARTRIDGE FUSES — Art. 240 Part VI
Class I, Division 2 Locations . . . . 501.115(B)(3)
  Lighting (Fixtures) Luminaires . . . . 501.115(B)(4)
Classification of 240.61
Disconnecting Means 240.40
Marking . . . . 240.60(C)
Maximum Voltage, 300-Volt Type . . . . 240.60(A)
Noninterchangeable . . . . 240.60(B)
Renewable Fuses . . . . 240.60(D)
Supplementary Overcurrent Protection, Fixed Electric Space Heating
. . . . 424.22(C)

CATV SYSTEMS . . . . Art. 820
See COMMUNITY ANTENNA TELEVISION AND RADIO DISTRIBUTION SYSTEMS . . . . Ferm's Finder

CAUTION SIGNAGE
Electric Deicing and Snow-Melting Equipment, Fixed Outdoor 426.13
Engineered Series Combination Systems . . . . . . 110.22(B)
High-Impedance Grounded Neutral AC System . . 408.3(F)(3)
High-Leg Identification . . . . 408.3(F)(1)
Lighting Systems Operating Over 1000 Volts 410.146
Pipeline and Vessels, Fixed Electric Heating 427.13
Requirements . . . . 110.21(B)
Resistively Grounded DC Systems . . . . 408.3(F)(5)
Tested Series Combination Systems . . . . 110.22(C)
Ungrounded AC Systems . . . . 408.3(F)(2)
Ungrounded DC Systems . . . . 408.3(F)(4)
Ungrounded Systems . . . . 250.21(C)
Unqualified Persons . . . . 110.31(B)(1)

CEILING GRID, LOW-VOLTAGE SUSPENDED POWER DISTRIBUTION SYSTEMS
Conductor Sizes and Types 393.104
Connections 393.57
Connectors . . . . 393.40(A)
Definitions 393.2
Disconnecting Means 393.21
Enclosures . . . . 393.40(B)
Grounding 393.60
Installation 393.14
Interconnection of Power Sources . . . . 393.45(B)
Listing Requirements 393.6
Overcurrent Protection . . . . 393.45(A)
Reverse Polarity . . . . 393.45(C)
Scope 393.1
Securing and Supporting 393.30
Splices 393.56
Uses Not Permitted 393.12
Uses Permitted 393.10

CEILING OUTLET (BOX) . . . . 314.27(A)(2)

CEILING-SUSPENDED (PADDLE) FANS
Bathtubs and Shower Areas, Above . . . . 410.10(D)
Outlet Box, Support of . . . . 314.27(C)
Spare Separately Switched Ungrounded Conductors . 314.27(C)
Spas and Hot Tubs, Indoor . . . . 680.43(B)(1)
Support of 422.18
Swimming Pools . . . . 680.22(B)

CELL
Cellular Concrete Floor Raceways
  Definition 372.2
Cellular Metal Floor Raceways
  Definition 374.2
Electrolytic Cells . . . . Art. 668
Fuel Cell Systems . . . . Art. 692
  See FUEL CELL SYSTEMS . . . . Ferm's Finder
Solar Cell, Definition of 690.2
Storage Batteries . . . . Art. 480

CELL LINE, ELECTROLYTIC CELLS . . . . Art. 668

CELLARS See BASEMENT . . . . Ferm's Finder

CELLULAR CONCRETE FLOOR RACEWAYS . . . Art. 372
Connection to Cabinets and Other Enclosures . . . 372.18(B)
Definitions 372.2
Discontinued Outlets 372.58
Header . . . . 372.18(A)
Inserts . . . . 372.18(D)
Installation 372.18
Junction Boxes . . . . 372.18(C)
Markers . . . . 372.18(E)
Number of Conductors 372.22
Size of Conductors 372.20
Splices and Taps 372.56
Uses Not Permitted 372.12
  Installation of Conductors with Other Systems 300.8

CELLULAR METAL FLOOR RACEWAYS . . . . . . **Art. 374**
  Construction Specifications 374.100
  Definitions 374.2
  Discontinued Outlets 374.58
  Inserts . . . . 374.18(C)
  Installation 374.18
  Junction Boxes . . . . 374.18(B)
  Listed 374.6
  Markers . . . . 374.18(D)
  Number of Conductors 374.22
  Size of Conductors 374.20
  Splices and Taps 374.56
  Uses Not Permitted 374.12
    Installation of Conductors with Other Systems 300.8

CENTER PIVOT IRRIGATION MACHINES
. . . . *See* **ELECTRICALLY DRIVEN**
OR CONTROLLED IRRIGATION MACHINES . . . .**Art. 675**

CENTIGRADE TO FAHRENHEIT Tables 310.16 through 310.21

CHAIR LIFTS . . . . **Art. 620**

CHILD CARE FACILITIES
  Definition 406.2
  Tamper-Resistant Receptacles . . . . 406.12(3)

CHURCHES . . . . **Art. 518**

CINDER FILL
  Electrical Metallic Tubing . . . . 358.10(C)
  Intermediate Metal Conduit . . . . 342.10(C)
  High Density Polyethylene Conduit Type HDPE . . . 353.10(3)
  Nonmetallic-Underground Conduit with Conductors 354.10(3)
  Reinforced Thermosetting Resin Conduit . . . . . . 355.10(C)
  Rigid Metal Conduit . . . . 344.10(C)
  Rigid PVC Conduit . . . . 352.10(C)

CIRCUIT BREAKERS . . . . **Art. 240 Part VII**
  Accessibility and Grouping, Where Used as Switches . . 404.8(A)
  Adjustable Trip . . . . 240.6(B)
    Restricted Access Adjustable-Trip . . . . 240.6(C)
  Applications, 1-Phase, 3-Phase, Slash Rating 240.85
  Ampere Ratings . . . . Table 240.6(A)
    Adjustable-Trip Circuit Breakers . . . . 240.6(B)
    Fuses and Fixed-Trip Circuit Breakers . . . . . . 240.6(A)
    Restricted Access Adjustable-Trip Circuit Breakers . . 240.6(C)
    Table for Standard Ampere Ratings . . . . . . Table 240.6(A)

Back Fed Breakers
  Overcurrent Protection . . . . 408.36(D)
  Solar Photovoltaic . . . . 710.15(E)
Back Fed due to Loss of Primary Source of Power
  Electric Vehicle Charging Systems 625.46
  Electrified Truck Parking Places 626.26
Circuits Over 1000 Volts . . . . 490.21(A)
Definition of . . . . Art. 100 Part I
Disconnection of Grounded Conductors . . . . 404.2(B)
  For Dispensing Equipment . . . . 514.11(A)
Enclosures . . . . 404.3(A)
Interrupting Rating 110.9
  Required Marking . . . . 240.83(C)
Marking 240.83
Nontamperable 240.82
Overcurrent Device, as . . . . 240.15(B)
Overcurrent Protection Over 1000 Volts . . . . Art. 240 Part IX
Parallel, Use in 240.8
Rating, Nonadjustable Trip . . . . 240.6(A)
  See BREAKERS AND FUSES . . . . Ferm's Finder
  See (DHJR) . . . . UL Product iQ
Series-Combination Ratings, Identification 110.22
Series Ratings 240.86
Short-Circuit Current Rating 110.10
Transformers, Circuit Breaker Ratings 450.3
  1000 Volts and Less . . . . Table 450.3(B)
  Over 1000 Volts . . . . Table 450.3(A)
Transient Voltage Surge Suppressor
  See (DIMV) . . . . UL Product iQ
Used As Switches, Lighting Loads
  HID . . . . 240.83(D)
  SWD . . . . 240.83(D)

CIRCUIT, CONTROL *See* **CONTROL CIRCUITS** . . . . . .
. . . . Ferm's Finder

CIRCUIT INTEGRITY CABLE
Definition
  Class 1, Class 2, and Class 3 Circuits 725.2
  Communication Systems 800.2
Fiber Optic Cables . . . . 770.179(E)(1)
Listing for PLTC . . . . 725.179(F)(1)

Mechanical Execution of Work . . . . 760.24(B)
Power Limited Fire Alarm Circuits . . . . 760.176(F)(1)
Survivability Characteristics, Communication Circuits . . . . . .
. . . . 800.179(G)(1)

CIRCUITS
Identification of 110.22
    In Switchboards and Panelboards (Field Required) 408.4
Low Voltage
    Class 1, 2 & 3 Systems . . . . Art. 725
    Communications Circuits . . . . Art. 805
    Fire Alarms . . . . Art. 760
    Less than 50 Volts . . . . Art. 720
    Low Voltage Lighting . . . . Art. 411
    Sound Recording. . . . . Art. 640
    See LOW VOLTAGE SYSTEMS. . . . . Ferm's Finder

Number Per Building
Allowed, General. 225.30
Required, Dwelling Unit Bathroom . . . . .210.11(C)(3)
Required, Dwelling Unit Laundry . . . . .210.11(C)(2)
Required, Dwelling Unit Small-Appliance . . . . .210.11(C)(1)
Required, General 210.11
Use of Demand Factors Not for Number of 220.42
See BRANCH CIRCUITS . . . . .Ferm's Finder

CIRCULAR MIL AREA OF CONDUCTORS . . . .
Chapter 9 Table 8
See BUSBARS. Ferm's Finder

CIRCUSES . . . . Art. 525
See CARNIVALS, CIRCUSES, FAIRS, AND SIMILAR EVENTS
. . . . Ferm's Finder

CLAMP FILL, BOXES . . . . 314.16(B)(2)

CLAMPS, GROUND 250.8
Electrodes, Connection to 250.70
Protection of Attachment. 250.10

CLASS I, II, AND III LOCATIONS (HAZARDOUS)
See HAZARDOUS (CLASSIFIED) LOCATIONS Ferm's Finder

CLASS 1, 2 & 3 REMOTE CONTROL CIRCUITS . Art. 725
Abandoned Cables 725.25
Applications in Buildings 725.154
Circuit Identification 725.30
Classification, General Definition 725.2
    Class 1 . . . . 725.41
    Class 2 & 3. . . . . 725.121
Conductors
    Copper Conductors for Control Circuit Devices . 430.9(B)
Definitions 725.2
Grounding. . . . . Art. 250
Physical Protection . . . . 725.31(B)
Power Source Limitations . . . .
Chapter 9 Tables 11(A)&(B)
Safety-Control Equipment 725.31
Class 1 Circuits . . . . .Art. 725 Part II
Circuits Beyond Building 725.52
    As Messenger Supported Wiring . . . . Art. 396
Different Systems in Same Enclosure or Raceway . . . 725.48
Mechanical Execution of Work. 725.24
Motor Control Circuits . . . . Art. 430 Part VI
    Copper Conductors . . . . 430.9(B)
    Mechanical Protection of Conductors Physical Damage 430.73
    Overcurrent Protection 430.72
    Ungrounded for Integrated Electrical Systems 685.14
    Where Not Required to Be Grounded 250.21
Number in Raceway, and Derating. 725.51
Overcurrent Protection 725.43
    Device Location 725.45
Size. . . . . 725.49(A)
Wiring Methods 725.46
Class 2 & 3 Circuits . . . . Art. 725 Part III
Applications of Listed Cables. 725.154
Circuit Marking 725.124
Different Systems in Same Enclosure. 725.139
Industrial Establishments . . . . 725.135(J)
Installation 725.133
Limits – Voltage & Current 725.121
Listing and Marking of. 725.179
Power Limitations (AC) . . . . Chapter 9 Table 11(A)
Power Limitations (DC) . . . . Chapter 9 Table 11(B)
Separation of Circuits 725.136
Uses Permitted and Substitutions. . . . . .725.154(A)
Vertical Support for Fire-Rated Cables and Cond. . . 725.3(I)
Wiring Methods, Load Side of Power Source 725.130
Wiring Methods, Supply Side of Power Source 725.127
Cable Routing Assemblies. . . . . .725.3(M)
Communications Raceways . . . . .725.3(N)
Identification of Circuits 725.30

Installation Requirements. 725.135
Limited Power (LP) Cables . . . . 725.179(I)
Listing Requirements . . . . Art. 725 Part IV
    Class 2, Class 3 and PLTC Cables 725.179
    Equipment (Powered Devices) 725.170
Mechanical Execution of Work 725.24
    See LOW VOLTAGE SYSTEMS. . . . . Ferm's Finder
Transmission of Power and Data 725.144

CLASS 2 & 3 TRANSFORMERS
See (XNWX) . . . . UL Product iQ

CLEAN SURFACES, GROUNDING 250.12

CLEARANCES
Clearance & Head Room about Equipment 110.26
    Switchboards and Panelboards 408.18
Clearance of Outside Branch Circuits. . . . . Art. 225
    From Buildings. 225.19
    For Overhead Conductors and Cables 225.18
Clearance of Service Drops
    Carnivals, Circuses, etc. 525.5
        Clearances to Portable Structures . . . . 525.5(B)
        Vertical Clearances. . . . . 525.5(A)
    From Building Openings. . . . . 230.9(C)
    Other Than Service Cables on Insulators . . . . . . 230.51(B)
    Overhead Service Drop Conductors 230.24
    Point of Attachment Below Service Head . . . . . 230.54(C)
    Recreational Vehicles 551.79
    Swimming Pools 680.9
        Communication Systems . . . . 680.9(B)
        Network-Powered Broadband Communication Systems. . .
        . . . . 680.9(C)
        Power, (Not in and Enclosed Raceway) . . . . . . 680.9(A)
Elevation & Clearance of Electric Signs 600.9
Fire Ladders . . . . 225.19(E)
High Voltage 490.34
    See Life Safety Code, NFPA 101 . . . . NFPA 101
Live Bare Parts
    Bus Enclosures. 408.5
    In Auxiliary Gutters . . . . .366.100(E)
    In Cabinets, Cutout Boxes and Meter Enclosures 312.11(A)(3)
    Over 1000 Volts 110.34
        Minimum Separation 490.24
    Under 1000 Volts 110.26

    Switchboards and Panelboards 408.56
Overhead Conductors
    Carnivals, Circuses, etc. 525.5
    In Recreational Vehicle Parks 551.79
    Clearance from Ground 225.18
    Railroad Tracks. . . . . 230.24(B)(5)
Overhead Open Conductors 225.14
    Carnivals, Circuses, etc. 525.5
    CATV Systems. . . . . .Art. 820 Part II
    Communications Systems . . . . .Art. 800 Part II
    From Buildings. 225.19
    From Ground 225.18
    Network-Powered Broadband Systems . . . . .Art. 830 Part II
    Overhead Feeder Conductors and Cables 225.18
    Overhead Service Drop Conductors 230.24
        Individual Open Conductors . . . . Table 230.51(C)
    Point of Attachment Below Service Head and Goosenecks . . .
    . . . . 230.54(C)
    Recreational Vehicles 551.79
    Radio and Television Equipment. . . . . . . .Art. 810 Part II
        Amateur Station Antenna Systems. . . . . Art. 810 Part III
    Swimming Pools 680.9
        Communication Systems . . . . 680.9(B)
        Network-Powered Broadband Communication Systems. . .
        . . . . 680.9(C)
        Power (Not in an Enclosed Raceway) . . . . . . 680.9(A)
Over 1000 Volts
    Entrance to Work Space 110.33
    From Non-Electrical Pipes and Ducts . . . . 110.34(F)
    Work Space (General) 110.34
    Work Space about Equipment 110.32
    See Life Safety Code, NFPA 101 . . . . NFPA 101
Underground Wiring from Swimming Pools . . . 680.11, 300.5

CLINICS . . . . Art. 517 Part II
Essential Electrical Systems, in 517.45
Example of (Health Care Facilities) . . 517.2 Informational Note
See HEALTH CARE FACILITIES . . . . Ferm's Finder
See Standard for Health Care Facilities, NFPA 99 . . . .NFPA 99
Tamper-Resistant Receptacles. . . . . 406.12(5)

CLOCK OUTLET . . . . 210.52(B)(2) Ex. 1

CLOTHES CLOSET
Definition. . . . . Art. 100

Luminaire (Fixture) Types Not Permitted . . . . 410.16(B)
Luminaire (Fixture) Types Permitted . . . . 410.16(A)
Location of Luminaires (Fixtures) . . . . 410.16(C)
Overcurrent Devices Prohibited. . . . . 240.24(D)
Storage Space, *See* Clearance from Different Types of Luminaires
. . . . Figure 410.2

CLOTHES DRYERS
Calculations for Dwelling Units. 220.54
Calculations for Other than Dwelling Units . . . . 220.14(B)
Feeder Demand Factors . . . . Table 220.54
Grounding the Frame of 250.140
Grounding by Grounded Conductor . . . . . 250.142(B) Ex 1
Mobile Homes, Wiring Methods . . . . 550.15(E)
Insulated Neutral. . . . . 550.16(A)
Park Trailers. . . . . 552.55(C)
Recreational Vehicles, Insulated Neutral . . . . . 551.54(C)

CLOTHING MANUFACTURING PLANTS
(Ignitable Fibers) . . . . 500.5(D)
Hazardous (Classified) Locations (Class III). . . . . . Art. 503

CO/ALR MARKING
Receptacles . . . . .406. 3(C)
See (RTRT). . . . . *UL Product iQ*
Switches . . . . 404.14(C)
See (WJQR) . . . . *UL Product iQ*

COAXIAL CABLE
CATV Systems . . . . Art. 820
Communications Systems . . . . Art. 805
Definition. . . . . Art. 100
Fire Alarm Systems . . . . 760.179(H)
Network-Powered Broadband Communication Systems Art. 830
Radio and Television Equipment . . . . Art. 810

CODE ARRANGEMENT . . . . 90.3 and FIGURE 90.3

COLOR CODING
Equipment Grounding Conductor . . . . 210.5(B)
Conductor Identification 200.6
Equipment Grounding Conductors. 250.119
For Branch Circuits . . . . 210.5(B)
Flexible Cords. Art. 400 Part II
Grounded Conductors .200.6
For Branch Circuits . . . . 210.5(A)
Means of Identifying .200.6

Heating Cables 424.35
Higher Voltage to Ground (Delta), General 110.15
Delta Service, Midpoint Grounded. 230.56
Intrinsically Safe Systems . . . . 504.80(C)
Mobile Home Power Supply
Cord (Power Supply). . . . . 550.10(B)
Feeders . . . . 550.33(A)
Mast Weatherhead or Raceway . . . . .550.10(I)
Multiwire and Individual Branch Circuits . . . . . . 210.5(C)
Park Trailer Power Supply
Cord (Power Supply). . . . . 552.43(B)
Feeders . . . . 552.43(A)
Mast Weatherhead or Raceway . . . . 552.43(C)
Sensitive Electronic Equipment . . . . 647.4(C)
Trailing Cable, Marina Hoists, Cranes, etc. . . . . . 555.34(B)(3)
Ungrounded Conductors . . . . 210.5(C)
Isolated Power Systems, Health Care Facilities . 517.160(A)(5)

COMBUSTIBLE DUSTS . . . . Art. 502
Definition. . . . . Art. 100
General . . . . 500.6(B)

COMMERCIAL GARAGES . . . . Art. 511
Battery Charging Equipment . . . . 511.10(A)
Class I Locations . . . . 500.5(B)
Area Classification 511.3
Electric Vehicle Charging . . . . 511.10(B)
Elevators & Escalators in 620.38
Equipment above Class I Locations . . . . 511.7(B)
Equipment in Class I Locations 511.4
Fuel Dispensing Units . . . . 511.4(B)(1)
Ground-Fault Circuit-Interrupter Protection 511.12
Grounding and Bonding . . . . Art. 250
Grounding 511.16
Hazardous (Classified) Locations. 250.100
Special Requirements. 501.30
Portable Hand Lamps . . . . 511.4(B)(2)
Sealing 501.15
At Horizontal and Vertical Boundaries 511.9
Process Sealing 501.17
Surge Protection Class I Locations 501.35
*See Standard for the Installation of Lightning Protection Systems,
NFPA 780* . . . . NFPA 780
Wiring Methods
Above Class I Locations . . . . 511.7(A)

In Class I Locations (General) 501.10
In Class I Locations . . . . 511.4(A)
Major Repair Garages . . . . 511.3(D)
Major and Minor Repair Garages . . . . 511.3(C)
Underground Wiring. 511.8
Modifications to Classification . . . . 511.3(E)
Parking Garages. 511.3(A)
Repair Garages with Dispensing . . . . 511.3(B)
Underground Wiring 511.8
Under Hazardous Areas. . . . . 511.4(A)
See HAZARDOUS (CLASSIFIED) LOCATIONS Ferm's Finder
See Code for Motor Fuel Dispensing Facilities and Repair Garages

COMMON AREA BRANCH CIRCUITS . . . . . . 210.25(B)

COMMON GROUNDING ELECTRODE 250.58
Agricultural Buildings . . . . 547.9(B)(3)
Bonding of Made and Other Electrodes. 250.50
Bonding to Other Systems 250.94
CATV Systems . . . . 800.100(D)
Communication Circuits . . . . 800.100(D)
Grounding Electrode System 250.52
Intersystem Bonding Termination, CATV Systems . . 800.100(B)
Intersystem Bonding Termination, Communications Circuits . . 800.100(B)
Intersystem Bonding Terminations, Network-Powered Broadband 800.100(B)
Lightning Protection Systems 250.106
Air Terminals (Strike Terminal Devices) 250.60
See also *Standard for the Installation of Lightning Protection Systems, NFPA 780* . . . . NFPA 780
Network-Powered Broadband Systems . . . . . . 800.100(D)
Radio and Television Equipment . . . . .810.21(J)
Separately Derived Systems . . . . .250.30(A)(5 and 6)
Size of Conductor. 250.66

COMMON NEUTRAL (Not Permitted)
Office Furnishings, Cord- and Plug-Connected . . . .605.9(D)

COMMON NEUTRAL CONDUCTOR (Permitted)
Continuity in Branch Circuits, Device Removal . . . 300.13(B)
Feeders 215.4
In Metal Raceways or Enclosures. . . . . 215.4(B)
Neutral Load. 220.61
Number of Sets Permitted . . . . 215.4(A)
Lighting Equipment Installed Outdoors . . . . 225.7(B)

Multiwire Branch Circuits 210.4
See NEUTRAL OR GROUNDED CONDUCTOR . . . . Ferm's Finder
Where Considered Current-Carrying. . . . . 310.15(E)
See NEUTRAL OR GROUNDED CONDUCTOR Ferm's Finder

COMMUNICATIONS CIRCUITS . . . . **Art. 805**
Cable Assemblies, see (DUNH). . . . . *UL Product iQ*
Cable Substitution 805.154
Circuit Integrity Cable . . . . 805.179(C)(1)
Circuits Requiring Primary Protectors 805.50
Conductors
Other Conductors with . . . . 805.133(A)
Definitions 805.2
Definition of . . . . Art. 100
Dwelling Unit Communications Outlet 805.156
Electrical Circuit Protective System, Defined . . . . . Art. 100
Fire Resistance 805.179
Fire-Resistive Cable. . . . . 805.179(C)(2)
Innerduct, Defined . . . . Art. 100
Listing Required (Equipment) 805.170
Listing Required (Wires and Cables) 805.179
Optical Fiber Cable. 770.110
Plenum Grade Cable Ties . . . . .805.170(C)
Protection . . . . Art. 805 Part III
Devices. 805.90
Grounding 805.93
Requirements
Application, Primary Protector . . . . 805.90(A)
Hazardous (Classified)Location, Primary Protector 805.90(C)
Location, Primary Protector . . . . 805.90(B)
Requirements, Secondary Protector . . . . 805.90(D)
Underground Circuits Entering Buildings 805.47

COMMUNICATIONS SYSTEMS . . . . **Art. 800**
Abandoned Cables 800.25
Access to Electrical Equipment 800.21
Bonding at Mobile Homes . . . . .800.106(B)
Bonding of Electrodes . . . . 800.100(D)
Cable Assemblies, see (DUNH). . . . . *UL Product iQ*
Cable Routing Assemblies,
Listed 800.182
Marked. 800.182
Support . . . . .800.110(C)
Conductors

Applications of Listed Conductors 800.154
  Overhead 800.44
  Temperature Limitations 800.27
  Within Buildings . . . . Art. 800 Part V
Definition of . . . . Art. 100
Electrical Circuit Protective System, Defined . . . . . . Art. 100
Grounding
  Cables 800.100
  Devices 800.180
  Mobile Homes 800.106
  Methods 800.100
Grounding of Entrance Conduits Containing Communication
Circuits 800.49
Innerduct, Defined . . . . Art. 100
Installation, Listing Required . . . . .800.113(A)
  Abandoned Cables 800.25
  Intersystem Bonding Termination . . . . 800.100(B)(1)
Lightning Conductors Separated by at least 6 Feet 800.53
Listing Required (Communications Raceways) 800.154
Listing Required (Grounding Devices) 800.180
Listing Required (Installed in Buildings) . . . . . 800.113(A)
Mechanical Execution of Work 800.24
Optical Fiber Cable 770.110
Primary Protector Grounding at Mobile Homes 800.106
Temperature Limitations 800.27
Types of Cable 800.179
Spread of Fire or Products of Combustion 800.26

COMMUNICATIONS RACEWAY, DEFINED . . . . Art. 100

**COMMUNITY ANTENNA TELEVISION AND RADIO
DISTRIBUTION SYSTEMS** . . . . Art. 820
Abandoned Coaxial Cable, Definition 800.2
Abandoned Cables 800.25
Access to Electrical Equipment 800.21
Cable Substitution 820.154
Cable Ties and Accessories, Nonmetallic 800.24
Cables Outside and Entering Buildings . . . . . Art. 820 Part II
Cables within Buildings . . . . Art. 820 Part V
  Cable Routing Assemblies . . . . 800.110(C)
  Innerduct for Coaxial Cables . . . . 800.110(A)(3)
  Raceway Fills . . . . 800.110(B)
  Types . . . . 800.110(A)
Definitions 800.2
Grounding Methods . . . . Art. 820 Part IV
Grounding Metallic Entrance Conduits 800.49
Intersystem Bonding Termination . . . . 800.100(B)(1)
Listing, Marking, and Installation of Cables 820.113
  Alternate Wiring Methods of Art. 830 Permitted . . . 820.3(B)
Mechanical Execution of Work 800.24
Power Limitations 820.15
Protection . . . . Art. 820 Part III
Wiring in Ducts . . . . 800.3(B)
See (DVCS) . . . . UL Product iQ
Unlisted Cable Entering Buildings 820.48

COMPACT CONDUCTORS . . . . Chapter 9 Table 5A
Compact Stranding, Definition of . . . . Notes to Tables
. . . . Informative Annex C
Wire-Bending Space (Terminals) . . . . . Table 312.6(A) and (B)

COMPRESSOR MOTORS
See AIR-CONDITIONING & REFRIGERATING
EQUIPMENT . . . . Ferm's Finder

COMPUTER ROOMS . . . . Art. 645
See INFORMATION TECHNOLOGY EQUIPMENT . . . . .
. . . . Ferm's Finder

CONCEALED (Definition of) . . . . Art. 100 Part I

CONCENTRIC KNOCKOUTS
Bond Around, Service . . . . 250.92(B)
Bonding for Over 250 Volts 250.97

CONCRETE
Concrete Encased Electrodes . . . . 250.52(A)(3)
  DC Systems, Sole Connection Size . . . . 250.166(D)
Concrete Walls Considered Grounded . Table 110.26(A)(1) Note
Metal Raceways, Equipment in . . . . 300.6(A)(3)
Use with Minimum Cover Requirements . . . . . Table 300.5
Wet Location, (Definition of) . . . . Art. 100 Part I

CONDUCTORS . . . . Art. 310
Aluminum Conductors, Material 310.3
Note: Use oxide inhibitor where required
  Compact Conductors . . . . Chapter 9 Table 5A
  Dwelling Unit Services and Feeders 310.12
  Not Permitted to Terminate Within 450 mm (18 in.) of Earth or
  Masonry Walls for Grounding . . . . 250.64(A)
  See Aluminum Conductor Terminations . . . .
  . . . . Ferm's Charts and Formulas

Aluminum to Copper, Limited Connections 110.14

Ampacity of, 0 – 2000 Volts (General) 310.14

    Calculated Under Engineering Supervision. . . . . 310.14(B)

        See Informative Annex B for Formula Applications Informative . . . . Annex B

    Determined from Tables . . . . . . Tables 310.16 through 21

    See AMPACITY . . . . Ferm's Finder

Ampacity of, 2001- 35,000 Volts (General) 311.60

    Calculated Under Engineering Supervision. . . . . 311.60(B)

    Determined from Tables . . . Tables 311.60(C)67 through 86

    See Informative Annex B for Formula Applications Informative Annex B

Ampacity of Bare or Covered Conductors. . . . . . Table 310.21

    Limits, Used with Insulated Conductors . . . . . 310.15(D)

Ampacity of Busways, Maximum Allowed . . . . . . 230.42(A)

Application & Insulation . . . . Table 310.4(A)

    Over 1000 Volts . . . . Tables 310.4(B)

Area in Circular Mils . . . . Chapter 9 Table 8

    See Busbars . . . . *Ferm's Charts and Formulas*

Area in Square Inches . . . . *Ferm's Charts and Formulas*

Backfill . . . . 300.5(F)

    Over 1000 Volts . . . . 300.50(E)

Bare Bus Bars in Sheet Metallic Auxiliary Gutters. . . 366.23(A)

Bare, Covered, Insulated Conductor (Definition of) . . . . Art. 100 Part I

Bending Space

    Auxiliary Gutters, in 366.58

    Bending Radius Over 1000 Volts. 300.34

    Enclosures for Motor Controllers and Disconnects, in 430.10(B)

    Examination of Equipment . . . . 110.3(A)(3)

    Manholes. 110.74

    1000 Volts, Nominal, or Less. . . . . 110.74(A)

    Over 1000 Volts, Nominal . . . . 110.74(B)

    Pull and Junction Boxes Not Over 600 Volts, in 314.28

    Pull and Junction Boxes Over 1000 Volts, in 314.71

    Switch or Circuit Breaker Enclosures, in . . . . 404.3(A)

        How Measured (wire-bending space) 404.28

    Switchboards and Panelboards, in — Wire Bending Space . . . . 408.3(G)

        Clearance Entering Bus Enclosures 408.5

        Provisions for 408.55

    Terminals of Cabinets, Cutout Boxes, Meter Sockets 312.6(B)

    Wireways, in Metal . . . . 376.23(A)

    Wireways, in Nonmetallic . . . . 378.23(A)

Bundled as to Ambient Temperature . . . . 310.15(B)(2)

Buried (General) 300.5

    Bushing on Open End of Conduit . . . . 300.5(H)

    Over 600 Volts (Shielding) 311.44

    Over 1000 Volts (Other Nonshielded Cables) . . 300.50(A)(3)

        See OVER 1000 VOLTS, NOMINAL . . . . Ferm's Finder

    Type Identified for the Purpose. . . . . 310.10(F)

Cable (In Cable Tray)

    Ampacity of, 2000 Volts or Less . . . . 392.22(A)

    Ampacity of, 2001 Volts and Over . . . . 392.22(C)

    Ampacity of Type TC Cable 336.80

    Number of, Cables 2001 Volts and Over. . . . . . 392.22(C)

    Number of Multiconductor Cables 2000 Volts of Less . . 392.22(A)

    Number of Single Conductor Cables 2000 Volts or Less . 392.22(B)

Circular Mil Area of . . . . Chapter 9 Table 8

    See COLOR CODING . . . . Ferm's Finder

Combination of, in Raceway

    Conductor Properties . . . . Chapter 9 Table 8

    Dimensions of Compact Aluminum Building Wire Chapter 9 Table 5A

    Dimensions of Insulated Conductors and Luminaire (Fixture) Wires. . . . . Chapter 9 Table 5

    Percent Area Conduit and Tubing . . . . . Chapter 9 Table 4

    Percent of Cross-Sectional Area Allowed . . Chapter 9 Table 1

Compact Conductors (Aluminum) . . . . . . Chapter 9 Table 5A

    Compact Stranding, Defined . . Table C.A(A) (Below Table)

Conductor Fill in Class 1 Sealing Fittings. . . . . .501.15(C)(6)

Connection to Terminals 110.14

    Motor Control Circuit Devices. 430.9

Continuous Loads (Definition) . . . . Art. 100 Part I

    Branch Circuit Conductors . . . . 210.19(A)(1)

    Feeder Conductors . . . . 215.2(A)(1)

    Fixed Electric Space-Heating . . . . 424.4(B)

    Individual Branch Circuit, Appliances . . . . 422.10(A)

    Pool Water Heaters. 680.10

    Storage-Type Water Heaters 422.13

Copper Conductors 110.5

    Agricultural Buildings

        Bonding and Equipotential Plane 547.10

        Equipment Grounding Conductor, Underground (Direct Buried) . . . . 547.5(F)

        Wiring Methods 547.5

    Ampacity of, 0 – 2000 Volts (General) 310.14

        Calculated Under Engineering Supervision . . . 310.14(B)

        Determined from Tables. . . . . Tables 310.16 through 21

        For Short-Time Rated Cranes and Hoists . . . . 610.14(A)

See Informative Annex B for Formula Applications Informative Annex B

See AMPACITY . . . . Ferm's Finder

Ampacity of, 2001- 35,000 Volts (General) 311.60

    Calculated Under Engineering Supervision . . . 311.60(B)

    Determined from Tables. . . . .

    . . . . Tables 311.60(C)(67) through 86

    See Informative Annex B for Formula Applications Informative

    . . . . Annex B

Class I, Zone 0, 1, and 2, Protection "e" Field Wiring 505.18(A)

Copper to Aluminum Permitted 110.14

Dimensions & Percent Area of . . . . Chapter 9 Table 5

    Conductor Properties . . . . Chapter 9 Table 8

Dwelling Unit Services and Feeders 310.12

Fire Alarm Systems

    NPLFA Circuit Conductors 760.49

    Health Care Facilities. . . . . .Art. 517 Part II

        Critical Care (Category 1) Spaces, Patient Bed Locations. . 517.19

        Equipment Grounding Conductor, Patient Care   517.13(B)

        General Care (Category 2) Spaces, Patient Bed Locations . .

        . . . . 517.18

        Panelboard Bonding Normal and Essential Circuits . 517.14

        Panelboard Grounding, Critical Care (Category 1) Spaces . .

        . . . . 517.19(E)

Low-Voltage Wiring, Park Trailers . . . . 552.10(B)(1)

Marinas and Boatyards, Equipment Grounding 555.37

Material Not Specified Means Copper 110.5

Motor Controllers . . . . 430.9(B)&(C)

Properties of . . . . Chapter 9 Table 8

Required for Type MI Cable Conductors 332.104

Swimming Pool, Fountains, and Similar Installations

    Bonding Conductor . . . . 680.26(B)

    Fountains . . . . .Art. 680 Part V

    Hydromassage Bathtubs, Bonding. 680.74

    Permanently Installed Pools . . . . .Art. 680 Part II

    Shall Be Long Enough (Replacement Pump Motor)   680.74

    Spas and Hot Tubs. . . . . Art. 680 Part IV

    Therapeutic Pools and Tubs . . . . Art. 680 Part VI

    Underwater Audio Equipment. 680.27

    Underwater Luminaires (Lighting Fixtures) 680.23

Correction Factors . . . . Tables 310.15(B)(2)

Crane & Hoist Conductors . . . . 610.14(A)

Deflection of, *See* CONDUCTORS, Bending Space . . . Ferm's Finder

Derating Allowable Ampacity

    Ambient Temperature . . . . Tables 310.15(B)(2)

Class 1 Circuit, Overcurrent Protection for 725.43

Class 1 Circuit Conductors 725.51

Grounding or Bonding Conductors Not Considered . 310.15(F)

Heated Ceilings 424.36

Neutral Conductor. . . . . 310.15(E)

Number of Current-Carrying Conductors, Adjustment

Factors . . . . 310.15(C)

    Metal Wireways 376.22

    Nonmetallic Wireways 378.22

Different Systems in Same Enclosure

    1000 Volts, Nominal or Less . . . . 300.3(C)(1)

    Cable Trays

        Cables Over 1000 Volts . . . . 392.20(B)

        Multiconductor Cables 1000 Volts or Less. . . . 392.20(A)

    CATV and Radio Distribution Systems 820.133

    Class 1 Remote Control, Signaling Circuits 725.48

    Class 2 & 3 Remote Control, Signaling Circuits . . . 725.136

    Communications Circuits 805.133

    Elevators, Dumbwaiters, etc. 620.36

    Emergency System Wiring . . . . 700.10(B)

    Essential Electrical System, Life Safety Branch . . . 517.42(D)

    Fire Pumps, Independent Routing . . . . 695.6(A)(2)(a)

    Generator Control Wiring . . . . 695.14(F)

    Grounded Conductors, Identification . . . . 200.6(D)

    Intrinsically Safe Systems 504.30

    Legally Required Standby Wiring 701.10

    Network-Powered Broadband Systems 830.133

    Non-Electrical Systems Prohibited 300.8

    Non-Power-Limited Fire Alarm (NPLFA) Circuits . . . 760.48

    Optical Fiber Cables 770.133

    Optional Standby Wiring 702.10

    Over 1000 Volts, Nominal . . . . 300.3(C)(2)

        Requirements for 300.32

    Power-Limited Fire Alarm (PLFA) Circuits. 760.133

Different PLFA Circuits

    Prohibition of with Service Conductors 230.7

    Solar Photovoltaic Systems . . . . 690.31(B)

    Surface Metal Raceways 386.70

    Surface Nonmetallic Raceways 388.70

Dimensions & Area of . . . . Chapter 9 Table 5

    Compact Aluminum Wire . . . . .Chapter 9 Table 5A

    Conductor Properties . . . . Chapter 9 Table 8

Dwelling Unit Services & Feeders 310.12

    Individual Unit Feeder Conductors 310.12

Equipment Grounding Conductors
　Box Fill. . . . . 314.16(5)
　Identification. 250.119
　Installation. 250.120
　Sectioned. . . . . 250.122 and 310.10(H)(5)
　Size. 250.122
　Increased Due to Voltage Drop . . . . .250.122(B)
Exposed to Direct Sunlight . . . . 310.10(D)
Feeder Identification
　Direct-Current Systems . . . . .215.12(C)(2)
　Equipment Grounding Conductor. . . . . 215.12(B)
　Grounded Conductor Identification . . . . 215.12(A)
　Ungrounded Conductor Identification
　　Alternating-Current Systems. . . . . .215.12(C)(1)
Feeder to Mobile Home. . . . . 550.33(A)
　Conductor Size . . . . 550.33(B), 310.12
　Minimum Capacity 550.33
　　Allowable Demand Factors 550.31
　　Fill within Boxes, Conductors 314.16
Fine Stranded Conductors 110.14
Fire Alarm Systems . . . . Art. 760
　Nonpower-Limited Circuits (NPLFA) . . . .Art. 760 Part II
　Power-Limited Circuits (PLFA) . . . . Art. 760 Part III
Fire Pumps . . . . Art. 695
Flexible Cords . . . . Art. 400
Free Length at Boxes 300.14
　Deicing and Snow-Melting Equipment . . . . . . 426.23(A)
　Fixed Electric Space Heating Cables 424.43
　Pipeline and Vessel Heating Equipment . . . . . 427.18(A)
Gasoline & Oil Resistant (General). . . . . 310.10(G)
　Class I Locations. 501.20
　See (ZLGR) . . . . UL Product iQ
High Voltage, Rated 2001 to 35,000 Volts 311.60
. . . . Tables 311.60(C)67 through 86
　Grounding Conductor Continuity. . . . . 200.2(B)
　Requirements for. . . . . Art. 300 Part II
　See OVER 1000 VOLTS, NOMINAL. . . . . . Ferm's Finder
Identification of
　Branch Circuit Conductors 210.5
　Equipment Grounding Conductors, General . . . . 210.5(B)
　Equipment Grounding Conductors 250.119
　Flexible Cords . . . . .Art. 400 Part II
　　Grounded Conductors. . . . . 210.5(A)
　　(Fixture) Luminaire Wires 402.8
　Means of Identifying. 200.6
　Use of White or Gray 200.7
Heating Cables. 424.35
Higher Voltage to Ground (Delta), General 110.15
　Service Conductors 230.56
Intrinsically Safe Systems 504.80
Isolated Power Systems . . . . 517.160(A)(5)
Mobile Home Power Supply
　Cords, Power Supply . . . . 550.10(B)
　Feeders . . . . 550.10(A)
　Mast Weatherhead or Raceway . . . . .550.10(I)
Multiwire Branch Circuits and Individual Branch Circuits . . 210.5(C)
Park Trailer Power Supply
　Cords, Power Supply . . . . 552.43(B)
　Feeder . . . . 552.43(A)
　Mast Weatherhead or Raceway . . . . 552.43(C)
　Ungrounded Conductors . . . . 210.5(C)
Induction, Arrange Wiring to Prevent. . . . . 300.3(B)
　In Metal Raceways and Enclosures 300.20
　　Over 1000 Volts. 300.35
　Installation in Cable Trays 392.46
　　Flanged Connections . . . . 392.46(B)
　　Through Bushed Conduit or Tubing. . . . . . 392.46(A)
　Single Cables of Type MI Cable 332.31
　Three-Way and Four-Way Switch Wiring . . . . . 404.2(A)
　Underground Installations . . . . 300.5(I)
Inserting in Raceways
　Completed Runs. 300.18
　General Rules 300.3
　Number and Size of (General) 300.17
　Requirements for Stranded Conductors . . . . . . 310.3(C)
Integrated Gas Spacer Cable 326.104
Low Voltage & High Voltage in Same Enclosure 300.3
　Cable Trays
　　Cables Over 1000 Volts . . . . 392.20(B)
　　Multiconductor Cables 1000 Volts or Less . . . 392.20(A)
　CATV and Radio Distribution Systems 820.133
　Class 1 Circuits 725.48
　Class 2 & 3 Circuits . . . . .725.136(A)
　Communications Circuits . . . . 805.133(A)(1)
　Network-Powered Broadband Systems . . . . . 830.133(A)(1)
　NPLFA Circuits 760.48
　Optical Fiber Cables 770.133

PLFA Circuits 760.136
Luminaire (Fixture) . . . . Art. 402
    Allowable Ampacity 402.5
    Ampacity of . . . . 240.4(A)
    Overcurrent, Protection of 240.5
    Supplementary Overcurrent Protection 240.10
    Taps . . . . .210.19(A)(4)
        To Branch Circuit Conductor . . . . 240.5(B)(2)
        To Luminaire (Fixture) Outlet Box . . . . . . .410.117(C)
        Manufactured Wiring Systems. . . . . 604.100(A)(2) Ex. 1
    Types. Table 402.3
    Whips .210.19(A)(4)
        Flexible Metal Conduit . . . . .348.20(A)(2)
        Flexible Metallic Tubing . . . . 360.20(A) Ex. 2
        Length and Size of Luminaire (Fixture) Wire . 240.5(B)(2)
        Liquidtight Flexible Metal Conduit . . 350.30(A) Excs. 3 and 4
        Liquidtight Flexible Nonmetallic Conduit . . . 356.30(2)
        Type AC Cable . . . . .320.30(D)(3)
        Type MC Cable . . . . .330.30(D)(2)
        Type NM Cable . . . . 334.30(B)(2)
        Wiring of . . . . Art. 410 Part VI
Machine Tool Wire
    Ampacity of . . . . .310.14(A)(1) Info. Notes 2
    Application and Insulation . . . . Table 310.4(A)
    See Electrical Standard for Industrial Machinery NFPA 79 NFPA 79
Material Not Specified Means Copper 110.5
Metal-Clad Cable . . . . Art. 330
    Ampacity of Conductors 330.80
    Conductor Material and Size 330.104
    Insulation 330.112
Mineral-Insulated, Metal-Sheathed Cable . . . . . . Art. 332
    Ampacity of Conductors 332.80
    Insulation within 332.112
    Type of Conductors 332.104
    See MINERAL-INSULATED, METAL-SHEATHED CABLE
. . . . Ferm's Finder
Minimum Size (General) . . . . 310.3(A)
Branch Circuits . . . . .Art. 210 Part II
Equipment Grounding Conductor 250.122
Feeders 215.2
    600 Volts and Less . . . . 215.2(A)
    Over 600 Volts . . . . 215.2(B)
    (Fixture) Luminaire Wire. 402.6
    Flexible Cords and Cables 400.21

Portable Cables Over 600 Volts . . Art. 400 Part III, 400.31
Grounding Electrode Conductor 250.66
Outside Branch Circuits and Feeders. 225.6
Parallel Conductors . . . . 310.10(G)
Services
    Minimum Permitted for Dwellings 310.12
    Overhead Service Drop 230.23
    Service Entrance 230.42
    Underground Service-Lateral . . . . 230.31(B)
Mobile Home Service or Feeder . . . . 550.33(A)
    Minimum Permitted Size . . . . 550.31(B), 310.12
Neutral, Definition of . . . . Art. 100
Number of, in
    Auxiliary Gutters. 366.22
    Boxes. 314.16
    Cable Trays
        Multiconductor Cables 2000 Volts or Less. . . . 392.22(A)
        Single Conductor Cables 2000 Volts or Less. . . 392.22(B)
        Type MV and MC Cables (2001 Volts and Over) 392.22(C)
    Cellular Concrete Floor Raceways 372.22
    Cellular Metal Floor Raceways 374.22
    Combination of, in Raceways . . . . . . Chapter 9 Table 1
        Areas of Conduit or Tubing for . . . . Chapter 9 Table 4
        Dimensions for . . . . Chapter 9 Table 5
        Dimensions for Compact Aluminum Building Wire . . . .
. . . . .Chapter 9 Table 5A
    Conduit Bodies . . . . 314.16(C)
    Electrical Metallic Tubing 358.22
        Conductors of the Same Size Informative Annex C, Table C1
        Conductors of the Same Size (Compact) . . . .
. . . . .Informative Annex C, Table C1A
    Electrical Nonmetallic Tubing 362.22
        Conductors of the Same Size Informative Annex C, Table C2
        Conductors of the Same Size (Compact) Informative Annex C, Table C2A
    Elevators . . . . Art. 620
        Conductors in Raceways 620.33
        Conductors in Wireways 620.32
    (Fixture) Luminaire Wires 402.7
    Flexible Metal Conduit. 348.22
        Conductors of the Same Size . . . .
. . . . Informative Annex C, Table C3
        Conductors of the Same Size (Compact)
. . . . .Informative Annex C, Table C3A
    3/8" Flexible Metal Conduit . . . . Table 348.22

General Installations 300.17
Intermediate Metal Conduit 342.22
    Conductors of the Same Size Informative Annex C, Table C4
    Conductors of the Same Size (Compact) . . . .
        . . . . .Informative Annex C, Table C4A
Liquidtight Flexible Metal Conduit 350.22
    Conductors of the Same Size Informative Annex C, Table C7
    Conductors of the Same Size (Compact)Informative Annex C, Table C7A
Liquidtight Flexible Nonmetallic Conduit 356.22
    Conductors of the Same Size (Compact) (FNMC-A)
        . . . . .Informative Annex C, Table C6A
    Conductors of the Same Size (Compact) (FNMC-B)
        . . . . .Informative Annex C, Table C5A
    Conductors of the Same Size (Type FNMC-A) . . . .
        . . . . Informative Annex C, Table C6
    Conductors of the Same Size (Type FNMC-B) . . . .
        . . . . Informative Annex C, Table C5
Nipples (60% Fill) . . . . Chapter 9, Note 4
Nonmetallic Underground Conduit with Conductors . 354.22
Outlet Boxes 314.16
Rigid Metal Conduit 344.22
    Conductors of the Same Size Informative Annex C, Table C8
    Conductors of the Same Size (Compact)
        . . . . .Informative Annex C, Table C8A
Rigid PVC (Type A) 352.22
    Conductors of the Same Size . . . .
        . . . . .Informative Annex C, Table C11
    Conductors of the same Size (Compact)
        . . . . Informative Annex C, Table C11A
Rigid PVC (Type EB) 352.22
    Conductors of the Same Size . . . .
        . . . . .Informative Annex C, Table C12
    Conductors of the Same Size (Compact)
        . . . . Informative Annex C, Table C12A
Rigid PVC (Schedule 40 & HDPE) 352.22
    Conductors of the Same Size . Informative Annex C, Table C10
    Conductors of the Same Size (Compact) . . . .
        . . . . Informative Annex C, Table C10A
Rigid PVC Conduit, Schedule 80 (PVC). 352.22
    Conductors of the Same Size Informative Annex C, Table C9
    Conductors of the Same Size (Compact) . . . .
        . . . . .Informative Annex C, Table C9A
Skeleton Tubing, Field-Installed . . . . 600.31(C)
Surface Raceways (Metal) 386.22
    Surface Raceways (Nonmetallic) 388.22
    Underfloor Raceways 390.22
    Wireways (Metal) 376.22
    Wireways (Nonmetallic). 378.22
Oil & Gasoline Resistant (General) . . . . 310.10(G)
    Class I Locations 501.20
    See (ZLGR) . . . . *UL Product iQ*
Operating Over 1000 Volts . . . . .Art. 300 Part II
    Conductor Construction and Applications 310.4
Overcurrent Protection of 240.4
    Over1000 Volts . . . . Art. 240 Part IX
    Protection of Flexible Cords and Luminaire (Fixture) Wires240.5
    Service Entrance Conductors 230.90
    See OVERCURRENT PROTECTION . . . . . Ferm's Finder
Parallel Conductors 1/0 AWG and Larger Permitted   310.10(G)
    Auxiliary Gutters (within and grouped) 366.20
    Control Power . . . . 310.10(G)(1) Ex. 1
    Existing Installations, 2 AWG Permitted . .310.10(G)(1) Ex. 2
    Grounding (Equipment) . . . . .250.122(F)
    Underground. 300.5(I)
Point, Neutral Point, Definition of . . . . Art. 100
Power and Control Tray Cable . . . . Art. 336
Properties of . . . . Chapter 9 Table 8
Protection of, Physical Damage (General) 300.4
    Cables in Accessible Attics 320.23
    Equipment Grounding Conductors Smaller Than 6 AWG . . .
    . . . . .250.120(C)
    Grounding Conductors
        Communications Circuits 805.93
        Community Antenna Television and Radio Distribution Systems . . . . 800.100(A)(6)
        Network-Powered Broadband Communication Systems. . . 800.100(A)(6)
        Radio and Television Equipment . . . . 810.21(D)
    Grounding Electrode Conductors . . . . 250.64(B)
    Multiconductor NPLFA Cables . . . . 760.53(A)
    Open Wiring on Insulators 398.15
        In Accessible Attics 398.23
    Over 1000 Volts
        Metal-Sheathed Cables 300.42
        Underground Installations . . . . 300.50(B)
    PLFA Circuits . . . . 760.130(B)
    Safety Control Circuits. . . . . 725.31(B)
    Service Entrance Conductors Aboveground 230.50
    Sleeves . . . . 300.15(C)

Types NM and NMC . . . . 334.15(B)
Types NM and NMC Cables (In Unfinished Basements) . . .
. . . . 334.15(C)
Underground Installations . . . . 300.5(D)
Underground Service-Laterals 230.32

Protection of, at Boxes
    Boxes, Conduit Bodies, or Fittings 314.17
    Cabinets, Cutout Boxes, and Meter Socket Enclosures 312.5
    Covers of Outlet Boxes and Conduit Bodies. 314.41
    Intermediate Metal Conduit. 342.46
    High Density Polyethylene Conduit, Type HDPE . 353.46
    Nonmetallic Underground Conduit with Conductors. . . 354.46
    Reinforced Thermosetting Resin Conduit, Type RTRC . . 355.46
    Rigid Metal Conduit 344.46
    Rigid PVC Conduit 352.46

Radius of Bends
    Manholes. 110.74
    Nonmetallic Underground Conduit with Conductors . 354.24
    Over 1000 Volts Nominal 300.34
    Pull and Junction Boxes 314.28
        Over 1000 Volts 314.71
    Service Entrance Cable: Type SE and USE 338.24
    Type AC Cable 320.24
    Type IGS Cable 326.24
    Type ITC Cable 727.10
    Type MC Cable 330.24
    Type MI Cable 332.24
    Type NM and NMC Cable 334.24
    Type TC Cable 336.24
    Type UF Cable. 340.24
    See Also, BENDS. . . . . Ferm's Finder

Resistance of . . . . Chapter 9 Table 8
    AC Resistance and Reactance. . . . . Chapter 9 Table 9

Rooftops, Sunlight and Ambient Temperature Adjustments. . . .
. . . . 310.15(B)(2)

Sealing Underground Raceways
    600 Volts and Under . . . . 300.5(G)
    Over 1000 Volts . . . . 300.50(F)
    Branch Circuits and Feeders 225.27
    Service Raceways. 230.8

Service Drop Conductors . . . . Art. 230 Part II
    Allowable Ampacity for
        Single Conductor Free Air 60 - 90°C . . . . Table 310.17
        Single Conductor Free Air 150 - 250°C . . . . Table 310.19
    Clearances 230.24
    Means of Attachment 230.27
    Size and Rating. 230.23

Shielding, Over 2000 Volts 311.44
    Direct Burial Conductors 311.60

Shielding, 2001 to 35,000 Volts . . . . 311.60(D)(1)
    Direct Burial Conductors . . . . 311.60(D)(2)
    See OVER 1000 VOLTS, NOMINAL . . . . . Ferm's Finder

Single Conductor . . . . 300.3(A)
    In Cable Trays . . . . 392.22(B)
    Nonheating Leads of Electric Space-Heating Cables . . 424.43
    Solar Photovoltaic Systems . . . . 690.31(C)
    Underground. . . . . 300.5(I)

Small Motors . . . . 430.22(G)

Splice Not Permitted of or in
    Flexible Cord Initial Installation 400.13
    Grounding Electrode Conductor . . . . 250.64(C)
    Raceways (General) . . . . 300.13(A)
    Raceways (See Specific Raceway Article)
    Sealing Fittings . . . . 501.15(C)(4)
    Swimming Pool Lighting Ground Wire . . . . 680.23(F)(2)

Splices of or in . . . . 110.14(B)
    Auxiliary Gutters. 366.56
    Boxes Required. 300.15
        See BOXLESS DEVICES . . . . Ferm's Finder
        Temporary Wiring. . . . . 590.4(G)
    Cable Trays. 392.56
    Cellular Concrete Floor Raceway. 372.56
    Cellular Metal Floor Raceway 374.56
    Conduit Bodies (General) 300.15
        When Permitted . . . . 314.16(C)(2)
    Direct Burial Cables . . . . 300.5(E)
    Enclosures for Switches and Overcurrent Devices . . 312.8(A)
    Flexible Cords
        For Repair Only 400.13
        Temporary Wiring. . . . . 590.4(G)
    Heating Cables
        For Pipelines and Vessels. . . . . 427.19(B)
        Outdoor Deicing and Snow-Melting . . . . . 426.24(B)
        Space-Heating Cables, Cannot Alter Length . . . 424.40
        Space-Heating Cables, Embedded . . . . 424.41(D)
    Insulation of . . . . 110.14(B)
    Service Entrance Conductors 230.46
    Underfloor Raceway 390.56

Wireways, Metal 376.56
Wireways, Nonmetallic 378.56
Stranded Required . . . . 310.3(C)
Suitable For Wet Locations . . . . 310.10(C)
    Weatherproof, Definition of . . . . Art. 100 Part I
    Wet Location, Definition of . . . . Art. 100 Part I
Supply Conductors, Described . . . . 250.102(C)(2) Info. Note
Supporting in or by
    Festoon Lighting . . . . 225.6(B)
    Messenger . . . . Art. 396
    Open Wiring on Insulators. 398.30
    Pull & Junction Boxes . . . . 314.28(B)
    Vertical Raceways 300.19
Swimming Pools
    Bonding 680.26
    Equipment Grounding. . . . . Art. 680 Part II
    Wet-Niche Luminaires (Fixtures) . . . . 680.26(B)(4)
Temperature Rating of Insulation . . . . 310.14(A)(3)
Allowable Ampacity for . . . . Tables 310.16 through 21
Conductor Application and Insulation . . . . . Table 310.4(A)
Fixture Wire . . . . .Table 402.3
Luminaire (Fixture) Outlet Boxes 410.21
Over 2000 Volts Rating
    Allowable Ampacity for . . . Table 311.60(C) 67 through 86
Terminations 110.14
    CO/ALR Marking
        Receptacles . . . . 406.3(C)
        Switches . . . . 404.14(C)
    Grounding Conductor Connections
    Connection to Electrodes 250.70
    Equipment Grounding Conductor at Receptacles . . 250.146
    Equipment Grounding Conductors at Boxes 250.148
    Grounding and Bonding Equipment. 250.8
    Suitable for the Purpose 110.14
    See (AALZ) . . . . UL Product iQ
    See (AALZ) . . . . UL Product iQ
Tray Cable . . . . Art. 336
    Ampacity of 336.80
    Installation of, in Cable Trays 392.20
UF, Type . . . . Art. 340
    See (YDUX) . . . . UL Product iQ
Under Buildings, In Raceway . . . . 300.5(C)
    When Considered Outside Building 230.6
Underground

Aircraft Hangars, Under Hangar Floor 513.8
Bulk Storage Plants
    Underground Wiring Below Class I Locations. 515.8
Dispenser, Sealing at 514.9
Motor Fuel Dispensing Facilities 514.3
Sealing of Conduits Under Hangar Floor 513.9
See BURIED CONDUCTORS . . . . Ferm's Finder
See HAZARDOUS (CLASSIFIED) LOCATIONS, Wiring
Under . . . . Ferm's Finder
See UNDERGROUND WIRING. . . . . Ferm's Finder
Under Swimming Pools. 680.11
Use Oxide Inhibitor Where Required 110.3
USE, Type . . . . Art. 338
    See (TYLZ) . . . . UL Product iQ
Voltage Drop of Branch Circuits . . . . 210.19(A) Info. Notes 4
    Conductors. . . . . .310.14(A)(1) Info. Notes 1
    Feeders . . . . 215.2(A)(1) Info. Notes 2
    Fire Pumps 695.7
    Sensitive Electronic Equipment . . . . 647.4(D)
    See VOLTAGE DROP . . . . Ferm's Finder
Volume Required Per . . . . 314.16(B)
Wet Locations . . . . 310.10(C)

CONDUIT
Aluminum
    Be Sure It Is Suitable for the Condition 300.6
    Corrosion Protection . . . . 344.10(B)
    Shall Not Be Used in Concrete or Earth Burial
        Without Supplementary Corrosion Protection
    See (DYWV) . . . . UL Product iQ
See BENDS . . . . Ferm's Finder
See CONDUIT BODIES. . . . . Ferm's Finder
Note: Boxes such as FS & FD or larger cast or sheet metal boxes are not classified as conduit bodies. See definition of conduit body in Article 100 Part I.
Burial Depth 600 Volts and Less 300.5
Burial Depths Over 1000 Volts 300.50
Continuity of Run
    Bonding in Hazardous Locations. 250.100
    Bonding Loosely Jointed Metal Raceways 250.98
    Bonding of Other Enclosures . . . . 250.96(A)
    Bonding of Service Raceways. 250.92
    Bonding Over 250 Volts 250.97
    Class I Locations . . . . 501.30(A)
        Zone 0, 1, & 2 Locations . . . . 505.25(A)

Zone 20, 21, & 22 Locations . . . . 506.25(A)
Class II Locations . . . . 502.30(A)
Class III Locations . . . . 503.30(A)
Complete Runs . . . . 300.18(A)
Electrical Continuity Metal Raceways 300.10
Electrical Metallic Tubing 358.42
Intermediate Metal Conduit 342.42
Isolated Grounding Circuits . . . . 250.96(B)
Mechanical Continuity 300.12
Method of Bonding at Service . . . . 250.92(B)
Rigid Metal Conduit 344.42
Threaded Conduit within Hazardous Locations . . . 500.8(E)
Dimensions & Percent Area . . . . Chapter 9 Table 4
    Conduit Fill Tables . . . . Informative Annex C
Enclosures with Devices or (Fixtures) Luminaires Supported by .
. . . . 314.23(F)
Enclosures without Devices or (Fixtures) Luminaires
Supported by . . . . 314.23(E)
Exposed to Sunlight on Rooftops . . . . 310.15(E)
Field-Cut Threads . . . . 300.6(A) Info. Notes
Fitting (Definition) . . . . Art. 100 Part I
    See (DWTT) . . . . UL Product iQ
Flexible Metal Conduit . . . . Art. 348
    See FLEXIBLE METAL CONDUIT . . . . . . Ferm's Finder
High Density Polyethylene Conduit, Type HDPE . . . Art. 353
Intermediate Metal Conduit . . . . Art. 342
    See INTERMEDIATE METAL CONDUIT . . Ferm's Finder
Liquidtight Flexible Metal Conduit . . . . Art. 350
    See LIQUIDTIGHT FLEXIBLE METAL CONDUIT Ferm's Finder
Liquidtight Flexible Nonmetallic Conduit . . . . . . Art. 356
    See LIQUIDTIGHT FLEXIBLE NONMETALLIC CONDUIT Ferm's Finder
Nonmetallic Underground Conduit with Conductors . Art. 354
    See NONMETALLIC UNDERGROUND CONDUIT WITH CONDUCTORS . . . . Ferm's Finder
Number of Bends Permitted in
    Flexible Metal Conduit 348.26
    High Density Polyethylene Conduit Type HDPE . . . 353.26
    Intermediate Metal Conduit 342.26
    Liquidtight Flexible Metal Conduit 350.26
    Liquidtight Flexible Nonmetallic Conduit 356.26
    Nonmetallic Underground Conduit with Conductors . 354.26
    Reinforced Thermosetting Resin Conduit, Type RTRC . . . .
. . . . 355.26
    Rigid Metal Conduit 344.26
    Rigid PVC Conduit 352.26
    See BENDS . . . . Ferm's Finder
Number of Conductors Permitted in
    See CONDUCTORS, Number of, in . . . . . . Ferm's Finder
    Note: Conduit fill tables apply only to complete conduit systems and are not intended to apply
    to sections of conduit used to protect exposed wiring from physical damage . . . . Chapter 9, Note 2
Radius of Bends
    Flexible Metal Conduit 348.24
    Flexible Metal Tubing 358.24
    High Density Polyethylene Conduit, Type HDPE . . . 353.24
    Intermediate Metal Conduit 342.24
    Liquidtight Flexible Metal Conduit 350.24
    Liquidtight Flexible Nonmetallic Conduit 356.24
    Nonmetallic Underground Conduit with Conductors . 354.24
    Reinforced Thermosetting Resin Conduit, Type RTRC 355.24
    Rigid Metal Conduit 344.24
    Rigid PVC Conduit 352.24
Reaming & Threading
    Electrical Metallic Tubing 358.28
    Intermediate Metal Conduit 342.28
    Rigid Metal Conduit 344.28
Rigid Metal Conduit . . . . Art. 344
    See RIGID METAL CONDUIT . . . . Ferm's Finder
Rigid PVC Conduit . . . . Art. 352
    See RIGID PVC CONDUIT . . . . Ferm's Finder
Risers Not Over 75 mm (3 in.) under Floor Standing Panels 408.5
Supports of or from
    Cables or Nonelectrical Equipment, Not Permitted 300.11(B)
    Flexible Metal Conduit 348.30
    Intermediate Metal Conduit 342.30
    Liquidtight Flexible Metal Conduit 350.30
    Liquidtight Flexible Nonmetallic Conduit 356.30
    Nonmetallic Rigid Conduit 352.30
        Support of (Fixtures) Luminaires or Other Equipment
        Prohibited . . . . 352.12(B)
        Reinforced Thermosetting Resin Conduit, Type RTRC 355.30
    Rigid Metal Conduit 344.30
    Welding Not Permitted . . . . 300.18(B)
Trimming
    Flexible Metal Conduit 348.28
    Liquidtight Flexible Nonmetallic Conduit 356.28
    Nonmetallic Underground Conduit with Conductors . 354.28

Rigid PVC Conduit 352.28
Wet Locations 300.6
   Rigid PVC Conduit . . . . 352.10(D)
      See RIGID PVC CONDUIT . . . . .Ferm's Finder
Electrical Metallic Tubing . . . . Art. 358
      See ELECTRICAL METALLIC TUBING . . Ferm's Finder
Electrical Nonmetallic Tubing . . . . Art. 362
      See ELECTRICAL NONMETALLIC TUBINGFerm's Finder
Flexible Metallic Tubing . . . . Art. 360
      See FLEXIBLE METALLIC TUBING . . . . Ferm's Finder

CONDUIT BODIES (Where Required) 300.15
Conductors Entering 314.17
Cross-Sectional Area . . . . 314.16(C)
Installation and Use (General) 314.1
Number of Conductors, General . . . . .314.16(C)(1)
   With Splices, Taps, or Devices . . . . .314.16(C)(2)
To Be Accessible 314.29
Unused Openings. . . . . 314.17(A)
Used As Pull or Junction Boxes 314.28

CONDUIT AND TUBING FILL TABLES . . INFORMATIVE ANNEX C

CONTINUITY
Auxiliary Gutters . . . . 366.100(A)
Cellular Metal Floor Raceways 374.100
Conductors, Mechanical and Electrical 300.13
Electrical Continuity Metal Raceways and Enclosures . . 300.10
Equipment Grounding Conductor 250.148
Ground Path Not to Rely on Water Meters or Filtering Equipment . . . . .250.53(D)(1)
Grounding Electrode Conductor . . . . 250.64(C)
Mechanical, Raceways and Cables 300.12
Neutral of Multiwire Branch Circuits . . . . 300.13(B)
Service Conductors 230.46
Service Equipment
   Bonding of 250.92
   Continuity of Enclosures 300.10
   Method of Bonding . . . . 250.92(B)
See CONDUIT, Continuity of Run. . . . . .Ferm's Finder
Connected to an Equipment Grounding Conductor . . 250.134
   Note: This phrase is used throughout the *Code* to give more specific direction as to how electrical equipment is to be grounded.

CONNECTIONS
Aluminum to Copper Conductors 110.14
Cables and Conductors . . . . 110.12(C)
Cellular Metal Floor Raceways to Cabinets and Extensions . . . . 374.18(A)
Essential Electrical Systems
   Automatic to Life Safety Branch 517.43
   Equipment Branch to Alternate Power Source 517.35
   For Clinics, Medical and Dental Offices 517.45
   To Equipment Branch 517.44
Feed-Through Connections of Neutral Conductors . 300.13(B)
Grounding and Bonding Conductors 250.8
   Equipment Grounding Conductor, Multiple Circuit Connections 250.144
   Grounding Electrode Conductor
      Method of Connection 250.70
      Size of Conductor 250.66
      Metallic Water Pipe and Structural Metal . . . . 250.68(C)
      To Grounding Electrode. 250.68
Grounding Connection AC Systems . . . . 250.24(A)
Grounding Connection DC Systems 250.164
Grounding Connection Ungrounded Systems . . . . 250.24(E)
High-Impedance Grounded Neutral System 250.36
Integrity of . . . . 110.12(B)
Multiple Circuit, See MULTIPLE ENCLOSURE SERVICES . . . . .Ferm's Finder
Point of Interconnected Power Production Sources 705.12
Splices . . . . 110.14(B)
Switches 404.2
Temperature Limitations . . . . 110.14(C)
Terminals, General Provisions . . . . 110.14(A)
   Location of in Switchboards and Panelboards . . . .408.3(D)
   See (AALZ). . . . . UL Product iQ
Torque (tightening connection). . . . . 110.14(D)
Underfloor Raceways to Cabinets and Wall Outlets. . . . 390.76
X-Ray Installations 517.71

CONNECTORS
Armored Cable (Type AC) 320.40
Cabinets and Cutout Boxes . . . . 312.5(C)
Cable, Theater 520.67
Electrical Metallic Tubing 358.42
Flexible Metal Conduit 348.42
General Requirement for . . . . 300.15(C)
   Listed for Specific Wiring Method 300.15
Intermediate Metal Conduit 342.42

Liquidtight Flexible Metal Conduit 350.42
Liquidtight Flexible Nonmetallic Conduit 356.42
Portable Switchboards on Stage . . . . Art. 520 Part IV
Pressure (Solderless) Connector, Definition of . . . . . Art. 100
Rigid Metal Conduit 344.42
Rigid PVC Conduit 352.48
Single-Pole Separable
    Definition of 530.2
    Portable Distribution or Termination Boxes, Carnivals . 525.22(D)
    Use of in Motion Picture and Television Studios 530.22
    Use of in Theaters . . . . 520.53(I)
    Solar Photovoltaic Systems . . . . Art 690
        Component Interconnections 690.32
        Connectors 690.33
        Flexible, Fine Stranded Cables . . . . 690.31(E)
        Grounding Connection Connectors 690.47
See (ECIS) . . . . UL Product iQ

CONNECTOR STRIPS . . . . 520.44(B)

CONSTRUCTION SITES
See TEMPORARY INSTALLATIONS . . . . Ferm's Finder

CONSTRUCTION, TYPES OF . . . . . . Informative Annex E

CONTINUOUS DUTY
Definition of . . . . Art. 100 Part I
Motor and Branch-Circuit Overload Protection Art. 430 Part III
Motor Circuit Conductors . . . . Art. 430 Part II

CONTINUOUS INDUSTRIAL PROCESS
Electrical System Coordination 240.12
Ground-Fault Protection of Equipment (Branch Circuits) 210.13 Ex. 1
Ground-Fault Protection of Equipment (Feeders) . . 215.10 Ex. 1
Ground-Fault Protection of Equipment (Services) 230.95
Integrated Electrical Systems . . . . Art. 685
Orderly Shutdown 430.44
Power Loss Hazard . . . . 240.4(A)

CONTINUOUS LOAD
Branch-Circuit Conductor Ratings . . . . 210.19(A)(1)
    Overcurrent Protection . . . . 210.20(A)
Branch-Circuit Ratings, Appliances . . . . 422.10(A)
Branch-Circuits, Electric Space Heating . . . . 424.4(B)
Definition of . . . . Art. 100

Electric Pool Water Heaters 680.10
Feeder Conductor Ratings . . . . 215.2(A)(1)
    Overcurrent Protection 215.3
Storage-Type Water Heaters 422.13

CONTROL CIRCUITS
Definition . . . . Art. 100
Fire Pumps 695.14
Industrial Control Panels . . . . Art. 100
Motors . . . . Art. 430 Part VI
    AC Systems 50-1000 Volts Not Required to be Grounded . . . 250.21(A)(3)a, b, c
    Arrangement of 430.74
    Class 1, 2, and 3 Remote Control Circuits . . . . . . Art. 725
    Copper Conductors Required . . . . 430.9(B)
    Definition of Control Circuit . . . . Art. 100
    Disconnecting Means 430.75
    Mechanical Protection 430.73
    Ungrounded 685.14

Overcurrent Protection
Conductors . . . . 430.72(A)&(B)
Control Circuit Transformers . . . . 430.72(C)
Class 1 Circuits 725.41
    Location of Overcurrent Device 725.45
Class 2 or 3 Power Source, Supply Source 725.121
    Class 2 or 3 Circuits . . . . Chapter 9 Tables 11(A) & (B)
Cranes 610.53
For Specific Conductor Applications . . . . 240.4(G)
Maximum Rating . . . . Table 430.72(B)
Motors 300% or 400% . . . . Table 430.72(B)
Orderly Shutdown, Motors 430.44
Torque Requirements . . . . 430.9(C)
See CLASS 1, 2 & 3 REMOTE CONTROL CIRCUITS . . . . Ferm's Finder
See DIFFERENT SYSTEMS IN SAME ENCLOSURE . . . . Ferm's Finder

CONTROL SYSTEMS FOR PERMANENT AMUSEMENT ATTRACTIONS . . . . Article 522

CONTROLLED RECEPTACLE
Building Automation . . . . 406.3(E)
Energy Management . . . . 406.3(E)
Marking . . . . Figure 406.3(E
Replacement . . . . 406.4(D)(7)

CONTROLLER, MOTOR . . . . **Art. 430 Part VII**
   See MOTOR CONTROLLERS . . . . Ferm's Finder

CONVERSION TABLE, AC CONDUCTOR   **Chapter 9 Table 9**

CONVERTERS
  Definition for Phase Converters 455.2
       See PHASE CONVERTERS. . . . . Ferm's Finder
  Definition of, for Recreational Vehicles and Park Trailers . . . . .
  . . . . 551.2 and 552.2
  Voltage Converters Park Trailers. 552.20
  Voltage Converters Recreational Vehicles 551.20

COOKING UNIT, COUNTER-MOUNTED
  Branch Circuits . . . . .210.19(A)(3)
  Definition of . . . . Art. 100 Part I
  Load Calculations 220.55

COOLING OF EQUIPMENT (General) . . . . . . **110.13(B)**
  Ventilation for Transformers 450.9

COORDINATION (SELECTIVE)
  Critical Operations Power Systems (COPS) 708.54
  Definition. . . . . .Article 100
  Elevators 620.62
  Emergency Systems. 700.32
  Fire Pumps . . . . 695.3(C)(3)
  Information Technology Equipment 645.27
  Legally Required Standby Systems 701.27

COORDINATION OF ELECTRICAL SYSTEMS
  Elevator Feeders, Selective Coordination 620.62
  Emergency Systems, Selective Coordination 700.32
  Essential Electric System . . . . 517.31(G)
  Health Care Facilities . . . . 517.17(C)
  Legally Required Standby Systems, Selective Coordination . . 701.27
  Orderly Shutdown 240.12

COPPER CONDUCTORS, *See* **CONDUCTORS, Copper** . .
. . . . Ferm's Finder

(COPS) CRITICAL OPERATIONS POWER SYSTEMS . . . .
Informative Annex F
  See CRITICAL OPERATIONS POWER SYSTEMS . Ferm's Finder

CORD SETS & POWER SUPPLY CORDS
  Green Outer Cover or Jacket Permitted. 400.23
  Ground-Fault Circuit-Interruption Protections . . . 590.6(A)(1)

Mobile and Manufactured Homes 550.10
Park Trailers. 552.43
Recreational Vehicles 551.46
Temporary Installations
  GFCI Protection for Personnel. 590.6
  Not on Floor or Ground . . . . 590.4(J)
See (ELBZ) . . . . UL Product iQ

CORDS . . . . **Art. 400**
Air Conditioners, Room 440.64
Ampacity of Flexible Cord
  Derating for More Than 3 Current-Carrying Conductors . . 400.5
  Engineering Supervision . . . . 400.5(C)
  Flexible Cables . . . . Table 400.5(A)(2)
  Flexible Cords and Cables . . . . Table 400.5(A)(1)
  General. 400.5
  Supply Cord of Listed Appliance or Portable Lamps . . .240.5
Appliances
  Cord- and Plug-Connected Subject to Immersion . . . 422.41
  Disconnection of Cord- and Plug-Connected 422.33
  General. . . . . 422.16(A)
  Polarity in Cord- and Plug-Connected 422.40
  Range Hoods. . . . . 422.16(B)(4)
  Specific Appliances . . . . 422.16(B)
Between an Existing Receptacle Outlet and an Inlet . . 400.10(A)(11)
Bushings Required 400.14
Busways, Branches from . . . . 368.56(B)
  Trolley-Type Busways, Branches from . . . . 368.56(C)
Carnivals, Circuses, Fairs, Wiring Method 525.20
Color Coding, Grounded Conductor. 400.22
Color Coding, Equipment Grounding Conductor 400.23
Counter-Mounted Cooking Units . . . . 422.16(B)(3)
Dishwashers . . . . .422.16(B)(2)
Electric Vehicle Charging System 625.17
Elevators, Not Required to be in Raceway . . . . . 620.21 Ex.
Emergency Unit Equipment . . . . 700.12(F)
  Range Hoods . . . . 422.16(B)(4)
Flexible Cord, Uses Permitted (General) . . . . 400.10(A)
  Adjustable (Fixtures) Luminaires . . . . 410.62(B)
  Attachment Plug Required . . . . 400.10(B)
  Branches from Busways . . . . 368.56(B)
  Elevators, Dumbwaiters, Escalators, etc.
    Car Wiring. .620.21(A)(2)
    Escalators. . . . . .620.21(B)(3)

Hoistways . . . . .620.21(A)(1)
Machine Room and Space, Equipment Components. . . .
. . . . .620.21(A)(3)
Electric Discharge (Fixtures) Luminaires Supported Independently
. . . . 410.24
Electric Discharge (Fixtures) and LED Luminaires   410.62(C)
Motor Disconnecting Means . . . . 430.109(F)
Motors on General-Purpose Branch Circuit . . . . 430.42(C)
Hazardous (Classified) Locations
    Aircraft Energizers, Ground Support Equipment. . 513.10(C)(3)
    Aircraft Mobile Servicing Equipment . . . . . .513.10(D)(2)
    Class I, Division 1 & 2 Locations, Basic Requirements . . . 501.140
    Class II, Division 1 & 2 Locations, Basic Requirements . . 502.140
    Class III, Division 1 Locations, Basic Requirements. 503.140
    Class I, Division 1 Locations, Flexible Connections . . 501.10(A)(2)
    Class I, Division 2 Locations, Flexible Connections . 501.10(B)(2)
    Class II, Division 1 Locations, Flexible Connections . 502.10(A)(2)
    Class II, Division 2 Locations, Flexible Connections . 502.10(B)(2)
    Class III, Division 1 Locations, Flexible Connections 503.10(A)(3)
    Class III, Division 2 Locations, Flexible Connections . .503.10(B)
    Pendant Luminaires (Fixtures) Class II, Division 1 . 502.130(A)(3)
    Pendant Luminaires (Fixtures) Luminaires) Class II, Division 2
    . . . . 502.130(B)(4)
    Pendants Above Class I Locations, Garages. . . . 511.7(A)(2)
    Pendants Not in Class I Locations, Hangars . . . . . 513.7(B)
    Portable Equipment Not in Class I Locations, Hangars. . 513.10(E)
    Portable Lighting Equipment used in Commercial
        Garage, Class I Locations . . . . 511.4(B)(2)
    Process Control Instruments, Class I, Division 2 Locations . .
    . . . . 501.105(B)(6)
Heater Cords & Other Heating Appliances . . 422.43(A), (B)
High Voltage Cable (Portable) . . . . .Art. 400 Part II
Immersion Heaters 422.44
Kitchen Waste Disposers . . . . .422.16(B)(1)
Legally Required Standby Unit Equipment . . . . 701.12(G)
Listed Assembly Type. . . . . 400.10(A)(11)
Manufactured Wiring Systems . . . . 604.100(A)
Markings
    Standard . . . . 400.6(A)
    Optional . . . . 400.6(B)
Mobile Home Feeder Supply . . . . 550.10(B)
Mobile Home Length of Cord . . . . 550.10(D)
Motion Picture and TV Studios . . . . 530 Part II
Motion Picture Projectors. 540.15
Motors

Disconnecting Means . . . . .430.109(F)
Overload Protection Cord- and Plug-Connected Motors. . . .
. . . . 430.42(C)
Office Furnishings
    Freestanding-Type Partitions 605.9
    Lighting Accessories . . . . 605.6(B)
    Partition Interconnections 605.5
Overcurrent Protection 240.5
    Allowable Ampacity Cables. . . . . Table 400.5(A)(2)
    Allowable Ampacity Flexible Cords and Cables Table 400.5(A)(1)
    Cords Approved for Specific Appliances 400.13
Park Trailers Connection and Length of Cord. 552.44
Park Trailers Feeder Supply . . . . 552.43(A)&(B)
Pendants
    Aircraft Hangars . . . . 513.7(B)
    Bathtubs and Shower Areas, Not within Zone . . . 410.10(D)
    Clothes Closets, Not Permitted in . . . . 410.16(B)
    Commercial Garages . . . . 511.7(A)(2)
    Conductors. 410.54
    Dressing Rooms, Lampholders Prohibited 520.71
    Hazardous (Classified) Locations
        Aircraft Hangars. 513.7(B)
        Class I, Division 1 Locations . . . . 501.130(A)(3)
        Class I, Division 2 Locations . . . . 501.130(B)(3)
        Class II, Division 1 Locations . . . . 502.130(A)(3)
        Class II, Division 2 Locations . . . . 502.130(B)(4)
        Class III, Division 1 & 2 Locations . . . . . . 503.130(C)
        Commercial Garages. . . . . 511.7(A)(2)
        Hospitals, Above Hazardous Location   .517.61(B)(3) Ex. 2
        Hospitals, Other-Than-Hazardous Location . . 517.61(C)(1) Ex.
        Makeup Rooms, Not permitted in. 520.71
Permission Required by AHJ (if not in Table 400.4) 400.4
Rating or Setting of Breakers for Cords (General). 240.5
    For Cables by Type. . . . . Table 400.5(A)(2)
    For Flexible Cords and Cables by Type . . . Table 400.5(A)(1)
Range Hoods . . . . .422.16(B)(4)
Recreation Vehicle Feeder Supply . . . . 551.46(A)
Recreation Vehicle Length of Cord . . . . 551.46(B)
Room Air Conditioners. . . . . .Art. 440 Part VII
Show Cases 410.59
Splices, on Construction Sites . . . . .590.4(G)
    Permitted for Repair Only 400.10
Support of, at Terminations for Busways . . . . . . 368.56(B)
Swimming Pools, Fountains, etc.

Equipment Other Than an Underwater Luminaire (Lighting Fixture) 680.8
Fountains . . . . .Art. 680 Part V
Luminaires (Lighting Fixtures) Other Than Underwater . . . . 680.22(B)(5)
Motors. 680.21
Spas and Hot Tubs . . . . Art. 680 Part IV
Storable Pools . . . . Art. 680 Part III
Strain Relief Required . . . . 680.24(E)
Wet-Niche Luminaires (Fixtures). . . . . 680.23(B)
Temporary Locations
   Assured Equipment Grounding Conductor Program . 590.6(B)(2)
   Ground-Fault Circuit-Interrupter Protection . . 590.6(A)(1)
   Physical Protection. . . . . .590.4(H)
   Support . . . . 590.4(J)
Theaters
   Border Lights. . . . . 520.44(C)
   Dressing Rooms, Not Permitted in. 520.71
   Luminaire Supply Cords 520.68
   Other Portable Stage Equipment . . . . . . .Art. 520 Part V
   Permitted for Portable Equipment . . . . 520.5(B)
Trash Compactors (Length of Cord) . . . . 422.16(B)(2)
Travel Trailer (Length of Cord) . . . . 551.46(B)
Types . . . . Table 400.4
Uses Not Permitted 400.12
Uses Permitted 400.10
   Application Use . . . . Table 400.4
   Class I, Division 1 & 2 . . . . 501.140
   From Busways in Industrial Establishments 400.10
   To Connect Lampholders and (Fixtures) Luminaires . . 410.62
   To Connect Motors . . . . 430.42(C)
      Attachment Plug As Disconnect for . . . . 430.109(F)
Water Resistant Cord 400.4
   For Fountain Equipment. 680.56
See Flexible Cords, (ZJCZ) . . . . UL Product iQ

CORRECTION FACTORS, AMBIENT TEMPERATURE
For Allowable Ampacity Values . . . Table 310.15(B)(1) and (2)

CORROSION AND DETERIORATION
Class 1, Class 2, and Class 3 Circuits . . . . 725.3(L)
Corrosive Environment-Defined 680.2
Deteriorating Agents 110.11
Protection, General . . . . 300.6(A) and (B)
   Bushings at Motors. 430.13

Cable Trays. . . . . 392.100(C)
Cellular Metal Floor Raceways . . . . 374.12(1)
Conductors . . . . 310.10(G)
Deicing and Snow-Melting Equipment
   Resistance Heating Elements 426.26
   Skin Effect Heating 426.43
   Electrical Metallic Tubing . . . . 358.10(B)
Equipment Enclosures, etc., Agricultural Buildings . . .547.5(C)(3)
Field-Cut Threads . . . . 300.6(A) Info. Notes
Intermediate Metal Conduit . . . . 342.10(B)
Luminaires (Light Fixtures) . . . . 410.10(B)
Metal Boxes, Conduit Bodies, and Fittings. . . . . 314.40(A)
Metal-Clad Cable 330.12
Metal Wireways . . . . 376.12(2)
Mineral-Insulated, Metal-Sheathed Cable 332.12
Mobile and Manufactured Homes, Power Supply . 550.10(H)
Nonmetallic Extensions . . . . 382.12(3)
Nonmetallic-Sheathed Cable, Type NMC . . . . 334.10(B)(1)
Nonmetallic Wireways . . . . 378.10(2)
Open Wiring on Insulators. . . . . 398.15(B)
Pools- Swimming. 680.14
Pull and Junction Boxes Over 1000 Volts . . . . . 314.72(A)
Rigid Metal Conduit . . . . 344.10(B)
Rigid PVC Conduit . . . . 352.10(B)
Strut-Type Channel Raceway. . . . . 384.10 & 12
Surface Metal Raceways . . . . 386.12(3)
Swimming Pools, Fountains and Similar Installations . 680.14
Underfloor Raceways. 390.12
See Electrical Metallic Tubing (FJMX) . . . . . UL Product iQ
See Intermediate Metal Conduit (DYBY). . . . UL Product iQ
See Rigid Ferrous Metal Conduit (DYIX) . . . UL Product iQ
See Rigid Nonferrous Metal Conduit (DYWV) . . . UL Product iQ
See Rigid PVC Conduit (DZYR). . . . . UL Product iQ

COUNTERTOP OUTLETS — Receptacles
Face-Up Position, Not Permitted . . . . .406.5(G)
Listed Receptacle Assemblies, Face-Up Permitted . . . .406.5(H)
Location and Placement, Dwelling Units
   Bathrooms . . . . 210.52(D)
   Countertop Receptacles not considered required by 210.52(A) . . . . 210.52(A)(4)
   General. 406.5(E)
   Kitchens . . . . 210.52(C)
On Small Appliance Branch Circuits . . . . .210.11(C)(1)

COVE LIGHTING **410.18**

COVER FOR DIRECT BURIAL WIRING METHODS
Definition of . . . . Table 300.5 Note 1
Minimum Cover . . . . Table 300.5
Over 1000 Volts, Nominal, Definition . . . .
. . . . Table 300.50 Superscript Note 1
   Industrial Establishments. . . . . .Table 300.50, Note 3
   Minimum Cover . . . . Table 300.50
   Protected by Concrete . . . . . . .Table 300.50, Footnote d
Swimming Pools, Fountains and Similar Installations. . . . . . .
. . . . 680.11 and Table 300.5

COVERS
Auxiliary Gutters . . . . 366.100(D)
Boxes
   Attachment Methods (Screws) 314.25
   Construction 314.41
   Over 1000 Volts . . . . 314.72(E)
   Required, General . . . . 314.28(C)
      Luminaire Outlets 410.22
Extensions From . . . . 314.22 Ex.
Handholes . . . . 314.30(D)
Manholes . . . . 110.75(D)

CRANES & HOISTS . . . . **Art. 610**
Ampacity of Conductors . . . . 610.14(A)
   Contact Conductors . . . . 610.14(D)
   Minimum Size . . . . 610.14(C)
   Secondary Resistor Conductors . . . . 610.14(B)
Class III Hazardous (Classified) Locations 503.155
Common Return 610.15
Contact Conductors . . . . Art. 610 Part III
Control Circuits . . . . Art. 610 Part VI
Demand Factors . . . . Table 610.14(E)
   Application of . . . . 610.14(E)
Disconnecting Means . . . . Art. 610 Part IV
   Runway Conductors . . . . 610.31 Ex.
Grounding and Bonding 610.61
Operating, Class III Locations 503.155
Overcurrent Protection . . . . .Art. 610 Part V
Runway Conductors, Disconnecting Means . . . . . 610.31 Ex.
Ungrounded Supply Circuit Over Class III Locations 250.22(1)
See (ELPX) . . . . UL Product iQ

CRAWL SPACES

Lighting Outlets
   GFCI Protection . . . . 210.8(C)
Receptacles
   GFCI Protection
      Dwelling Units . . . . 210.8(A)(4)
      Other Than Dwelling Units . . . . 210.8(B)(9)

CRIMP TOOLS, CLASSIFIED FOR USE WITH SPECIFIC WIRE CONNECTORS
See (ZMLS) . . . . UL Product iQ

CRITICAL OPERATIONS POWER SYSTEM (COPS) Art. 708
Accessible to Authorized Persons Only 708.50
Availability and Reliability/Development and
Implementation/Functional
Performance Testing . . . . Informative Annex F
Branch Circuit (Circuit Wiring) 708.10
Branch Circuits Supplied by COPS 708.30
Capacity of Power Sources 708.22
Commissioning 708.8
   Definitions 708.2
Emergency Operations Plan 708.64
   Equipment 708.11
Feeder (Circuit Wiring) 708.10
Ground-Fault Protection of Equipment 708.52
Identification of Boxes and Enclosures . . . . 708.10(A)
Generators
   Outdoor, Permanently Installed . . . . . . 708.20(F)(5)(a)
   Outdoor, Portable . . . . 708.20(F)(5)(b)
Overcurrent Protection . . . . Art. 708 Part IV
Physical Security 708.5
Risk Assessment 708.4
Scope 708.1
Selective Coordination 708.54
Selectivity . . . . 708.52(D)
Sources of Power 708.20
Surge Protection . . . . 708.20(D)
Testing and Maintenance 708.6
Transfer Equipment 708.24
Ventilation 708.21
Wiring of Feeders and Branch Circuits 708.10
See . . . . Informative Annex F

CROSS-SECTIONAL AREAS
Compact Aluminum . . . . Chapter 9 Table 5A

Conductors . . . . Chapter 9 Table 5
   Bare . . . . Chapter 9 Table 8
Conduit . . . . Chapter 9 Table 4

CURRENT LIMITING

Cartridge Fuses, Marking . . . . 240.60(C)
Circuit Breakers, Marking 240.83
   See (DIRW) . . . . UL Product iQ
Circuit Impedance, Short Circuit Current Ratings and Other Characteristics 110.10
Interrupting Rating .110.9
Overcurrent Device, Definition of 240.2
Series Ratings 240.86
See Cartridge Fuses, Nonrenewable(JDDZ) . . . . UL Product iQ
See Cartridge Fuses, Renewable (JDRX) . . . . . UL Product iQ

D

DAMP OR WET LOCATIONS

Agricultural Buildings . . . . 547.5(C)(2)
Receptacles . . . . .547.5(G)
Boxes and Fittings. 314.15
Cabinets, Cutout Boxes, and Meter Socket Enclosure . . .312.2
Class 1, Class 2, and Class 3 Circuits . . . . 725.3(L)
Definition of . . . . Art. 100
Deteriorating Agents 110.11
Drainage Openings, Field Installed (Boxes, etc.) 314.15
Electric Signs and Outline Lighting. . . . . .600.9(D)
 Electrode Connections . . . . 600.42(G)
Equipment Connected by Cord and Plug. . . . . 250.114(4)(F)
Fixed Electric Space Heating Equipment . . . . 424.12(B)
Grounding Equipment in . . . . .250.110(2)
Lampholders 410.96
Luminaires in Specific Locations (Fixtures) . . . . . . 410.10(A)
Marinas and Boatyards . . . . .Art. 555 Part II
Materials to Be Identified for the Use 110.11
Open Wiring 398.10
Overcurrent Protection 240.32
Panelboards 408.37
Receptacles, Weather Resistant 406.9
Signs, Portable and Mobile . . . . .600.9(D)
Switches. 404.4

Temporary Installations . . . . 590.6(A)(3)
Switchboards 408.16

DANCE HALLS (ASSEMBLY OCCUPANCIES) . . . Art. 518

DANGER SIGNS 300.45

DATA PROCESSING (INFORMATION TECHNOLOGY) EQUIPMENT . . . . Art. 645
See INFORMATION TECHNOLOGY EQUIPMENT. Ferm's Finder

D.C. SYSTEMS . . . . Art. 250 Part VIII

Branch Circuit Identification of Ungrounded Conductor. . . . . 210.5(C)(2)
Combination Electrical Systems
   Park Trailers 552.20
   Recreational Vehicles 551.20
Concrete Encased Electrodes . . . . 250.166(D)
Equipment Grounding Conductor . . . . .250.134(B)
   Grounded System Required 250.162
   Sizing. 250.166
Ground-Fault Detection
   Grounded Systems . . . . .250.167(B)
   Marking . . . . .250.167(C)
   Ungrounded Systems. . . . . 250.167(A)
Integrated Electrical Systems 685.12
Overcurrent Protection Generator Systems 445.12
Point of Connection 250.164
Signage, Caution
   Resistively Grounded DC Systems . . . . .408.3(F)(5)
   Ungrounded DC Systems . . . . .408.3(F)(4)
Size of Bonding Jumper. 250.168
Size of Grounding Electrode Conductor 250.166
Storage Batteries . . . . Art. 480
Theaters, Motion Picture and Television Studios, and Similar Locations . . . .
   Over 1000 Volts, Direct Current Switchboards. 530.64
Solar Photovoltaic (PV) Systems . . . . Art. 690
   AFCI Protection 690.11
   Circuit Requirement 690.7
   Circuit Sizing and Current 690.8
   Disconnecting Means 690.13
   Grounding 690.41
   Grounding Electrode System . . . . 690.47(A)
   Marking . . . . Art. 690 Part VI

Overcurrent Protection. . . . . 690.9(B)
Over 1000 Volts . . . . Art. 690 Part IX
Self-Regulated PV Charge Control (Battery) 690.72
Wiring Method . . . . 690.31(G)

Ungrounded Conductor Identification . . . . 210.5(C)(2)
Ungrounded Separately Derived Systems 250.169
Use of Grounded Conductor for Load Side Equipment. . . . . .
. . . . 250.142(B) Ex. 3
Wind Electric Systems . . . . Art. 694
   Charge Control (Battery). . . . . Art. 480
   Disconnecting Means 694.20
   Overcurrent Protection. 694.15
   Storage Batteries . . . . Art. 480
   Wiring Method 694.30

DEAD ENDS
Busways 368.58
Cablebus . . . . 370.42(2)
Flat Cable Assemblies . . . . 322.40(A)
Metal Wireways 376.58
Nonmetallic Wireways 378.58

DEAD FRONT
Attachment Plugs . . . . 406.7(A)
Definition. . . . . Art. 100 Part I
Informative Annex A, Enclosed and Dead Front Switched . . UL 98
Mobile and Manufactured Homes, Distribution Panelboards . 550.11
Panelboards 408.38
Park Trailers, Distribution Panelboards . . . . 552.45(C)
Recreational Vehicles
   Definition 551.2
   Distribution Panelboard . . . . 551.45(C)
Theaters, Motion Picture and Television Studios, and
Similar Locations
   Stage Switchboards, Fixed 520.21
   Over 1000 Volts, Substations. 530.64

DEDICATED
Branch Circuits
   Bathrooms (Dwellings). . . . . 210.11(C)(3)
   Central Heating Equipment Other Than Electric . . . 422.12
   Fire Alarm or Burglar System. . . . . 210.8(A)(5) Ex.
   Laundry (Dwellings) . . . . 210.11(C)(2)
   Receptacle for Deicing or Snow-Melting Equipment,
      Exempt from GFCI. . . . . 210.8(A)(3) Ex.

   Small Appliance . . . . 210.11(C)(1)
      Disconnect for HVAC Equipment 645.10
      Fire Pump Transformers, Sizing of Dedicated . . 695.5(A)
   Signs . . . . 600.5(A)
   Unfinished Basements Appliances That Occupy . . . .
. . . . 210.8(A)(5)
   Wind Electric Systems . . . . 694.7(D)
Equipment Space
   Indoor. 110.26(E)(1)
   Outdoor . . . . 110.26(E)(2)
Feeder
   Fire Pump . . . . 695.3(A)(3)

DEFINITIONS
NOTE 1: Many of the categories listed below have additional definitions that are specific to that general subject. These additional definitions are also found in the respective ".2" sections.

NOTE 2: Article 100 includes only those definitions that are considered to be essential to the proper application of the *NEC*. In general, only those terms used in two or more articles are defined in Article 100. The definitions included here are provided in the article in which they are used. In some cases, these may be referenced in Article 100.

Air-Conditioning and Refrigerating Equipment 440.2
Aircraft Painting Hangar . . . . Art. 100
Arc-Fault Circuit-Interrupter (AFCI) . . . . Art. 100
Armored Cable 320.2
Audio Signal Processing, Amplification and Reproduction
   Equipment. 640.2
Busways. 368.2
Cable Tray 392.2
Cablebus 370.2
Cellular Concrete Floor Raceways (Cell) 372.2
Cellular Metal Floor Raceways 374.2
Class I, II, & III Locations (General Area Classifications) . .500.5
Class I, Zone 0, 1, & 2 Locations. . . . . Art. 100
   General Area Classifications 505.5
Communication Circuits 800.2
Community Antenna TV and Radio Distribution Systems .800.2
Concealed Knob-and-Tube Wiring 394.2
Control Circuit, Motor . . . . Art. 100
Controller, Motor. 430.2
Cover (Direct Burial) . . . . Table 300.5, Note 1

Over 1000 Volts . . . . Table 300.50, Superscript Note a
Electric-Discharge Lighting . . . . Art. 100
Electric Signs . . . . Art. 100
Electrical Metallic Tubing 358.2
Electrical Nonmetallic Tubing 362.2
Electrically Driven or Controlled Irrigation Machines . . . 675.2
Electrolytic Cells 668.2
Elevators, Dumbwaiters, Escalators, Etc. 620.2
Fire Alarm Circuit 760.2
Fire Alarm Circuit Integrity Cable 760.2
Fire Alarm Systems 760.2
Fixed Outdoor Electric Deicing and Snow-Melting Equipment. 426.2
Fixed Electric Heating Equipment for Pipelines and Vessels . . 427.10
Flat Cable Assemblies 322.2
Flat Conductor Cable 324.2
Flexible Metal Conduit 348.2
Flexible Metallic Tubing 360.2
Floating Buildings 555.2
Fuel Cell Systems 692.2
General, Not Over 1000 Volts, Nominal . . . . . Art. 100 Part II
Grounding . . . . Art. 100
Hazardous (Classified) Locations . . . . Art. 100
Health Care Facilities 517.2
High Density Polyethylene Conduit, Type HDPE 353.2
Induction and Dielectric Heating Equipment. 665.2
Industrial Control Panels . . . . Art. 100
Industrial Machinery . . . . Art. 100
Instrumentation Tray Cable. 727.2
Integrated Gas Spacer Cable 326.2
Interconnected Electric Power Production Sources 705.2
Intermediate Metal Conduit 342.2
Intrinsically Safe Systems . . . . Art. 100
Inverter (Interactive) . . . . Art. 100
Liquidtight Flexible Metal Conduit. 350.2
Liquidtight Flexible Nonmetallic Conduit 356.2
Manufactured Buildings 545.2
Manufactured Wiring Systems 604.2
Medium Voltage Cable 311.2
Messenger Supported Wiring 396.2
Metal Wireways. 376.2
Metal-Clad Cable 330.2
Mineral-Insulated, Metal-Sheathed Cable. 332.2
Mobile Equipment . . . . Art. 100
Mobile Homes, Manufactured and Mobile Home Parks . . 550.2

Motion Picture and TV Studio & Similar Locations 530.2
Motion Picture Projectors. . . . . Art. 100
Motor Fuel Dispensing Facilities . . . . Art. 100
Natural and Artificially Made Bodies of Water 682.2
Network-Powered Broadband Communication Systems . . 830.2
Nonmetallic Extensions 382.2
Nonmetallic Sheathed Cable 334.2
Nonmetallic Underground Conduit with Conductors . . . 354.2
Nonmetallic Wireways 378.2
Non-Power-Limited Fire Alarm Circuit (NPLFA) 760.2
Open Wiring On Insulators. 398.2
Optical Fiber Cables and Raceways 770.2
Over 1000 Volts, Nominal (General) . . . . . . Art. 100 Part II
   High Voltage 490.2
P Cable 337.2
Park Trailers. 552.2
Phase Converters 455.2
Power and Control Tray Cable 336.2
Power-Limited Fire Alarm Circuit (PLFA) 760.2
Radiant Heating Panels 424.2
Radiant Heating Panel Sets 424.2
Radio and TV Equipment 810.2
Recreational Vehicles and Recreation Vehicle Parks 551.2
Remote Control Signaling, and Power Limited Circuits
Class I, 2 & 3 . . . . 725.2
Retrofit Kit . . . . Art. 100
Service Entrance Cable 338.2
Sign Body 600.2
Section Sign. 600.2
Skeleton Tubing 600.2
Solar Photovoltaic Systems 690.2
Storage Batteries 480.2
Surge Arresters . . . . Art. 100
Surge-Protective Device (SPD) . . . . Art. 100
Swimming Pools, Fountains, etc. 680.2
Tap Conductor 240.2
Theaters,
Motion Picture and Television Studios, and Similar Locations 520.2
Transformer. 450.2
Underfloor Raceway 390.2
Wind Electric Systems 694.2
X-Ray Equipment,
   Medical 517.2
   Nonmedical 660.2

DEFLECTION FITTINGS . . . . **300.7(B)**

DEFLECTION OF CONDUCTORS
   See CONDUCTORS, Bending Space . . . . Ferm's Finder

DEICING & SNOW-MELTING . . . . **Art. 426**
   Ampacity and Size . . . . 210.19(A)(4), Ex. 1(e)
   Branch-Circuit Sizing 426.4
   Control & Protection . . . . Art. 426 Part VI
   Controllers 426.51
   Corrosion Protection
      Resistance Heating Elements 426.26
      Skin-Effect Heating 426.43
   Disconnecting Means 426.50
   Expansion Joints
      Embedded Nonheating Leads . . . . 426.22(D)
      Embedded Resistance Elements . . . . 426.20(E)
      Exposed Resistance Elements. . . . . .426.21(C) & (D)
   Free Conductors at Boxes
      Resistance Heating Elements, Embedded. . . . . . 426.22(E)
      Resistance Heating Elements, Exposed . . . . . . 426.23(A)
   Grounding
      Resistance Heating Elements 426.27
      Skin-Effect Heating 426.44
   Ground-Fault Protection of Equipment, Resistance Units 426.28
   Impedance Heating . . . . Art. 426 Part IV
   Installation . . . . .Art. 426 Part II
   Overcurrent Protection . . . . 240.4(B)
      Setting of (Continuous Load) 426.4
   Resistance Heating Elements . . . . Art. 426 Part III
   Skin-Effect Heating . . . . .Art. 426 Part V
   Tap Conductors. . . . . 210.19(A)(4), Ex. 1(e)

DELTA BREAKERS (Prohibited in Panelboards) . . . .
. . . . **408.36(C)**

DELTA CONNECTED (HIGH-LEG)
   High-Leg Marking, General 110.15
   Phase Arrangement in Switchboards and Panelboards. . 408.3(E)
   Service Conductor Identification 230.56

DEMAND FACTORS
   See CALCULATIONS, Demand Factors . . . . . . Ferm's Finder

DENTAL CLINICS AND OFFICES
   Defined, Health Care Facilities 517.2
   Essential Electrical System 517.45
   Wiring and Protection . . . . .Art. 517 Part II
   See HEALTH CARE FACILITIES . . . . Ferm's Finder
   See Standard for Health Care Facilities, NFPA 99 .NFPA 99-2012

DERATING AMPACITY
   Auxiliary Gutters . . . . 366.22(A)
      When Applicable. 366.23
   Cable Trays
      Conductors Rated 2000 Volts or Less 392.22
      Conductors Rated 2001 Volts or Over . . . . 392.22(C)
   Fire Alarm Systems
      NPLFA Circuits . . . . .Art. 760 Part II
      PLFA Circuits Using NPLFA Methods. . . 760.130(A) Ex. 3
   Parallel Conductors . . . . 310.10(G)(4)
   Remote Control, Signaling Circuits, Class 1 . . . . . . 725.51
   Strut-Type Channel Raceway 384.22
   Surface Metal Raceway 386.22
   See Also AMPACITY OF, Derating Allowable Ampacity . . . . .
. . . . Ferm's Finder
   See Also CONDUCTORS, Derating *Allowable Ampacity*. . . . .
. . . . Ferm's Finder

DETERIORATING AGENTS **110.11**
   Integrity of Equipment and Connections . . . . . . 110.12(B)

DEVICE
   Defined. . . . . Art. 100
   Nonmetallic-Sheathed Cable Interconnector . . . . . 334.40(B)
   Self-Contained . . . . 334.40(B)

DEVICE OR EQUIPMENT FILL FOR BOXES . **314.16(B)(4)**

DIAGRAMS (FIGURES)
   Adjustable Speed Drive Control. . . .Informative Annex D, D10
   Autotransformer Overcurrent Protection 450.4
   Bulk Storage (Marine Terminal). 515.3
   Branch Circuit, Feeder, and Service Calculation Methods. .220.1
   Cable Installation Dimension Details Over 2001 Volts . . 311.60
      For Use under Engineering Supervision   Informative Annex B
. . . . .Table B.2 (1,3,5,6,7,8,9,10,11)
   Cable TV Substitutions (Coaxial Type Cables Only) . . 820.179
   Class I, Zone 0, 1, and 2 Locations, Marking. . . . 505.9(C)(2)
   Class 2 and Class 3 Cable Substitutions . . . . . . .725.154(A)
   Class 2 and Class 3 Circuits 725.121
   Closet Storage Space 410.2
   Code Arrangement 90.3

Control Drawings, Required for Intrinsically Safe System Equipment 504.10(A)

Elevators, Dumbwaiters, Escalators, etc. 620.13

    Control System 620.2

Feeders, Diagrams If Required 215.5

Fire Alarm System Cable Substitutions . . . . .760.154(A)

Generator Field Control . . . . Informative Annex D, D9

Grounding and Bonding 250.1

Grounding Symbol 250.126

    For Receptacles. . . . . .406.10(B)(4)

Hospital Essential Electrical Systems IN Figure 517.31 No.1 & 2

Mobile Home Power Supply Configurations . . . . . 550.10(C)

Motors, Article 430 Contents. 430.1

Motor Fuel Dispensing Facilities 514.3

Nursing Homes and Limited Care Facilities. . . . .
. . . . Info. Note Figure 517.42(B) No. 1 & 2

Optical Fiber Cable Substitutions . . . . . . Table 770.154(b)

Park Trailer Power Supply Configurations. . . . . . 552.44(C)

Recreational Vehicle Power Supply Configurations . . 551.46(C)

Service 230.1

Solar Photovoltaic Systems

    Identification of System Components . . . . . Figure 690.1(a)

    Identification of System Components in Common System Configurations. . . . . .Figure 690.1(b)

Spray Applications . . . . Figure 516.4

Swimming Pools, Conductor Clearances . . . . Figure 680.8(A)

Wind Electric Systems

    Components, Interactive System . . .IN Figure 694.1 No. 1(a)

    Components, Stand-Alone System . IN Figure 694.1 No. 2(b)

DIFFERENT SYSTEMS IN SAME ENCLOSURE
See CONDUCTORS, Different Systems in Same Enclosure
. . . . Ferm's Finder

DIMENSIONS & PERCENT AREA OF CONDUCTORS
See CONDUCTORS, Dimensions & Area of . . . Ferm's Finder

DIMENSIONS & PERCENT AREA OF CONDUIT   Chapter 9 Table 4
See . . . . Conduit Fill Tables, Chapter 9 Table 5
See . . . . .For Compact Conductors, Chapter 9 Table 5A

DIMMERS
Device Fill for Boxes . . . . .314.16(B)(4)
Grounding of . . . . .404.9(B)
In Fixed Stage Switchboards. 520.25

DIP TANKS (PAINT) . . . . **Art. 516**
See SPRAY APPLICATIONS, DIPPING, COATING & PROCESSES. . . . . Ferm's Finder

DIRECT BURIAL
See BURIED CONDUCTORS . . . . Ferm's Finder
See CABLES, Underground. . . . . Ferm's Finder
See CONDUCTORS, Underground . . . . Ferm's Finder

DIRECT CURRENT MICROGRIDS. . . . . .Art. 712
Building Directory . . . . 712.10(B)
Circuit Requirements . . . . .Art. 712 Part II
Definitions 712.2
Directory 712.10
Disconnecting Means . . . . Art. 712 Part III
Identification of Circuit Conductors 712.25
Interrupting and Short-Circuit Ratings 712.72
Listing and Labeling 712.4
Marking. .Art. 712 Part V
Overcurrent Protection 712.70
Protection. Art. 712 Part VI
Systems over 1000 Volts . . . . .Art. 712 Part VII
System Voltage 712.30
Wiring Methods . . . . Art. 712 Part IV

DIRECT CURRENT SYSTEMS
See D.C. Systems . . . . Ferm's Finder

DIRECTLY CONTROLLED LUMINAIRES 700.24

DIRECTORY SEE LABELING . . . . .Ferm's Finder
See MARKING . . . . Ferm's Finder

DISCHARGE STORED ENERGY
Capacitors Under 1000 Volts 460.6
Capacitors Over 1000 Volts. 460.28

DISCONNECTING MEANS
Agricultural Buildings. . . . . 547.9(B)(C)
Air Conditioning & Refrigeration . . . . .Art. 440 Part II
    For Cord-Connected Equipment. 440.13
    For Dedicated HVAC in Information Technology Room . 645.10
    Location of. 440.14
    Room Air Conditioners 440.63
Appliances . . . . Art. 422 Part III
    Cord-Connected 422.33
    Motor-Driven . . . . 422.31(C)

Permanently Connected 422.31
Unit Switch(es) 422.34
Batteries 480.6
Capacitors . . . . 460.8(C)
Over 1000 Volts, Nominal 460.24
Carnivals, Circuses, Fairs and Similar Events . . . . . 525.21(A)
Circuits, Motor Fuel Dispensing Facilities 514.11
Computers 645.10
See INFORMATION TECHNOLOGY EQUIPMENT . . . . Ferm's Finder
Cord-and-Plug, Exemption for Lockable . . . . . . 110.25 Ex.
Cranes and Hoists . . . . Art. 610 Part IV
Definition of . . . . Art. 100 Part I
Deicing and Snow-Melting Equipment . . . . . Art. 426 Part VI
Duct Heaters 424.65
Electric Vehicle Charging System 625.43
Branch Circuit 625.40
Disconnecting Means 625.43
Overcurrent Protection 625.41
Rating 625.42
Electrolytic Cells 668.13
Electroplating 669.8
Elevators . . . . Art. 620 Part VI
Car HVAC 620.54
Car Lights, Receptacles, and Ventilation 620.53
Power from More Than One Source 620.52
Single Disconnecting Means Each Unit 620.51
To Disconnect Normal and Emergency Power . . . 620.91(C)
Utilization Equipment 620.55
Fire Pumps . . . . 695.4(B)(1)(3)
Fixed Electric Heating Equipment for Pipelines and Vessels . . 426.50
Fixed Electric Space-Heating Equipment 424.19
Fuel Cell Systems . . . . Art. 692 Part III
Fuses & Cutouts 240.40
Readily Accessible 240.24
Generators 445.18
Heating Equipment (Induction & Dielectric) 665.12
Heating Equipment (Space) . . . . Art. 424 Part III
Location for Duct Heater 424.65
Required 424.19
Thermostats as 424.20
Heating Equipment for Pipelines and Vessels . . Art. 427 Part VII
High Voltage Systems
Mobile and Portable Equipment . . . . 490.51(D)

Overcurrent Device as 230.206
Service Disconnecting Means 230.205
See OVER 1000 VOLTS, NOMINAL . . . . . . Ferm's Finder
Hot Tubs & Spas 680.13
Emergency Switch for Nondwelling Units 680.41
Identification of (General) 110.22
Circuits in Panelboards 408.4
Industrial Machinery . . . . 670.4(B)
Information Technology Equipment 645.10
See INFORMATION TECHNOLOGY EQUIPMENT . . . . Ferm's Finder
Interconnected Electric Power Production Sources
Disconnect Device 705.22
Disconnecting Means, Equipment 705.21
Disconnecting Means, Sources 705.20
Irrigation Machines 675.8
Lockable 110.25
Air-Conditioning and Refrigerating Equipment . . . 440.14
Appliances . . . . 422.31(B)
Carnivals, Circuses, Fairs and Similar Events
Rides, Tents, Concessions 525.21
Services 525.10
Cranes and Hoists . . . . 610.31 and 610.32
Electric-Discharge Lighting, More than 1000 Volts . 410.141(B)
Electric Vehicle Charging System 625.43
Electrically Driven or Controlled Irrigation Machines . . 675.8(B)
Electrified Truck Parking Spaces
Parking Space . . . . 626.24(C)
Supply Wiring . . . . 626.22(D)
Transport Refrigerated Units 626.31
Elevators, Etc.
Car Light, Receptacle(s), and Ventilation Disconnecting Means 620.53
Disconnects, General 620.51
Heating and Air-Conditioning Disconnecting Means . . 620.54
Utilization Equipment Disconnecting Means 620.55
Equipment Over 1000 Volts, Circuit Breakers 490.46
Equipment Over 1000 Volts, Interrupter Switches . . . 490.44
Feeder Disconnecting Means . . . . 225.52(C)
Fire Pumps 695.4
Fixed Electric Space-Heating Equipment . . 424.19(A) and (B)
Fixed Outdoor Electric Deicing and Snow-Melting Equipment . . . . 426.51(A) and (D)
Generators 445.18
Induction and Dielectric Heating Equipment 665.12

Motors	430.102(A) and (B)
Motors, Over 1000 Volts	430.227
Motors with More Than One Source of Power	430.113
Outdoor Lamps	225.25
Sensitive Electronic Equipment, Lighting Equipment	647.8(A)
Signs	600.6(A)
Transfer Equipment	702.5
Transformers	450.14

Marinas and Boatyards, Shore Power Connection . . . 555.33(A)
Mobile and Manufactured Homes 550.11
   Interconnection . . . . 550.19(B)
   Service Equipment 550.32
Motors & Controllers . . . . Art. 430 Part IX
   All Disconnects in Circuit to Comply 430.108
   Ampere Rating and Interrupting Capacity 430.110
   Both Controller and Disconnecting Means 430.111
   Energy from More Than One Source 430.113
   Grounded Conductors 430.105
   Location of 430.102
   Motors Served by Single Disconnecting Means 430.112
   Operation 430.103
   Over 1000 Volts 430.227
   Readily Accessible 430.107
   To Be Indicating 430.104
   Type of 430.109
   See MOTORS . . . . Ferm's Finder
Motor Control Circuit 430.75
Motor Fuel Dispensers
   Circuits 514.11
   Provisions for Maintenance and Servicing 514.13
Multiple-Occupancy Building
   Grouping of 230.72
   Location in or on Premises . . . . 240.24(A)
   Locked Service Overcurrent Devices 230.92
   Maximum Number of 230.71
   Separate Service Disconnects Grouped One Location
   . . . . 230.40 Ex. 2
Outside Branch Circuits and Feeders
   Access to Occupants 225.35
   Construction of 225.38
   Grouping of 225.34
   Identification of 225.37
   Location of 225.32
   Maximum Number of 225.33

   Over 1000 Volts . . . . Art. 225 Part III
   Isolating Switches 225.51
   Location of . . . . 225.52(A)
   Not Readily Accessible . . . . 225.52(A)
   Type of . . . . 225.52(B)
   Rating of 225.39
   Supplying More Than One Building or Structure . . . 225.31
   Type 225.36
   To Be Suitable for Service Equipment 225.36
Park Trailers . . . . 552.45(C)
Phase Converters 455.8
Readily Accessible
   Air-Conditioning and Refrigerating Equipment 440.14
   Carnivals, Circuses, Fairs and Similar Events
      For Rides, Tents, and Concession 525.21
   Cranes and Hoists
      Cranes and Monorail Hoists 610.32
      Rating of 610.33
      Runway Contact Conductor . . . . 610.31(1)
   Definition of . . . . Art. 100 Part I
   Electric Vehicle Supply Equipment 625.43
   For Motors and Controllers 430.107
   In or On Premises 240.24
   Induction and Dielectric Heating 665.12
   Information Technology Equipment 645.10
   More than One Building or Structure Supplied 225.32
   Service Disconnecting Means . . . . 230.70(A)(1)
   Solar Photovoltaic Systems . . . . 690.13(A)
   Spas and Hot Tub Emergency Switch 680.41
   Switches and Circuit Breakers . . . . 404.8(A)
Recreational Vehicles . . . . 551.45(C)
   Site Supply Equipment . . . . 551.77(B)
Sensitive Electronic Equipment, Luminaires . . . . . . 647.8(A)
Sensitive Electronic Equipment, Three Phase Systems . . 647.5
Separate Buildings
   Supplied by Branch Circuit or Feeder
      See DISCONNECTING MEANS, *Outside Branch* . . . .
      . . . . Ferm's Finder
   Supplied by Service 230.70
      See DISCONNECTING MEANS, Service Equipment Ferm's
      Finder
Service Equipment . . . . Art. 230 Part VI
   Available Short-Circuit Current 110.9
   Cartridge Fuses and Fuseholders 240.60

Circuit Breakers 240.83
Circuit Impedance and Other Characteristics 110.10
Combined Rating 230.80
Connection to Terminals. 230.81
Equipment Connected to Supply Side 230.82
Grouping of 230.72
    Additional Service Disconnecting Means . . . . 230.72(B)
Labeled, Definition of . . . . Art. 100
Listed, Definition of . . . . Art. 100
Location of. . . . . 230.70(A)
Marked as Suitable for Use as Service Equipment . . . 230.66
Maximum Number of 230.71
Minimum Size and Rating 230.79
Mobile Home Disconnecting Means . . . . 550.32(F)
Mounting Height Switches and Circuit Breakers 404.8
    Multiple Branch Circuits (Simultaneous Disconnection) . 210.7
Multiple Occupancy Building
    Each Occupant to Have Access. . . . . 240.24(B)
    Grouping of 230.72
    Location, Relative to Overcurrent Protection. 230.94
    Locked Service Overcurrent Devices 230.92
    Two to Six Separate Enclosures, Grouped . . . . 230.40 Ex. 2
Number of Permitted. . . . . 230.71(A)
Provisions for Disconnecting the Grounded Conductor . 230.75
Provisions to Bond the Grounded Conductor to the Main
    Service Equipment . . . . 250.24(B)
    Load Side Grounding Connections . . . . 250.24(A)(5)
Rating of 230.79
Simultaneous Opening of Poles. 230.74
Supply-Side, Marking 230.82
See SERVICE EQUIPMENT, Main Disconnect . . Ferm's Finder
Signs and Outline Lighting 600.6
Solar Photovoltaic Systems . . . . Art. 690 Part III
    Energy Storage Systems 706.7
    Type 690.15
    Storage Batteries 480.7
Spas & Hot Tubs 680.13
    Emergency Switch for 680.41
Swimming Pool Equipment 680.13
Transformers 450.14
Welders
    Arc Welders 630.13
    Resistance Welders 630.33
Wet Locations

Cabinets and Cutout Boxes 312.2
Switch and Circuit Breaker Enclosures 404.4
Switchboards 408.16
Wind Electric Systems . . . . Art. 694 Part III
Storage Batteries 480.6
X-Ray Equipment
    In Health Care Facilities 517.72
    Independent Control, Nonmedical. 660.24
    Other Uses, Nonmedical 660.5

DISCONTINUED OUTLETS
Cellular Concrete Floor Raceways 372.58
Cellular Metal Floor Raceways 374.58
Underfloor Raceways 390.57

DISHWASHERS
Arc-Fault Circuit-Interrupter Protection . . . . 210.12(A)
Booster Heaters, Nondwelling Load Demand Factors . . 220.56
Cord Connected . . . . 422.16(B)(2)
Disconnecting Means . . . . Art. 422 Part III
Ground-Fault Circuit-Interrupter Protection . . . . 422.5(A)(7)
Grounding of, Cord- and Plug-Connected 250.114
Grounding of, Permanent Wiring Methods 250.110
Maximum Load. 220.18
Minimum Branch Circuit 422.10
Mobile Homes (Portable Appliance) . . . . . . 550.2 Info. Note
Permissible Load 210.23

DISPOSER, KITCHEN WASTE
Arc-Fault Circuit-Interrupter Protection . . . . 210.12(A)
Cord Connected . . . . 422.16(B)(1)
Disconnecting Means . . . . Art. 422 Part III
Ground-Fault Circuit-Interrupter Protection, Within 6 feet of Sink
. . . . 210.8(A)(7)
Grounding of, Cord- and Plug-Connected 250.114
Grounding of, Permanent Wiring Methods 250.110
Maximum Load. 220.18
Minimum Branch Circuit. 422.10
Permissible Load 210.23

DISSIMILAR METALS (General) 110.14
Electrical Metallic Tubing. 358.14
Ground Clamps Listed for Electrode & Conductor Materials 250.70
Intermediate Metal Conduit 342.14
Rigid Metal Conduit 344.14
    Stainless Steel (and fittings). 344.14

Tower Grounding Connections . . . . 694.40(B)

DOCUMENTATION
AC and Refrigerating Equipment . . . . 440.10(B)
AHJ (Waive Requirements) . . . . Informative Annex H
Arc Energy Reduction. 240.87
    Circuit Breakers . . . . 240.87(A)
    Fuses . . . . 240.67(A)
Critical Operations Power Systems 708.1
    Transfer Equipment . . . . 708.24(E)
Emergency Systems . . . . 700.5(E)
Hazardous (Classified) Locations . . . . 500.5(A)
Industrial Control Panels . . . . 409.22(B)
Integrated Electrical System 685.1
Intrinsically Safe Systems . . . . 504.10(C)
Large-Scale Photovoltaic Electric Power Production Facility . . .
    Arc-Fault Mitigation 691.10
    Conformance to Engineered Design 691.7
    DC Operating Voltage 691.8
    Engineered Design 691.6
    Fence Grounding 691.11
Legally Required Standby Systems . . . . 701.5(D)
Manholes . . . . 110.70 Ex.
Manufactured Buildings, Building System 545.2
Motor Fuel Dispensing Facilities . . . . . 514.3(D) Ex. 1 and 2
Outdoor Overhead Conductors Over 1000 Volts 399.30
Series Ratings . . . . 240.86(A)
Substations 490.48
Zone 0, 1, and 2 Locations . . . . 505.4(A)
Zone 20, 21, and 22 Locations . . . . 506.4(A)

DORMITORIES
Arc-Fault Circuit Interrupter Protection . . . . 210.12(C)
Definition . . . . Art. 100
Overcurrent Protection, Not in Bathrooms . . . . . 240.24(E)
Receptacle Outlet Placement Requirements . . . . . 210.60(B)

DOORS
Clearances from, Outside Branch Circuits and Feeders 225.19(C)
Clearances From, Services . . . . 230.9(A)
Door Stops 490.38
Enclosures, Over 1000 Volts . . . . 110.31(A)(3)
Flexible Cords Prohibited . . . . 400.12(3)
Garage Vehicle Door . . . . 210.70(A)(2)
Locks, Over 1000 Volts . . . . 110.31(A)(4)

Panic Hardware, Listed
    Over 1000 Volts . . . . 110.33(A)(3)
    Battery Support Systems 480.9
    Under 1000 Volts . . . . 110.26(C)(3)
Personnel, Electrical Vaults, Over 1000 Volts . . . . 110.31(A)
Personnel, Over 1000 Volts . . . . 110.33(A)(3)
Pools . . . . 680.26(B)(7)
Transformer and Transformer Vaults 450.43
Unqualified Access, Over 1000 Volts 490.35
Vaults and Tunnels, Over 1000 Volts . . . . 110.76(B)
Working Space . . . . 110.26(A)(2)

DOUBLE LOCKNUTS
Bonding Over 250 Volts to Ground . . . . 250.97 Ex
Mobile Homes Wiring Methods . . . . 550.15(F)
Not for Hazardous Areas (Bonding) 250.100
    Class I, Division 1 & 2 . . . . 501.30(A)
    Class I, Zone 0, 1, & 2 . . . . 505.25(A)
    Class II, Division 1 & 2 . . . . 502.30(A)
    Class III, Division 1 & 2 . . . . 503.30(A)
Not Recognized for Service Raceway Bonding . . . . 250.92(B)
Park Trailers . . . . 552.48(B)
Recreation Vehicles Wiring Methods . . . . 551.47(B)
Under Wholly Insulating Bushings Not Used to Secure Raceway
. . . . 300.4(G)

DRAINAGE
Boxes, Outlet, Devices and Junction (Approved Openings) . . 314.15
Capacitor Charge, see DISCHARGE STORED ENERGY . . . .
. . . . Ferm's Finder
Conduit Bodies (Approved Openings) 314.15
Equipment in Hazardous (Classified) Locations
    Class I, Division 1 & 2 Locations . . . . 501.15(F)
    Class I, Zone 0, 1, & 2 Locations . . . . 505.16(E)
Oil-Insulated Transformers, Outdoors 450.27
Raceways on Exterior Surfaces 225.22
Service Raceways 230.53
Swimming Pool Equipment Rooms & Pits 680.12
Transformer Vaults 450.46

DRESSING ROOMS
Motion Picture and Television Studios 530.31
Theaters and Similar Locations
    Lamp Guards 520.72
    Pendant Lampholders Prohibited 520.71

Switches Required 520.73

DRIP LOOPS AT WEATHERHEAD
Lowest Point for Conductor Clearance . . . . 230.24(B)(1)
Outside Branch Circuits and Feeders 225.11
Overhead Service Locations . . . . 230.54(F)
    Individual Conductors 230.52

DRIVEWAYS
Clearance of Conductors 225.18
Clearance of Service Drop . . . . 230.24(B)
Cover Requirements . . . . Table 300.5
Overhead Aerial Cables . . . . 830.44(C)

DROP BOXES
Definition 520.2
Listed . . . . 520.44(B)

DRY LOCATION (Definition of) . . . . **Art. 100**
Conductors . . . . 310.10(A)
Deteriorating Agents 110.11
Switchboards 408.20

DRYERS, CLOTHES
Disconnecting Means . . . . Art. 422 Part III
Flexible Cord and Cables Ampacity . . . . 400.5(A)
Grounding of 250.140
    By Connection to Grounded Conductor . . . 250.142(B) Ex. 1
    In Mobile Homes 550.16
    In Park Trailers 552.55
    In Recreational Vehicles 551.55
Minimum Load Branch Circuits (5000 VA), Dwelling Units . . 220.54
Receptacle Location . . . . 210.50(C)
Wiring Methods, Mobile and Manufactured Homes . . 550.15(E)

DRYWALL SCREWS
Covers and Canopies, Not to be Used 314.25
Grounding and Bonding Connection, Not to be Used . 250.8(5)
Receptacles, Not to be Used 406.5
Switches, Not to be Used . . . . 404.10(B)

DRY-TYPE TRANSFORMERS . . . . **Art. 450**
Accessibility 450.13
Grounding 450.10
In Vaults (Over 35,000 Volts) . . . . 450.21(C)
Installed Indoors 450.21

Installed Outdoors 450.22
Marking of 450.11
Not Covered by Article 450 (for Specific Applications) . . . 450.1
Exceptions
Overcurrent Protection 450.3
Ventilation 450.9
See (XQNX) . . . . *UL Product iQ*

DUCT HEATERS . . . . **424 Part VI**
Equipment and Wiring in Ducts . . . . 300.22(B)
Limited Access (Working Space) . . . . 110.26(A)(4)
Working Space Requirements 424.66
See (KOHZ) . . . . *UL Product iQ*

DUCTS & HOODS
Bonded
    Recreational Vehicles . . . . 551.56(F)
    Park Trailers . . . . 552.57(F)
Cable Ties and Other Cable Accessories, Nonmetallic . . 300.22(C)(1)
Heaters in . . . . Art. 424 Part VI
Luminaires (Lighting Fixtures) in . . . . 410.10(C)
Network-Powered Broadband Communications Systems . 800.24
Not to Infringe on Dedicated Space . . . . 110.26(E)(1)
Wiring In
Audio Signaling Process, Amplification and Reproduction Equipment . . . . 640.3(B)
CATV Systems . . . . 800.3(B)
Class 1, 2, and 3 Remote Control Circuits . . . . 725.3(C)
Classification for Spray Applications 516.5
Communications Circuits 805.154
Community Antenna Television and Radio Distribution Systems
. . . . 800.3(B)
Fire Alarm Systems . . . . 760.3(B)
General 300.22
Network-Powered Broadband Systems . . . . 830.3(B)
Premises-Powered Broadband Communications Systems 840.3(B)
Optical Fiber Cables and Raceways . . . . 770.3(B)
Spread of Fire 300.21
    To Be Suitable for the Environment 110.11

Note: Cable type wiring methods are required to be listed for plenum application. Refer to the Listing,

Marking, and Application sections within the specific system articles referenced above.

DUMBWAITERS . . . . **Art. 620**
See ELEVATORS, etc. . . . . *Ferm's Finder*

DUST-IGNITIONPROOF . . . . .Art. 502
Definition. . . . . Art. 100
Enclosures for Control Transformers and Resistors . . . 502.120
Enclosures for Switches, Circuit Breakers, etc. 502.115
Protection Technique, Class II Locations . . . . 500.7(B)

DUSTTIGHT
Class II Locations . . . . Art. 502
Definition. . . . . Art. 100
Enclosures for Control Transformers and Resistors . . . 503.120
Enclosures for Switches, Circuit Breakers, etc. 503.115
Protection Technique, Class II and Class III . . . . . . 500.7(C)
Wiring Methods Class III Locations 503.10

DUTY CYCLE
Arc Welders 630.11
Elevators
   Duty Rating . . . . 620.61(B)
   Feeder Demand Factors . . . . Table 620.14 Note
Explanation of (Term) . . . . 630.31(B) Info Note 3
Motors . . . . Table 430.22(E)
Resistance Welders . . . . 630.31(A)

DWELLING UNIT
AFCI Requirements. 210.12
   Branch Circuit Extensions and Modifications . . . 210.12(B)
   Locations Required. . . . . 210.12(A)

Branch Circuit Requirements
   Bathroom . . . . .210.11(C)(3)
   Central Heating Equipment Other Than Electric . . . 422.12
   Garage . . . . .210.11(C)(4)
   General Lighting 220.12
   Kitchens . . . . .210.11(C)(1)
      2 or More Branch Circuits. . . . . 210.52(B)(1)
      Small Appliance Branch Circuits . . . . . . .210.11(C)(1)
   Laundry . . . . .210.11(C)(2)
   Small Appliances . . . . .210.11(C)(1)
      See BRANCH CIRCUITS . . . . Ferm's Finder
Emergency Disconnects. 230.85

GFCI Requirements
   Accessory Buildings . . . . 210.8(A)(2)
   Basements . . . . 210.8(A)(5)
   Bathrooms . . . . 210.8(A)(1)
   Boat Hoists. 555.9
   Boathouses . . . . 210.8(A)(8)
   Crawl Spaces. 210.8(A)(4)
      Lighting Outlets. . . . . 210.8(C)
   Dishwashers . . . . 422.5(A)(7)
   Garages . . . . 210.8(A)(2)
   Kitchens . . . . 210.8(A)(6)
   Outdoors. . . . . 210.8(A)(3)
   Sinks, Other than Kitchens. . . . . 210.8(A)(7)

Lighting Requirements
   Bathrooms . . . . 210.70(A)(1)
   Closets 410.16
   Equipment Space. . . . . 210.70(C)
   Garages (Attached) . . . . 210.70(A)(2)
   Garages (Detached with Electric Power) . . . . 210.70(A)(2)
   Guest Rooms and Guest Suites. . . . . 210.70(B)
   Hallways . . . . 210.70(A)(2)
   Habitable Rooms. . . . . 210.70(A)(1)
   Stairways . . . . 210.70(A)(2)
   Storage Space. 210.70(C)

Receptacle Location Requirements
   Accessory Buildings . . . . 210.52(G)(2)
   Balconies. . . . . .210.52(E)(3)
   Basements .210.52(G)(3)
   Bathrooms . . . . 210.52(D)
   Countertop Receptacles
      Not Considered . . . . 210.52(A)(4)
   Decks . . . . 210.52(E)(3)
   Floor Receptacles. . . . . 210.52(A)(3)
   Foyers . . . . .210.52(I)
   Garage . . . . .210.52(G)(1)
   Guest Rooms, Guest Suites, Dormitories and
   Similar Locations . . . . 210.60(B)
   Hallways . . . . 210.52(H)
   Kitchens . . . . .210.52(B)(3)
      Island Spaces. . . . . .210.52(C)(2)
      Peninsular Spaces . . . . .210.52(C)(2)
      Receptacle Outlet Locations . . . . .210.52(C)(3)
   General Provisions . . . . 210.52(A)
   Laundry Areas . . . . 210.52(F)
   Outdoor Outlets
      Balconies, Decks, and Porches . . . . 210.52(E)(3)
      Multifamily Dwellings. . . . . .210.52(E)(2)
      One and Two-Family . . . . 210.52(E)(1)
   Porches. . . . . .210.52(E)(3)
   Spacing. . . . . .210.52(A)(1)

Wall Space . . . . 210.52(A)(2)
Receptacle Replacement Requirements . . . . 406.4(D)
   Arc Fault Circuit Interrupter Protection . . . . . 406.4(D)(4)
   Controlled Receptacles . . . . 406.4(D)(7)
   Ground-Fault Circuit Interrupters . . . . 406.4(D)(3)
   Grounding-Type Receptacles . . . . 406.4(D)(1)
   Non-Grounding Type Receptacles . . . . 406.4(D)(2)
   Tamper-Resistant Receptacles . . . . 406.4(D)(5)
   Weather-Resistant Receptacles . . . . 406.4(D)(6)
Surge Protection 230.67
Tamper-Resistant Receptacles
   Dormitories . . . . 406.12(7)
   Dwellings . . . . 406.12(1)
   Guest Rooms and Suites . . . . 406.12(2)
Voltage Limitations . . . . 210.6(A)

E

ECCENTRIC KNOCKOUTS
Bonding Over 250 Volts to Ground . . . . 250.97 Ex.
Bonding Service Equipment . . . . 250.92(B)

EFFECTIVE GROUND-FAULT CURRENT PATH . . ART. 100

ELBOWS, METAL, PROTECTION FROM CORROSION .300.6

ELECTRIC DISCHARGE LAMP CONTROL
See (FKOT) . . . . UL Product iQ

ELECTRIC DISCHARGE LIGHTING
1000 Volts or Less . . . . Art. 410 Part XII
Auxiliary Equipment, Lamps 410.104
Conductors within 75 mm (3 in.) of a Ballast Branch Circuit and Feeder 410.68
   Note: See Table 310.4(A) on pages 152-154 in NEC for Special Application of Type THW
Connection of Electric-Discharge and LED Luminaires (Fixtures) 410.24
Disconnecting Means for Ballasted Luminaires (Fixtures) 410.130(G)
   Metal Halide Lamp Containment . . . . 410.130(F)(5)
Definition . . . . Art. 100
Dwellings
   Open-Circuit Voltage Exceeding 1000 Volts Prohibited . . . . 410.140(B)
   Open-Circuit Voltage Over 300 Volts 410.135

Flush and Recessed Luminaires (Fixtures) . . . . Art. 410 Part XI
Luminaire (Fixture) Mounting 410.136
Marking Required for More Than 1000 Volts 410.146
Maximum Load for Branch Circuits . . . . 220.18(B)
More Than 1000 Volts . . . . Art. 410 Part XIII
Voltage Limitations 210.6
   Outdoor Equipment 225.7
See Categories under Luminaires (IEXT) . . . . . UL Product iQ

ELECTRIC HEAT (SPACE) . . . . Art. 424
Accessible . . . . 424.20(A)
Baseboard (Receptacle Outlet in) 210.52
   In Lieu of Required Wall Receptacles 424.9
Boilers
   Electrode Type . . . . Art. 424 Part VIII
   Resistance Type . . . . Art. 424 Part VII
Branch-Circuit Conductors 424.4
   Disconnect, More than One Supplying Equipment . . 424.19
   Sizing of . . . . 424.4(B)
   Supplying Supplementary Overcurrent Devices . . 424.22(D)
Cables . . . . Art. 424 Part V
   Area Restrictions 424.38
   Clearance from Other Objects and Openings 424.39
   Color Coding 424.35
   Derating and Clearances of Branch-Circuit Conductors . 424.36
   Finished Ceiling Materials 424.42
   Installation in Concrete or Poured Masonry Floors 424.44
      GFCI Protection . . . . 424.44(E)
   Installation on Dry Board, in Plaster and on Concrete
      Ceilings 424.41
   Listing Required 424.6
Low-Voltage Fixed Electric Space-Heating Equipment
   Branch Circuits 424.104
   Energy Sources 424.101
   Installation 424.103
   Listed Equipment 424.102
   Scope 424.100
Non-heating Leads 424.43
Not Permitted in or below Pool Decks . . . . . . 680.27(C)(3)
Space-Heating (Installation) 424.45
   Concrete or Poured Masonry 424.44
   Expansion Joints . . . . 424.45(B)
   GFCI Protection . . . . 424.45(E)
   Concrete or Masonry . . . . 424.44(E)

Under Flooring. . . . . 424.45(E)
Under Floor Coverings . . . . 424.45(A)
Splices in, Length Cannot Be Altered 424.40
Calculations of Load on Feeder 220.51
Dwelling Unit (Optional Method) 220.82
Existing Dwelling, Additional Loads (Optional Method) . .
. . . . 220.83
Multifamily Dwelling (Optional Method). 220.84
Control & Protection . . . . Art. 424 Part III
Deck Area Heating (Pools) . . . . 680.27(C)
Disconnecting Means 424.19
Lockable Disconnect 424.19
Rating of Disconnect, 125% of total load 424.19
Thermostat Serving as 424.20
Unit Switch(es) . . . . 424.19(C)
Duct Heaters . . . . Art. 424 Part VI
Limited Access . . . . .110.26(A)(4)
With Heat Pumps and Air Conditioners 424.61
Working Space Requirements 424.66
See (KOHZ) . . . . UL Product iQ
Feeder
More than One Supplying Equipment 424.19
Sizing 220.51
Heating Panels . . . . Art. 424 Part IX
Infrared Lamp Industrial Heating Appliances 425.14
Branch Circuits . . . . 210.23(C)
Overcurrent Protection. . . . . 422.11(C)
Inspection and Tests. 424.46
Labels 424.47
Low-Voltage Fixed Electric Space-Heating Equipment . . . .Art.
. . . . 424 Part X
Energy Source 424.101
Installation 424.103
Listed Equipment 424.102
Scope 424.100
Overcurrent Protection 424.22
48 Amperes Maximum Load . . . . 422.11(F)
Boilers Not Contained in ASME Rated Equipment . .424.72(B)
Resistance Elements . . . . 424.22(B)
120 Amperes Maximum Load
Boilers ASME Rated. . . . . 424.72(A)
Permanently Wired Radiant Heaters at Pools . . . .680.27(C)(2)
Radiant Heating Panels and Sets . . . . Art. 424 Part IX
Derating of Branch-Circuit Conductors

In Ceilings 424.94
In Interior Walls 424.95
Expansion Joints . . . . 424.98(B)
Nonheating Leads 424.97
Space Heating Cables, Marking and Color Code 424.35
Supplementary Overcurrent Protective Device
Definition of . . . . Art. 100
General 424.22
Not As Substitute for Branch-Circuit Devices 240.10
Resistance-type Boilers . . . . 424.72(C)
Thermostats for 424.20
Unit Heaters Pool Areas. . . . . .680.27(C)(1)
Unit Switch(es) as Disconnecting Means . . . . . . 424.19(C)
See Heating & Cooling Summary . *Ferm's Charts and Formulas*
See (LZLZ) . . . . UL Product iQ
See (LZFE) . . . . UL Product iQ

ELECTRIC POWER PRODUCTION AND DISTRIBUTION NETWORK . . . . Art. 100

ELECTRIC POWER PRODUCTION SOURCES
Fuel Cell Systems . . . . Art. 692
Generators . . . . Art. 445
Interconnected Electric Power Production Sources . . . Art. 705
Permitted for Fire Pumps . . . . 695.3(A)(2)
Solar Photovoltaic Systems . . . . Art. 690
Wind Electric Systems . . . . Art. 694

ELECTRIC SIGNS AND OUTLINE LIGHTING . . Art. 600
See SIGNS, ELECTRIC AND OUTLINE LIGHTING . . . . .
. . . . .Ferm's Finder

ELECTRIC SPACE HEATING CABLES. . . . . . . Art. 424 V
Clearances
From Objects and Openings 424.39
From Wiring (Ceilings) 424.36
Construction 424.34
Inspection and Tests. 424.46
Installation (Concrete or Masonry) 424.44
Installation (Under Floor Coverings) 424.45
Installation (Walls and Ceilings) 424.41
GFCI Protection Required
Concrete and Masonry . . . . 424.44(E)
Under Floor Covering . . . . 424.45(E)
Label Requirements. 424.47
Marking. 424.35

Restrictions (Area) 424.38

ELECTRIC TRUCK SPACE PARKING EQUIPMENT Art. 626

ELECTRIC UTILITIES
Connections, Meter Enclosures 230.82
Fire Pump. . . . . 695.3(A)(1)
Ground-Fault Protection for Personnel 590.6
Installations Covered by *NEC*. . . . . 90.2(A)
Installation Not Covered by *NEC*. . . . . 90.2(B)
Special Permission . . . . 90.2(C)
Transfer Equipment. 702.5

ELECTRIC VEHICLE CHARGING SYSTEM
Branch Circuit, For the Purpose of Charging 625.40
    Additional Feeders or Branch Circuits . . . . 225.30(A)
Cords and Cables 625.17
Definitions 625.2
Individual Branch Circuit. 625.40
Overcurrent Protection 625.41
Premise Wiring, Cord- and Plug-Connected 625.44
Ventilation 625.52
Wireless Power Transfer Equipment. . . . . Art. 625 Part IV
    Installation 625.102
    Grounding 625.101

ELECTRIC VEHICLE POWER TRANSFER SYSTEMS Art. 625
Branch Circuit 625.40
Definitions 625.2
Disconnecting Means. 625.43
Ground-Fault Circuit-Interrupter 625.54
Ground-Fault Circuit-Interrupter (On-Board Receptacle) 625.60
Installation . . . . Art. 625 Part III
Interactive System. 625.48
Listed 625.5
Load Calculation . . . . Table 220.3
Overcurrent Protection 625.41
Personnel Protection System 625.22
Plug-In Hybrid Electric Vehicle (Defined) 625.2
Rating. 625.42
Receptacle, On-Board (GFCI Protected) 625.60
Receptacle Enclosures. 625.56
Rechargeable Energy Storage System (Defined) 625.2
Scope, Power Export 625.1
Supply Equipment

Equipment Construction . . . . Art. 625 Part II
Locations 625.50
Ventilation Not Required . . . . 625.52(A)
Ventilation Required . . . . 625.52(B)
Voltages (Supply Equipment) 625.4
Wireless Power Transfer . . . . Art. 625 Part IV
    Construction 625.102
    Grounding. 625.101
Wiring Methods 625.44

ELECTRIC WELDERS
See WELDERS . . . . Ferm's Finder

ELECTRICAL EQUIPMENT
Dedicated Space . . . . 110.26(E)
Definition. . . . . Art. 100
Hazard Markings . . . . 110.21(A)(2)
Illumination . . . . 110.26(D)
Limited Access . . . . 110.26(A)(4)
Marking. . . . . 110.21(A)
Reconditioned . . . . 110.21(A)(2)
Torquing Requirements . . . . 110.14(D)
Working Space . . . . 110.26(A)

ELECTRICAL CIRCUIT PROTECTIVE SYSTEM
Communication Circuits (Defined). 800.2

ELECTRICAL METALLIC TUBING . . . . Art. 358
Bends, How Made 358.24
Bends, Number in One Run 358.26
Construction of 358.100
Corrosion Protection . . . . 358.10(B)
Couplings and Connectors 358.42
Dimensions and Percent Area of . . . . Chapter 9 Table 4
Dissimilar Metals 358.14
Installation in Hazardous Locations. . . . . 358.10(A)
Made of. 358.100
Marked 358.120
Minimum and Maximum Sizes 358.20
Not as a Means of Support for Cables or
Nonelectrical Equipment (General) . . . . 300.11(B)
    CATV Wiring . . . . 820.133(B)
    Class 2 or 3 Remote Control Circuits 725.143
    Communications Wires and Cables . . . . 805.133(B)
    Fire Alarm Circuits, PLFA 760.143

Network-Powered Broadband Communication Cables . . . .
. . . . .830.133(B)

Number of Conductors. 358.22
    Combination of Conductors (General). . . Chapter 9 Table 1
    Conductors and Fixture Wires, Same Size Tables C.1 & C.1(A)

Physical Damage . . . . 358.10(E)
Reaming of . . . . 358.28(A)
Securing and Supporting 358.30
Splices and Taps. 358.56
Threads on, Prohibited . . . . 358.28(B)
Uses Not Permitted 358.12
Uses Permitted 358.10
Wet Locations, in . . . . 358.10(C)
See (FJMX) . . . . UL Product iQ

ELECTRICAL NOISE (ELECTROMAGNETIC INTERFERENCE)

Currents That Introduce, Not Considered Objectionable . . 250.6(D)
Grounding
    Isolated Grounding Circuits . . . . 250.96(B)
    Isolated Grounding Receptacles . . . . 250.146(D)
    Isolated Ground Receptacles . . . . .406.3(D)
        For Audio Systems. . . . . 640.7(C)
        For Sensitive Electronic Equipment . . . . 647.7(B)
    Isolated Receptacles . . . . 250.146(D)
    Lighting Equipment 647.8
    Panelboards . . . . 408.40 Ex.
Sensitive Electronic Equipment
    Use of Separately Derived System 647.3
    Use of With Audio Systems . . . . 640.7(B)

ELECTRICAL NONMETALLIC TUBING: TYPE ENT Art. 362

Bends, How Made 362.24
Bends, Number in One Run 362.26
Bushings 362.46
Construction 362.100
Dimensions and Percent Area of . . . . Chapter 9 Table 4
Finish Grade Use Requirements 362.10
    Uses of in Places of Assembly 518.4(C) & (Informational Note)
Fished, Securing Not Required . . . . 362.30(A) Ex. (3)
Joints 362.48
Marking. 362.120
Maximum and Minimum Size 362.20
Number of Conductors 362.22
    Combination of Conductors (General). . . Chapter 9 Table 1
    Conductors and Luminaire (Fixture) Wires, Same Size Tables C.2 & C.2(A)

Securing and Supporting 362.30
Splices and Taps 362.56
Trimming. 362.28
Uses Not Permitted 362.12
Uses Permitted 362.10
See (FKHU). UL Product iQ

ELECTRICAL SERVICE AREAS. . . . . 210.63(B)

ELECTRICALLY ACTUATED FUSE

Feeders, Additional Requirements. 240.101
Over 1000 Volts (Defined) . . . . Art. 100, Part II
Transformers, Overcurrent Protection. . . Table 450.3(A) Note 4
Under 600 Volts (Defined) . . . . Art. 100, Part I

ELECTRICAL CIRCUIT PROTECTIVE SYSTEM

Class 1, Class 2, and Class 3 Circuits . . . . 725.179(F)(2)
Critical Operations Power Systems . . . . .708.10(C)(2)
Definition
    Communication Circuits. 800.2
    Fiber Optic Cables 770.2
Emergency Systems . . . . .700.10(B)(1)
Fiber Optic Cables and Raceways . . . . .770.179(E)
Fire-Rated Cables and Conductors . . . . 300.19(B)
Fire Alarm Systems
    Listing . . . . 760.176(F)(2)
    Other Article. 760.3(I)
Fire Pumps . . . . .695.6(H)
Fire-Resistive Cable Systems Considered This Protection . . . . .
. . . . 728.4 Info. Note 2
See (FHIT) . . . . UL Product iQ

ELECTRICALLY DRIVEN OR CONTROLLED IRRIGATION MACHINES . . . . Art. 675

ELECTRICALLY OPERATED POOL COVERS

Bonding of Metal Parts of. . . . . .680.26(B)(3)
Definition. 680.2
Ground-Fault Circuit Interrupter Required . . . . . 680.27(B)(2)
Motors and Controllers . . . . .680.27(B)(1)

ELECTRICALLY POWERED POOL LIFTS Art. 680 Part VIII

Bonding 680.83
Definition. 680.2
Equipment Approval 680.81

Nameplate Marking. 680.85

Protection. 680.82

Switching Devices. 680.84

ELECTRIFIED TRUCK PARKING SPACES . . . . . Art. 626
Definitions of. 626.2

Electrical Wiring Systems. . . . . .Art. 626 Part II

General Requirements . . . . Art. 626 Part I

Supply Equipment . . . . Art. 626 Part III

Transport Refrigerated Units . . . . .Art. 626 Part IV

ELECTRODES, GROUNDING . . . . Art. 250 Part III
At Agricultural Buildings . . . . 547.9(A)(5)

CATV and Radio Distribution Systems. . . . . . . .820.100(B)

Common Electrode for Building Services 250.58

Communications Circuits. . . . . 805.93(B)

Direct-Current Sizing. 250.166

Electrodes Permitted, Types 250.52

Electrode System Installation 250.53

For Separately Derived Systems 250.30

Fuel Cell Systems 692.47

Grounding Electrode System, Elements of 250.50

Intersystem Bonding Termination, Communications . 805.93(B)

Intersystem Bonding Termination, CATV . . . . . .820.100(B)

Intersystem Bonding Termination, Network Powered Broadband . . . . .800.100(B)

Irrigation Machines. 675.15

Lightning Protection Systems Bonded to 250.106

Network-Powered Broadband Systems . . . . . . .800.100(B)

Permanently Installed Generators 250.35

Radio and Television Equipment/Intersystem Bonding Termination. . . . . 810.21(F)

Restrictions on Use of Water Piping. . . . . .250.52(A)(1)

Rod, Pipe, or Plates, Augment If Resistance to Ground over 25 Ohms. . . . . 250.53(A)(2) Ex.

Sensitive Electronic Equipment . . . . 647.6(A)

Solar Photovoltaic Systems 690.47

Supplementary Electrodes. 250.54

Two or More Buildings from Common Service 250.32

Use of Strike Termination Devices in Lieu of, Prohibited. 250.60

Water Pipe Prohibited for Electrolytic Cells 668.15

Water Piping System Used to Ground a System Must Be Supplemented .250.53(D)(2)

See GROUNDING ELECTRODE CONDUCTORS Ferm's Finder

See (KDER). UL Product iQ

ELECTROLYTIC CELLS . . . . Art. 668
Application of Other Articles Limited. 668.3

Cell Line Working Zone (Area Encompassed) 668.10

Circuit Protection, Cell Line Working Zone . . . . . 668.30(D)

Cranes and Hoists in Working Zone 668.32

Definitions 668.2

Fixed and Portable Not Required to Be Grounded 668.30

Isolated Circuits Required . . . . 668.21(A)

Portable Equipment Not to Be Grounded . . . . . . 668.20(A)

Isolating Transformers to Supply Branch Circuits . 668.20(B)

Runway Conductors, Disconnecting Means . . . . . 610.31 Ex.

ELECTRONIC COMPUTER . . . . Art. 645
See INFORMATION TECHNOLOGY EQUIPMENT. . . . . Ferm's Finder

ELECTROPLATING . . . . Art. 669
Branch-Circuit Conductor Ampacity 669.5

Disconnecting Means 669.8

Overcurrent Protection 669.9

Warning Signs 669.7

Wiring Methods 669.6

ELEVATOR EQUIPMENT
See (FQKR). . . . . UL Product iQ

ELEVATORS, DUMBWAITERS, ESCALATORS, MOVING WALKS, PLATFORM LIFTS, AND STAIRWAY CHAIR LIFTS
. . . . .Art. 620

Adjustable Speed Drive Systems . . . . Article 430 Part X

Available Short-Circuit Current

Field Marking . . . . .620.51(D)(2)

Branch Circuits Required

Car Air-Conditioning and Heating Source . . . . . 620.22(B)

Car Lights, Receptacles, Ventilation . . . . 620.22(A)

Hoistway Pit Lighting and Receptacles 620.24

Machine Room/Space Lighting, Receptacles 620.23

Calculation of Feeder and Branch Circuit 620.13

Emergency and Standby . . . . 620.91(C)

Multiple Disconnecting Means 620.52

Conductors . . . . .Art. 620 Part II

Ampacity of 620.13

Installation . . . . Art. 620 Part IV

Minimum Size of. 620.12

Control . . . . Art. 620 Part VI

Control Room, Definition of 620.2

Control Space, Definition of 620.2
Cords and Cables, Not Required in a Raceway . . . . 620.21 Ex.
Definitions of Special Elevator Terms 620.2
Demand Factors 620.14
   Other Than Continuous Load . . . . 430.22(E)
   Feeders 430.26
   Single Motor 430.22
Different Systems in Cable or Raceway 620.36
Disconnecting Means 620.51
   Car Air-Conditioning and Heating 620.54
   Car Light, Receptacle(s) and Ventilation 620.53
   Utilization Equipment 620.55
Emergency and Standby Power Systems. . . . . .Art. 620 Part X
   General Requirements for Systems . . . . Art. 700
   Hospitals
      Connection of Equipment to Life Safety Branch 517.33(G)
      Equipment for Delayed Automatic or Manual
         Connection to Alternate Power Source . . . . 517.35(B)(2)
   Nursing Homes & Limited Care Facilities
      Automatic Connection to Life Safety Branch 517.43
      Connection to Equipment Branch. 517.44
Flexible Cords and Cables . . . . Art. 620 Part III
GFCI Protected Receptacles 620.6
   Pit Lighting, Machine Space Lighting Not on Load Side
   . . . . 620.23 & 24
Grounding and Bonding . . . . Art. 620 Part IX
Hoistway Wiring Methods . . . . .620.21(A)(1)
Machine Room . . . . Art. 620 Part VIII
Machine Space, Definition of 620.2
Overcurrent Protection . . . . .Art. 620 Part VII
   Selective Coordination 620.62
Power from More than One Source 620.52
Selective Coordination 620.62
   Signage. 620.65
Short-Circuit Current Rating 620.16
Surge Protection . . . . 620.51(E)
Traveling Cables . . . . .Art. 620 Part V
Voltage Limitations . . . . 620.5(A)
Wiring Methods . . . . Art. 620 Part III
See GRAIN HANDLING & STORAGE AREAS . Ferm's Finder

EMERGENCY DISCONNECTS
   Additional Service Allowed for 230.85

EMERGENCY SYSTEMS . . . . **Art. 700**

Additional Service Allowed for . . . . 230.2(A)(2)
   Grouping of Disconnects. . . . . 230.72(A) Ex.
   Location of Disconnects . . . . 230.72(A)
   Service Installed Sufficiently Remote . . . . 230.72(B)
      Permitted As Source of Power . . . . 700.12(D)
Automatic Load Control Relay, Defined 700.2
Boxes and Enclosures for (Identification) . . . . 700.10(A)
Branch Circuit Emergency Lighting Transfer Switch . . . 700.25
Capacity of, Load Shedding & Peak Load Shaving 700.4
Circuits for Lighting . . . . Art. 700 Part IV
   Control. .Art. 700 Part V
   See Life Safety Code NFPA 101 . . . . NFPA 101
Clinics 517.45
Cord-Connected Unit Equipment . . . . 700.12(F)
DC Microgrid Systems . . . . 700.12(H)
Definitions 700.2
Dimmer and Relay Systems 700.23
Fire Protection of
   Equipment for Feeder Circuits . . . . .700.10(D)(2)
   Feeder Circuit Wiring . . . . .700.10(D)(1)
   Sources of Power 700.12
Fuel Cells as Source . . . . 700.12(E)
Ground-Fault Indication . . . . .700.6(D)
Ground-Fault Protection of Equipment. 700.31
Identification
   Grounded Circuit Conductor Connection . . . . . 700.7(B)
   Permanently Marked (Components) . . . . 700.10(A)
   Receptacles (Distinctive Color). . . . . 700.10(B)
Illumination 700.16
Independent of Other Wiring . . . . 700.10(B)
Luminaires, Directly Controlled 700.24
Maintenance . . . . 700.3(C)
Multiwire Branch Circuits 700.19
Places of Assembly . . . . 518.3(C)
Selective Coordination 700.32
Signs
   Emergency Source . . . . 700.7(A)
   Grounding . . . . 700.7(B)
   Illumination and Exit Required 700.16
Surge Protection 700.8
Switch Locations 700.21
Maintenance 700.3
Transfer Switch(es)
   Branch Circuit Emergency Lighting 700.25

    General 700.5
    Listing Required . . . . 700.5(C)
    Short-Circuit Current Rating . . . . 700.5(E)
    With Generator Set . . . . 700.12(B)(1)
    See (WPWR) . . . . UL Product iQ
  Unit Equipment . . . . 700.12(F)
    Components . . . . 700.12(F)(1)
    Installation . . . . 700.12(F)(2)
    Remote Heads . . . . 700.12(F)(2)(6)
  Wiring Design and Location . . . . 700.10(C)
  See Life Safety Code, NFPA 101 . . . . NFPA 101

ENCLOSURES
Arcing Parts 110.18
Bonding, Service . . . . 250.92(A)(2)
Bonding Other Enclosures 250.96
Busways . . . . Art. 368
Cabinets, Cutout Boxes, and Meter Socket Enclosures . Art. 312
Circuits in, Number of . . . . 90.8(B)
Definition of . . . . Art. 100 Part I
Elevators, Dumbwaiters, Escalators, Moving Walks,
    Wheelchair Lifts, and Stairway Chair Lifts   Art. 620 Part VIII
For Grounding Electrode Conductors . . . . 250.64(E)
Grounding of . . . . 250 Part VI
High-Intensity Discharge Lamp Auxiliary Equipment 410.104(A)
Induction and Dielectric Heating 665.20
Installations Over 1000 Volts, Nominal 110.31
    Equipment Over 1000 Volts – Specific Provisions Art. 490 Part II
Intrinsically Safe Conductors in . . . . 504.30(A)(2)
Manholes and Other Enclosures for Personnel Entry Art. 110 Part V
Panelboards 408.38
Radio Equipment . . . . 810.71(A)
Signs and Outline Lighting (General) 600.8
  Enclosures Used as Pull Boxes . . . . 600.5(C)(2)
  Neon Tubing Electrode, Listed for the Purpose . . . 600.42(C)
Subsurface . . . . 110.12(B)
Switches 404.3
Types of Enclosures, NEMA . . . . Table 110.28
Underground 110.28

ENERGY CODE
Lighting Load, Design and Constructed . . . . 220.12(B)

ENERGY MANAGEMENT SYSTEM
Alternate Power Source 750.20
Building Automation . . . . 406.3(E)
Controlled Receptacle . . . . 406.3(E)
Definitions 750.2
Field Marking 750.50
Load Management 750.30
Marking . . . . Figure 406.3(E)
Recreational Vehicles . . . . 551.42(C) Ex. 2
Scope 750.1

ENERGY STORAGE SYSTEMS (ESS) . . . . .Art. 706
Charge Control 706.33
Circuit Requirements . . . . Art. 706 Part IV
Connection to Other Energy Sources 706.16
Definitions 706.2
Directory 706.21
Disconnecting Means . . . . .Art. 706 Part II
Dwelling Units . . . . 706.20(B)
Flow Battery . . . . .Art. 706 Part V
Installation Requirements . . . . Art. 706 Part III
Listed 706.5
Locations 706.10
Maintenance 706.7
Maximum Voltage .706.9
Multiple Systems 706.6
Nameplates 706.4
Other Energy Storage Technologies . . . . Art. 706 Part VI
Overcurrent Protection 706.31
Qualified Person 706.3
Storage Batteries 706.8
System Requirements 706.4
Ventilation . . . . 706.20(A)
Working Space . . . . 706.20(C)

ENFORCEMENT OF CODE 90.4
Administration and Enforcement . . . . . .Informative Annex H
Authority Having Jurisdiction (AHJ) . . . . Art. 100- Definition
  Note: See Informative Annex H, 80.5 for specific adoption Informative . . . . Annex H Section 80.5

EQUIPMENT
Acceptable Only If Approved 110.2
Approval of 90.4
Cooling of . . . . 110.13(B)
Definition of . . . . Art. 100 Part I
Examination of 90.7

Examination, Identification, Installation, and Use 110.3
Grounding . . . . Art. 250 Part VI
Hazard Markings . . . . 110.21(B)
Identification of Engineered Combination Systems . . 110.22(B)
    Engineered Combination Systems . . . . 240.86(A)
Installation, General Provisions . . . . Art. 110
Installation, General Provisions, Over 1000 Volts . . . Art. 490
Listing of . . . . 110.3(C)
Marking. 110.21
Mounting of . . . . 110.13(A)
Product Safety Standards . . . . Informative Annex A
Reconditioned . . . . .110.21(A)(2)
Requiring Service 210.63
Wireless Power Transfer Equipment. 625.101
    Construction. 625.102
    Definitions. 625.2
    Grounding 625.101

EQUIPMENT GROUNDING CONDUCTOR

Air Conditioning Equipment on Rooftops 440.9
Agricultural Buildings, Separate Conductor, Copper
or Aluminum . . . . 547.5(F)
    Underground to, Insulated, Copper or Aluminum . . 547.5(F)
Boatyards and Marinas 555.37
Bonded to Well Casing . . . . 250.112(M)
Bonding Metal Piping and Exposed Structural Steel to . 250.104
Box, Conductor Fill . . . . .314.16(B)(5)
Building- Metal Frames (Restricted) . . . . .250.121(B)
Cable Trays 392.60
Carnivals, Circuses, Fairs and Similar 525.31
    Continuity Assurance 525.32
Conductors of Same Circuit, with, General. . . . . . . 300.3(B)
Conductors of Same Circuit, with, Metal Raceways. . 300.20(A)
Conductors of Same Circuit, with, Underground . 300.5(I) Ex. 1
Connections 250.130
Continuity of 250.124
Cord- and Plug-Connected Equipment (General) . . . 250.138
Cord- and Plug-Connected Equipment (Specific). . . . 250.114
Definition of . . . . Art. 100 Part I
Earth Not Used As Effective Ground-Fault Current Path 250.4(A)(5)
Earth Not Used As Effective Ground-Fault Current Path
for Supplementary Electrodes. 250.54
Electrical Nonmetallic Tubing, in 362.60
Equipment Fastened in Place 250.134
Feeder Identification . . . . 215.12(B)

Flexible Metal Conduit 348.60
Floating Buildings . . . . 555.54(B)
Identification of, General . . . . 310.6(B)
    Feeder . . . . 215.12(B)
    Identification Methods. 250.119
Impedance Grounded Neutral Systems Over 1 kV . 250.187(D)
Increased In Size . . . . .250.122(B)
Installation of 250.120
Insulated Conductor for Isolated Ground Receptacles . . .250.146(D)
Isolated Grounding Circuits . . . . 250.96(B)
Luminaires (Lighting Fixtures)
    Exposed Conductive Parts 410.42
    Means for Attaching Equipment Grounding Conductor 410.46
    Methods of Grounding. 410.44
Liquidtight Flexible Metal Conduit. 350.60
Liquidtight Flexible Nonmetallic Conduit, in 356.60
Marinas and Boatyards 555.37
Messenger Supported Wiring . . . . 396.30(C)
Motor Terminal Housings, Means of Attachment . . 430.12(E)
Panelboards 408.40
    Bonding in Patient Vicinity 517.14
Parallel Conductors . . . . .250.122(F)
Patient Care Spaces . . . . 517.13(A)
    Additional Requirements for . . . . 517.13(B)
Rigid PVC Conduit, in 352.60
Signs and Metal Parts of Outline Lighting 600.7
Size of 250.122
Sizing . . . . Table 250.122
    Increased in Size . . . . .250.122(B)
Spliced or Joined in Boxes 250.148
Surface Metal Raceways 386.60
Surface Nonmetallic Raceways 388.60
Swimming Pools, Equipment to Be Grounded 680.6
    Methods of Grounding
        Cord and Plug-Connected Equipment 680.8
        Feeders 680.25
        Fountains 680.55
        Motors, Permanently Installed. . . . . 680.21(A)(1)
        Pumps, Storable Pools 680.31
        Underwater Luminaires (Lighting Fixtures) . . 680.23(F)(2)
Switches, Connected to EGC. . . . . 404.9(B)
Switchgear, Connected to EGC . . . . 250.112(A)
Transformers 450.10
Tunnel Installations Over 1000 Volts . . . . 110.54(B)

Types of. 250.118

Wireways

    Nonmetallic 378.60

    See BONDING . . . . Ferm's Finder

    See GROUNDING . . . . Ferm's Finder

EQUIPMENT GROUNDING CONDUCTOR FILL FOR BOXES 314.16(B)(5)

EQUIPMENT SPACE, DEDICATED

Indoor . . . . 110.26(E)(1)

Outdoor . . . . 110.26(E)(2)

EQUIPOTENTIAL BONDING

Class 1, Division 2. . . . . 501.125(B) Info. Note 2

Hot Tubs and Spas, Indoor Not Required . 680.43 Ex. 2 and Ex. 3

Hydromassage Bathtubs. 680.74

Performance, Pools . . . . 680.26(A)

Perimeter Surface . . . . 680.26(B)(2)

Pool Water . . . . 680.26(C)

Swimming Pools . . . . 680.26(B)

Zone 0, 1 and 2 Locations . . . . . . 505.20(C) Info. Note 2

EQUIPOTENTIAL PLANE

Bonding . . . . 547.10(B)

Definition, Agricultural Buildings. 547.2

Definition, Natural and Artificially Made Bodies of Water . 682.2

Installation 547.10

Natural and Artificially Made Bodies of Water 682.33

Receptacle Locations within . . . . 547.5(G)(1)

Required

    Indoors. . . . . 547.10(A)(1)

    Outdoors. 547.10(A)(2)

ESCALATORS . . . . **Art. 620**

    See ELEVATORS, ETC. . . . . Ferm's Finder

ESSENTIAL ELECTRICAL SYSTEM

Branches for Hospitals 517.30

    Independent Power Sources . . . . 517.30(A)

    Location of Components. . . . . 517.30(C)

    Types of Power Sources. . . . . 517.30(B)

Branches for Nursing Homes 517.41

    Independent Power Sources . . . . 517.41(A)

    Location of Components. . . . . 517.41(C)

    Types of Power Sources. . . . . 517.41(B)

Defined. 517.2

Health Care Facilities . . . . Art. 517 Part III

Transfer Switched

    Hospitals. . . . . 517.31(B)

    Nursing Homes . . . . 517.42(B)

See HEALTH CARE FACILITIES . . . . Ferm's Finder

EXAMPLES . . . . **Informative Annex D**

Cable Tray Calculations. . . . . Example D13

    Multiconductor Cables 4/0 and Larger . . . Example D13(a)

    Multiconductor Cables Smaller than 4/0 . . Example D13(b)

    Single Conductor Cables 1/0 through 4/0 . . Example D13(c)

    Single Conductor Cables 250 through 900 kcmil . . Example D13(d)

Feeder Ampacity for Adjustable Speed Drive Control Example D10

Feeder Ampacity for Generator Field Control. . . . Example D9

Industrial Feeders in a Common Raceway . . . . Example D3(a)

Mobile Home. . . . . Example D11

Motor Circuit Conductors, Overload Protection, etc. Example D8

Multi-Family (Optional, at 208Y/120, Volts, Three-Phase . . . . Example D5(b)

Multi-Family Dwelling (Optional) . . . . Example D4(b)

Multi-Family Dwelling . . . . Example D4(a)

Multi-Family Served at 208/Y/120 Volts, Three-Phase Example D5(a)

One-Family (Optional, Air-Conditioning Larger Than Heat). . . . . Example D2(b)

One-Family (Optional, Heat Larger Than Air-Conditioning). . . . . Example D2(a)

One-Family (Optional, with Heat Pump). . . . . Example D2(c)

One-Family Dwelling (Basic) . . . . Example D1(a)

One-Family Dwelling (Plus Air-Conditioning & Appliances) . . . . Example D1(b)

Park Trailer . . . . Example D12

Range Loads . . . . Example D6

Sizing of Service Conductors for Dwellings. . . . . Example D7

Store Building. . . . . Example D3

EXIT LIGHTS

Emergency Illumination 700.16

Places of Assembly, Emergency Systems . . 700.2 Informational Note

See EMERGENCY SYSTEMS . . . . Ferm's Finder

See Life Safety Code, NFPA 101 . . . . NFPA 101

EXOTHERMIC WELDING (Splicing Method)

Connection of Grounding and Bonding Equipment 250.8

Installation of Grounding Electrode Conductor, Continuous . .

.... 250.64(C)
Methods of Connection to Electrodes. 250.70
Not Required to be Accessible when Fireproofed. ...250.68(A) Ex. 2
Swimming Pool Bonding (Equipotential Bonding Grid)680.26(B)

EXPANSION JOINTS

Bonding Around 250.98
Busways, Over 1000 Volts 368.244
Deicing and Snow-Melting
    Protected from Expansion, Nonheating Leads, Embedded ...
    .... 426.22(D)
    Protected from Expansion, Resistance Heating Elements ...
    .... 426.20(E)

Heating Cables
    Under Floors..................... 424.45(B)
Within Concrete or Masonry............. 424.98(B)
Heating Panels and Heating Panel Sets
    Under Floors..................424.99(B)(1)
    Within Concrete or Masonry........... 424.98(B)
Pipelines and Vessels 427.16
    Flexural Capability................... 427.17
Raceways .... 300.7(B)
Rigid PVC Conduit 352.44
    Expansion Characteristics .... Tables 352.44
Structural Joints....................... 300.4(H)
Type RTRC Conduit .... 355.44, Table 355.44

EXPLANATORY MATERIAL (Informational Notes) . . 90.5(C)

EXPLOSIONPROOF EQUIPMENT

Class I, Zone 0, 1, and 2 Locations, Sealing..... 505.16(B)(2)
Class I, II, and III, Zone 20, 21, and 22 Locations, Sealing . . 506.16
Class II Locations 502.1
Definition of .... Art. 100 Part I
Protection Technique, Class I, Division 1 & 2 ..... 500.7(A)

EXPLOSIONPROOF

Explosionproof (Equipment) .... Art. 100
See HAZARDOUS (CLASSIFIED) LOCATIONS . . Ferm's Finder
See (FTRV) .... UL Product iQ

EXPOSED

Clearance for Live Parts 110.26
Definitions of (as applied to live parts) ...... Art. 100 Part I
Extensions, Boxes and Fittings 314.22
Structural Steel, Bonding .... 250.104(C)

Work Space and Guarding, Over 1000 Volts 110.34

EXTENSION CORD SETS

Assured Equipment Grounding Conductor Program . . 590.6(B)
On Construction Sites Utilizing GFCI Protection . . . 590.6(A)
Overcurrent Protection of 240.5
Temporary Installations, Support Not Required .... 590.4(J)

EXTENSION RINGS

Exposed Surface Extensions................ 314.22
For Box Volume Calculations.............. 314.16(A)
See (QCIT) .... UL Product iQ

EXTENSIONS

Auxiliary Gutters .... 366.12(2)
Boxes and Fittings, Exposed 314.22
Cellular Concrete Floor Raceways (uses not permitted) . 372.12
Cellular Metal Floor Raceways (uses not permitted) . . . 374.12
Flat Cable Assemblies, Type FC............ 322.40(D)
Nonmetallic Extensions................... Art. 382
Surface Metal Raceways .... 386.10(4)
Surface Nonmetallic Raceways .... 388.10(2)
Wireways
    From Metal 376.70
    From Nonmetallic................... 378.70
    Metal, Extension through Walls .... 376.10(4)
    Nonmetallic, Extension through Walls....... 378.10(4)

EXTRA-DUTY OUTLET BOXES .... .406.9(B)(1)

EXTRACTS FROM OTHER NFPA CODES AND STANDARDS

See **Informational Notes** that immediately follow the listed sections or titles of the following articles: Art. 500; Art. 505; Art. 511; Section 513.1; Art. 514; Art. 515; Art. 516; Art. 517; Art. 695.

See also NEC *Style Manual* and NFPA *Manual of Style* for further information regarding extracts.

F

FACEPLATES

Covers and Canopies................... 314.25
Flush-Mounted Installations 314.20
Flush Mounting, Receptacle Enclosure ... 406.5(B), (C), & (E)

Grounding Required 250.110
    Grounding Provisions . . . . 404.9(B)
    In Patient Care Spaces . . . . 517.13(B)(1) Ex. 1 to (2)
    Insulated Equipment Grounding Conductor . . . 517.13(B)
    Metal Covers in Completed Installations . . . . . 314.25(A)
    Metal or Non-Metallic, for Receptacles — Thickness 406.6(A)
Integral Night Light and/or USB Charger . . . . 406.6(D)
Isolated Ground Receptacles, for . . . . 406.3(D)(2) Ex.
Position of, for Receptacle . . . . 406.5(D)
Receptacles, for 406.6
Repair of Noncombustible Surface 314.21
Snap Switches, for 404.9
Switches (Snap) . . . . 404.9(A)
USB Charging . . . . 406.6(D)

FAHRENHEIT TO CENTIGRADE . . . .
. . . . Tables 310.16 through 310.21

FAIRS, See **CARNIVALS, CIRCUSES, FAIRS & SIMILAR EVENTS** . . . . Ferm's Finder

FANS, CEILING-SUSPENDED (PADDLE)
  See CEILING-SUSPENDED (PADDLE) FANS . Ferm's Finder

FARM BUILDINGS
Agricultural Buildings . . . . Art. 547
  See AGRICULTURAL BUILDINGS . . . . . Ferm's Finder
    Wiring Methods 547.5
Bonding and Equipotential Plane 547.10
    GFCI Protection for Receptacles . . . . 547.5(G)
Clearance of Service Conductors from Building Openings . . . 230.9
Calculating Loads . . . . Art. 220 Part V
Disconnecting Means
    And Overcurrent Protection (Site Isolation), Where Located .
    . . . . 547.9
    Grouping of 230.72
    Location, General 230.91
    Maximum Number of 230.71
    More than One Building or Structure . . . . . Art. 225 Part II
    Service Equipment, General 230.90
    Service, Location . . . . 230.70(A)
Grounding . . . . 547.9(B)
    Common Grounding Electrode 250.58
    Grounding Electrode System 250.50
    Two or More Buildings from Common Service 250.32
Outside Branch Circuits and Feeders . . . . Art. 225

Service Drop Conductors 230.22
See Flammable and Combustible Liquids at Farms and Isolated Sites Ferm's Finder
See Motor Fuel Dispensing Facilities . . . . Ferm's Finder

FAULT CURRENT, AVAILABLE 110.24
Definition . . . . Art. 100
Field Marking . . . . 110.24(A)
Modifications . . . . 110.24(B)
See SHORT-CIRCUIT CURRENT AVAILABLE   Ferm's Finder

FEEDERS . . . . Art. 215
Additional Loads on (Standard Calculation) 220.14
    (Optional Calculation) 220.87
Attached to Buildings or Structures 225.11
Busways, Installation 368.10
    Taps . . . . 240.21(E)
    Uses Not Permitted of As Feeder 368.12
    Uses Permitted of As Feeder 368.10
Calculations . . . . Art. 220
  See CALCULATIONS . . . . Ferm's Finder
Clearance for Overhead Feeder Conductors 225.18
Continuous and Noncontinuous Loads . . . . 215.2(A)
Conductor Ampacity (Over 600 Volts) . . . . 215.2(B)
Definition of (Feeder) . . . . Art. 100
Direct-Current Systems, Supplied From . . . . 215.12(C)(2)
Dwellings, Size of 310.12
    Number of Feeders 225.30
    Permitted Size of Feeders 310.12
Entering a Building or Structure 225.11
Essential Electrical Systems, Alternate Power Source . 517.31(F)
Exiting a Building or Structure 225.11
Farm Buildings . . . . Art. 220 Part V
Floating Buildings 553.7
Ground-Fault Protection of Equipment 215.10
Ground-Fault Protection of Equipment- Temporary . . . . . .
. . . . 215.10, Ex. No. 3
Ground-Fault Protection (Health Care) 517.17
Ground-Fault Circuit Interrupter (Personnel) 215.9
Grounding Means 215.6
    In Panelboards 408.40
    Marinas and Boatyards 555.54
    Swimming Pools 680.25
  See GROUNDING, Fixed Equipment . . . . . . Ferm's Finder
High-Leg (Identification) 110.15

Alternating-Current Systems, Ungrounded Conductors . . . .
. . . . 215.12(C)(1)
Direct-Current Systems, Ungrounded Conductors . . . .
. . . . 215.12(C)(2)
Equipment Grounding Conductor . . . . 215.12(B)
Grounded Conductor . . . . 215.12(A)
In Same Metallic Enclosure . . . . 215.4(B)
Kitchen Equipment, Commercial 220.56
Kitchen Equipment, Dwellings 220.55
Marinas and Boatyards 555.52
Minimum Size & Rating . . . . 215.2(A)
Minimum Size & Rating (Over 600 Volts) . . . . 215.2(B)
Calculated Load . . . . Art. 220
Mobile Home (Four Insulated Conductors) 550.10
Conductor Size 310.12
Installation and Capacity 550.33
Permitted Size of Conductors 310.12
See MOBILE HOMES . . . . Ferm's Finder
More than One Building or Structure . . . . Art. 225 Part II
Number of Supplies 225.30
Motors
Circuit Conductors . . . . Art. 430 Part II
Short-Circuit and Ground-Fault Protection . Art. 430 Part IV
See MOTORS . . . . Ferm's Finder
Neutral (Feeder Load) 220.61
Common Neutral 215.4
Considered Current-Carrying . . . . 310.15(E)
Supplying Outside Lighting . . . . 225.7(B)
Noncoincident Loads 220.60
Outside . . . . Art. 225
Ampacities of Conductors 310.14
General . . . . Art. 240
Over 1000 Volts . . . . Art. 240 Part IX
Overcurrent Protection 225.3
See OUTSIDE BRANCH CIRCUITS AND FEEDERS . . . . .
. . . . Ferm's Finder
Recreational Vehicles 551.73
Restaurants, New (Optional) 220.88
Sizing (General) 215.2
Ampacities of Conductors 310.14
Calculations . . . . Art. 220
Mobile Homes, See FEEDERS, Mobile Homes . Ferm's Finder
Motion Picture and TV Studios 530.19
Recreational Vehicles, Minimum Size 551.73
See CALCULATIONS . . . . Ferm's Finder

Supervised Industrial Installations . . . . . . Art. 240 Part VIII
Swimming Pools 680.25
Taps, Overcurrent Protection and Definition . . . . . 240.4(E)
Location of Overcurrent Protection . . . . 240.21(B)
Motors 430.28
Transformer Secondary Conductors . . . . . . .240.21(C)(6)
See TAPS . . . . Ferm's Finder
Ungrounded Identification
Alternating-Current Systems . . . . 215.12(C)(1)
Direct-Current Systems . . . . 215.12(C)(2)
Use of SE Cable 338.10

FEED-THROUGH CONNECTIONS OF NEUTRAL — PIGTAILING . . . . **300.13(B)**

FENCES (Metal)
Grounding and Bonding, Over 1000 Volts . . . . . 250.194(A)
Around the Vicinity of Transformers . . . . 450.10(B)

FERROMAGNETIC ENCLOSURE . . . . **300.20(A) Ex.2**

FERROMAGNETIC ENCLOSURE, SKIN EFFECT . . . **426.2**

FERROUS METALS
Agricultural Buildings . . . . 547.5(C)(3) Ex.2
Cable Tray Systems . . . . .392.100(C)
Electrically Continuous . . . . 250.64(E)
Electrical Metallic Tubing 358.2
Electrical Metallic Tubing, Corrosion Protection . . . 358.10(B)
Enclosures for Grounding Electrode Conductors . . . 250.64(E)
Faceplates, Receptacles . . . . 406.6(A)
Faceplates, Switches . . . . 404.9(C)
Field-Cut Threads . . . . 300.6(A) Info. Notes
Fixed Outdoor Electric Deicing and Snow-Melting Equipment . .
. . . . 426.26
Induced Currents 300.20
Metal Equipment . . . . 300.6(A) Info. Notes
Protection from Corrosion 300.6
Rigid Metal Conduit 344.2
Solely Protected by Enamel, Rigid Metal Conduit . . 344.10(4)
Solely Protected by Enamel, Strut-Type Channel Raceway . . 384.10
Strut-Type Channel Raceway, Uses Not Permitted . . 384.12(2)
Underfloor Raceways, Uses Not Permitted 390.12

FESTOON LIGHTING
Conductor Covering 225.4
Conductor Size and Support . . . . 225.6(B)

Definition. . . . . Art. 100
Outdoor Lampholders 225.24
Theaters and Audience Areas 520.65

FIBER BUSHINGS
AC Cable Ends 320.40
Raceways . . . . 300.4(G)

FIBER OPTICAL CABLES & RACEWAYS
See Optical Fiber Cables and Raceways . . . . Art. 770
See OPTICAL FIBER CABLES AND RACEWAYS Ferm's Finder

FIBERS, LINT, FLYINGS, IGNITIBLE . . . . Art. 506

FIELD APPLIED HAZARD MARKINGS. . . . . . 110.21(B)
See . . . . ANSI Z535.4-2011, Product Safety Signs and Labels

FIELD-CUT THREADS . . . . 300.6(A) Info. Notes

FIELD EVALUATION BODY (FEB)
Definition. . . . . Art. 100
Field Labeled (Evaluation Reports) . . . . Art. 100

FIELD IDENTIFICATION REQUIRED (Switchboards and Panelboards) 408.4

FINE PRINT NOTES (Now Informational Notes) (Info. Note)
. . . . 90.5(C)

FINELY STRANDED CABLES AND CONDUCTORS, TERMINATING 110.14

FINISH RATING
Definition of . . . . 362.10 Ex to (2) Info. Note
ENT, Uses Permitted, More Than Three Floors . . . . 362.10(2)
Places of Assembly 518.4

FINISHING PROCESSES . . . . Art. 516
Class I and II Locations 516.4
Special Precautions . . . . 500.5(B) Info. Note
Specific Occupancies 500.9
Electrostatic Equipment . . . . 516.6(E)
Hand Spraying Equipment . . . . 516.10(B)
Use of . . . . 516.6(E)
Grounding 516.16
Grounding & Bonding (General) . . . . Art. 250
In Class I Locations 501.30
In Class II Locations 502.30
In Hazardous (Classified) Locations 250.100

Sealing
In Class I Locations 501.15
In Class II Locations 502.15
Wiring Methods
Above Class I and II Locations 516.7
Equipment in Class I Locations 516.4
In Class I Locations 501.10
In Class II Locations 502.10
See Standard for Dipping and Coating Processes Using Flammable or Combustible Liquids
See HAZARDOUS (CLASSIFIED) LOCATIONS Ferm's Finder
See . . . . . . Standard for Spray Application Using Flammable and Combustible Materials

FIRE ALARM SYSTEMS . . . . Art. 760
AFCI Protection Prohibited. . . . . 760.41(B)
Abandoned Cables 760.25
Access to Electrical Equipment 760.21
Ahead of Mains, Service Connection . . . . 230.82(5)
For Emergency System, Prohibited 700.12
Relative Location of Service Overcurrent Device 230.94 Ex. 4
Branch Circuit (PLFA Circuits) . . . . 760.121(B)
Cable Routing Assemblies. . . . . 760.3(L)
Circuit Integrity (CI) Cable, Definition. 760.2
Circuits Extending Beyond One Building 760.32
Fire Conditions of Survivability . . . . 760.179(F)
Listing and Marking, NPLFA . . . . 760.176(D)
Listing and Marking, PLFA 760.179
Mechanical Execution of Work . . . . 760.24(B)
Circuit Requirements 760.35
Communication Raceways . . . . 760.3(M)
Conductors
Non-Power-Limited (NPLFA) . . . . Art. 760 Part II
Power-Limited (PLFA) . . . . Art. 760 Part III
Ducts . . . . 760.3(B)
Electrical Circuit Protective System, NPLFA . . . 760.176(F)(2)
GFCI Protection Prohibited . . . . 760.41(B)
Grounding . . . . 250.112(I)
Health Care Facilities . . . . Art. 517 Part VI
See Standard for Health Care Facilities, NFPA 99 . . . NFPA 99
See Life Safety Code, NFPA 101 . . . . NFPA 101
Identification of Circuits 760.30
Installation in Buildings 760.135
Installation with Conductors of Other Systems . . . . 760.3(G)
Listing of Cables 760.176, 760.179

  Non-Power Limited Fire Alarm Cables. 760.176

  Power Limited Fire Alarm Cables 760.154

Location, Hazardous . . . . 760.3(C)

Non-Power-Limited Circuits . . . . .Art. 760 Part II

  Cable Marking. 760.176

  Conductors 760.49

  Conductors of Different Circuits in Same Raceway . . 760.48

  Multiconductor Cable 760.53

  Number of Conductors and Cables in Raceways . . . 760.3(J)

  Number of Conductors in Raceway, Derating 760.51

  Overcurrent Protection 760.43

   Location of 760.45

  Plenums .760.53(B)(2)

  Wiring Methods . . . . 760.53(A)

   Application of Listed Cables . . . . 760.53(B)

   As Multiconductor Cable 760.53

Not Covered by Article 640. . . . . 640.1(B)

Plenums. . . . . 760.3(B)

Power-Limited Circuits . . . . Art. 760 Part III

  Cable Markings 760.154

  Cable Substitutions . . . . .Table 760.154(A)

  Circuit Markings 760.124

  Installation 760.133

  Line-Type Fire Detectors 760.179

  Plenums . . . . .760.135(C)

  Power Sources 760.121

  Separation of Conductors 760.136

  Substitute Cables. 760.154

  Wiring Methods and Materials

   Load Side 760.130

   Supply Side 760.127

Raceways and Sleeves Exposed to Different Temperatures . . 760.3(H)

Raceway Bushings . . . . 760.3(K)

Vertical Support for Fire Rated Cables and Conductors   760.3(I)

See Life Safety Code, NFPA 101 . . . . NFPA 101

See National Fire Alarm Code, NFPA 72. . . . . NFPA 72

FIRE LADDERS

 Clearance for Conductors Not Over 1000 Volts . . . 225.19(E)

 Clearance for Service Conductors. 230.9

 Over 1000 Volts, *See Life Safety Code, NFPA 101* . . . NFPA 101

FIRE PROTECTIVE SIGNALING CABLE

 See (HNGV) . . . . UL Product iQ

FIRE PUMPS . . . . **Art. 695**

 Connection Ahead of Mains . . . . 695.3(A)(1)

  Grouping of Disconnects. . . . . 230.72(B)

  Overload Protection . . . . 230.90 Ex. 4

  Permitted on Supply Side of Service Disconnect . . 230.82(5)

  Prohibited If Emergency System 700.12

  Relative Location of Overcurrent Device. . . . . 230.94 Ex. 4

 Continuity of Power 695.4

 Control Wiring. 695.14

  Arrangement. . . . . 695.14(A)

  Engine Drive Control . . . . 695.14(D)

  Wiring Methods . . . . 695.14(E)

 Disconnecting Means. . . . . 695.4(B)

  Marking Requirements, Controller. . . . . . . 695.4(B)(3)(d)

  Marking Requirements, Fire Pump. . . . . . . 695.4(B)(3)(c)

  Supervision . . . . 695.4(B)(3)(e)

 Emergency Power Supply . . . . Art. 700

 Equipment Location 695.12

 Feeder Sources . . . . 695.5(C)

  Overcurrent Protection. . . . . 695.5(B)

 Fire Rated Assembly (2 hour) to Protect Pump Circuit . . . . . .
. . . . 695.6(A)(2)(d)(3)

 Ground-Fault Protection of Equipment, Service . . . .
. . . . 230.95(C) Info. Note 2

  Building or Structural, Not Required. . . . . . . 240.13(3)

 On-Site Power Production Facility . . . . 695.3(A)(2)

 Overcurrent Protection

  Conductors . . . . 430.72(B) Ex. 1

  Control Circuit Transformer . . . . 430.72(C) Ex.

  Motor 430.31

 Power Sources to 695.3

 Power Wiring 695.6

  Circuit Conductors. . . . . 695.6(B)

  Fire Pump Motors and Other Equipment . . . . 695.6(B)(1)

  Fire Pump Motors Only . . . . 695.6(B)(2)

  Ground Fault Protection . . . . 695.6(G)

  Junction Boxes . . . . 695.6(I)

  Listed Electrical Circuit Protective Systems. . . . . 695.6(H)

  Loads Supplied by Controller and Transfer Switch . . 695.6(E)

  Mechanical Protection . . . . 695.6(F)

  Overload Protection . . . . 695.6(C)

  Pump Wiring. . . . . 695.6(D)

  Raceway Terminations . . . . 695.6(J)

  Service Conductors. . . . . 695.6(A)

 Voltage Drop. 695.7

Remote Control Circuits	430.72
Separate Service	230.2(A)(1)
For Emergency Systems	700.12(D)
Sufficiently Remote	230.72(B)
Utility Service Connection	695.3(A)(1)
Service Equipment Overcurrent Protection	230.90(A) Ex. 4
Surge Protection	694.15
Transfer of Power within Pump Room	695.3(F)
Transformers	695.5

See . Standard for the Installation of Stationary Fire Pumps for Fire Protection

FIRE RATED

Assemblies	300.11(A)(1)
Cables and Conductors	300.19(B)
Circuit Integrity Cables	725.3(I)
Circuit Protective Systems, see (FHIT)	UL Product iQ
Critical Operations Power Systems (COPS)	708.10(C)(2)(3)
Emergency Systems	700.10(D)(2)(4)
Non-Fire-Rated Assemblies	300.11(A)(2)

FIRE-RESISTIVE CABLE SYSTEMS

Boxes	728.5(E)
Cable Trays	728.5(D)
Communication Circuits	800.179(G)(2)
Definitions	728.2
General	728.4
Grounding	728.60
Installations	728.5
Marking	728.120
Mounting	728.5(A)
Optical Fiber Cables	770.179(E)(2)
Other Articles	728.3
Pulling Lubricants	728.5(F)
Raceway Fill	728.5(C)
Scope	728.1
Splices	728.5(H)
Support	728.5(B)
Vertical	728.5(G)

FIRE SPREAD, PREVENTION OF

CATV and Radio Systems	800.26
Installation of Cables	800.110
Class 1, 2 & 3 Remote Control Systems	725.3(B)
Communication Circuits	800.110
Fire Alarm Systems	760.3(A)
Network-Powered Broadband Systems	800.26
Installation of Cables	800.110
Optical Fiber Cables and Raceways	770.26
Wiring Methods, Spread of Fire or Products of Combustion	300.21

FIRE STOPS 300.21

Busways Barriers and Seals	368.234(B)
See FIRE SPREAD, PREVENTION OF	Ferm's Finder

FIREWALLS, WIRING THROUGH

See FIRE SPREAD, PREVENTION OF	Ferm's Finder
See (BXRH)	UL Product iQ

FITTINGS . . . . Art. 314

Conduit Bodies	314.16
Definition of	Art. 100 Part I
Drainage Openings, Field Installed	314.15
Expansion	
For Rigid PVC Conduit	352.44
Underground Subject to Ground Movement	300.5(J) Info. Note
Insulated Fittings Required	300.4(G)
Separable Attachment Fitting	314.27

FIXED ELECTRIC HEATING EQUIPMENT FOR PIPELINES & VESSELS . . . . Art. 427

Branch-Circuit Sizing	427.4
Control & Protection	427 Part VII
Disconnecting Means	427.55
Disconnect, Locked in Open Position	427.55(A)
Expansion Joints	427.16
Free Conductor at Boxes	427.18(A)
Ground-Fault Protection of Equipment	427.22
Grounding, Induction Heating	427.29
Grounding, Resistance Heating Elements	427.23
Grounding, Skin-Effect Heating	427.48
Identification — Caution Signs	427.13
Impedance Heating	Art. 427 Part IV
Induction Heating	Art. 427 Part V
Installation	Art. 427 Part II
Resistance Heating Elements	Art. 427 Part III
Skin-Effect Heating	Art. 427 Part VI

FIXED ELECTRIC SPACE HEATING EQUIPMENT Art. 424

See ELECTRIC HEAT (SPACE)	Ferm's Finder

FIXED EQUIPMENT
Bulk Storage Plants . . . . 515.7(A)
Electric Vehicle Charging System . . . . 625.44(C)
Health Care Facilities . . . . 517.20(A)
Motor Fuel Dispensing Facility . . . . Art. 100
Sensitive Electronic Equipment . . . . 647.4(D)(1)
Swimming Pools 680.2
See GROUNDING Fixed Equipment . . . . Ferm's Finder

FIXED OUTDOOR ELECTRIC DEICING AND SNOW MELTING EQUIPMENT . . . . **Art. 426**
See DEICING AND SNOW MELTING . . . . . Ferm's Finder

FIXED RESISTANCE AND ELECTRODE INDUSTRIAL PROCESS HEATING EQUIPMENT . . . . **Art. 425**
Boilers . . . . Art 425 Part VI and VII
Branch Circuits . . . . .425.4(B), 425.22
Control and Protection . . . . Art. 425 Part III
Disconnecting Means 425.19
Fixed Industrial Process Duct Heaters . . . . . . Art 425 Part V
Fixed Industrial Process Electrode-Type Boilers . . . Art 425 Part VII
Fixed Industrial Process Resistance-Type Boilers . . . Art 425 Part VI
Listed Equipment 425.6
Locations 425.12
Infrared Lamp Industrial Heating Equipment 425.14
Marking of Heating Equipment . . . . Art. 425 Part IV
Overcurrent Protection 425.72
Overpressure Limit Control 425.74
Overtemperature Limit Control 425.73
Special Permission 425.10

(FIXTURE) LUMINAIRE WHIPS
Armored Cable (Type AC) . . . . 320.30(D)
Conductor Sizes and Ampacity . . . . 210.19(A)(4)
Flexible Metal Conduit . . . . 348.30(A) Ex. 3
Flexible Metallic Tubing . . . . 360.20(A) Ex. 2
Length of . . . . 410.117(C)
Liquidtight Flexible Metal or Conduit . . . . . 350.30(A) Ex. 3
Liquidtight Flexible Nonmetallic Conduit . . . . . . 356.30(2)
Manufactured Wiring Systems . . . . . . 604.100(A)(2) and (3)
Metal-Clad Cable (Type MC) . . . . 330.30(D)(2)
Nonmetallic-Sheathed Cable . . . . 334.30(B)(2)
Protection . . . . 240.4(B)(2)

(FIXTURE) LUMINAIRE WIRE . . . . **Art. 402**
Ampacity of, Tapped to Branch Circuits . . . . . . 240.4(B)(2)
Allowable . . . . 402.5 & Table
Overcurrent Protection of 240.4
Overcurrent Protection (General) 402.14
Supplementary Protection 240.10
Taps . . . . 210.19(A)(4)
Length . . . . .410.117(C)
Manufactured Wiring Systems . . . . 604.100(A)(2) and (3)
Types . . . . Table 402.3
Wiring of . . . . Art. 410 Part VI
Whips, see (FIXTURE) Luminaire WHIPS . . . Ferm's Finder
See (ZIPR) . . . . UL Product iQ

FIXTURES (LUMINAIRES) FOR POOLS . . . . . . **Art. 680**
Storable Pools 680.33
See SWIMMING POOLS . . . . Ferm's Finder
See (WBDT) . . . . UL Product iQ

FIXTURES (LUMINAIRES) LIGHTING . . . . **Art. 410**
As a Raceway 410.64
Ballast Type (Electric Discharge)
1000 Volts or Less . . . . Art. 410 Part XIII
Calculations, Inductive and LED Lighting Loads . 220.18(B)
Conductors within 3" of a Ballast Must Be Rated Not Lower Than 90°C 410.68
Note: For Special Application of Type THW, See Application Provision Table 310.4(A) 310.4
Cord-Connected . . . . 410.62(B) & (C)
Fixture (Luminaire) Mounting 410.136
For Signs and Outline Lighting . . . . Art. 600
More than 1000 Volts . . . . Art. 410 Part XIII
Thermal Protection . . . . 410.130(E)
High-Intensity Discharge (Fixtures) Luminaires . . . 410.130(F)
Voltage Limitations
Branch Circuits 210.6
Dwellings, Open-Circuit Voltage Exceeding 300 Volts 410.135
Lighting Equipment Installed Outdoors 225.7
Operating 1000 Volts or Less . . . . .410.130(A)
Operating More than 1000 Volts 410.140
Bathtub and Shower Areas . . . . 410.10(D)
Branch-Circuit Ratings to (Fixtures) Luminaires . . . .
. . . . 210.19(A)(4) Ex. 1&2
Maximum Load 220.18
Overcurrent Protection 210.20
Permissible Loads 210.23
To Lampholders . . . . 210.21(A)

Breaker Rated "SWD" for Switching Fluorescent . . . 240.83(D)
Breaker Rated "HID" for Switching High Intensity Discharge . . . . 240.83(D)
See (DIVQ) . . . . *UL Product iQ*
Calculations, Recessed Luminaires . . . . 220.14(D)
Canopies . . . . Art. 410 Part III
Clearance Required
    Bathtubs & Showers . . . . 410.10(D)
    Clothes Closet Light 410.16
    Hot Tubs and Spas, Indoors . . . . 680.43(B)
    Hot Tubs and Spas, Outdoors . . . . 680.22(B)
    Over Combustible Materials 410.12
    Recessed (Fixtures) Luminaires, From Combustible Materials . . . . 410.116(A)(1)&(2)
    Recessed (Fixtures) Luminaires, From Thermal Insulation . . . 410.116(B)
Clothes Closet (Restrictions) 410.16
Cord-Connected . . . . 410.24(A)
    Adjustable Luminaires (Fixtures) . . . . 410.62(B)
    Listed Electric-Discharge and LED Luminaires (Fixtures) . . . . 410.62(C)
    Unit Equipment Emergency Systems . . . . 700.12(F)
    See PENDANTS . . . . Ferm's Finder
Covering of Combustible Material at Outlet Boxes 410.23
Disconnecting Means Required . . . . 410.130(G)
Decorative Lighting and Similar Accessories 410.160
Electric-Discharge Lighting
    See FIXTURES (LUMINAIRES) LIGHTING, Ballast Type. . . . . Ferm's Finder
Festoon Lighting
    See FESTOON LIGHTING . . . . Ferm's Finder
(Fixture) Luminaire Wires
    Taps
        See (FIXTURE) Luminaire WHIPS . . . . . . Ferm's Finder
    Whips
        See (FIXTURE) Luminaire WHIPS . . . . . . Ferm's Finder
        See(ZIPR) . . . . *UL Product iQ*
        See (FIXTURE) Luminaire WIRE . . . . Ferm's Finder
Flush and Recessed Lighting Luminaires (Fixtures) . . . Art. 410 Part XI
    Clearance Required 410.116
    Supports to 410.36
    Taps to Junction Box (Whip)
        See (FIXTURE) Luminaire WHIPS . . . . . . Ferm's Finder
    Temperature Limits
        Combustible Materials . . . . 410.115(A)

Conductors in Outlet Boxes 410.21
Construction of (Fixtures) Luminaires 410.119
Near Combustible Materials 410.11
Where Recessed in Fire-Resistant Material . . . 410.115(B)
Grounded (Neutral) Conductor Must Be Connected
    Polarization of Luminaires 410.23
    Screw Shell Lampholders 410.90
    Shell of Screw Shell Lampholders . . . . . 200.10(C) & (D)
Grounding . . . . Art. 250 Part VI
    Connected to an Equipment Grounding Conductor . 410.42
    (Fixtures) Luminaires and Lighting Equipment Art. 410 Part V
    Health Care, Patient Care Spaces . . . . 517.13(B)(1)(4)
    Methods of Equipment Grounding . . Art. 250 Part VII and 410.44
Lampholders . . . . Art. 410 Parts VIII & IX
    Circuits and Equipment, at Less Than 50 Volts 720.5
    Double-Pole Switched 410.93
    Heavy Duty (Rating of) . . . . 210.21(A)
    Permissible on 30-ampere Branch Circuits . . . 210.23(B)
    Permissible on 40- and 50-ampere Branch Circuits . 210.23(C)
    Voltage Limitations
        See FIXTURES (LUMINAIRES) LIGHTING, Voltage Limitations . . . . Ferm's Finder
Infrared Heating, Construction 422.48
    Branch-Circuits, 40- and 50-ampere Rating . . . 210.23(C)
    Branch-Circuit Requirements . . . . 424.4(A)
    Overcurrent Protection . . . . 422.11(C)
Infrared Lamp Industrial Heating Appliances 425.14
Lamps Only (No Plug Fuses) 410.90
Mechanical Strength of Luminaires and Parts . . . Art. 410 Part VII
Medium Base
    Branch Circuits . . . . 210.23(A)
    Not Permitted in Clothes Closets . . . . 410.16(B)
Multiwire Branch Circuits, Disconnects Required 410.130(G)(2)
Overcurrent Protection
    Ampacity of (Fixture) Luminaire Wires . . . . Table 402.5
    Branch-Circuit Ratings . . . . 210.19(A)(4)
    Outlet Device Ratings, Lampholders . . . . . . 210.21(A)
    Outlet Devices . . . . 210.20(D)
    Permissible Loads 210.23
    Protection of Conductors 240.5
Polarization
    Identification of Terminals . . . . 200.10(C) & (D)
    Installation of Luminaires (Lampholders) Art. 410 Part VIII
    Of (Fixtures) and Luminaires (Lampholders) . . . 410.50

Signs & Outline Lighting	600.4(B)
Supports	Art. 410 Part IV
Wet & Damp Locations	410.10(A)
To Be Weatherproof Type	410.96
LED Luminaires, Closets	410.16
Lighting Track	Art. 410 Part XIV
Calculated Load for	220.43(B)
Connected Load	410.151(B)
Definition	Article 100, Part 1
Fastening	410.154
Grounding	410.155(B)
Heavy-Duty Track	410.153
Installation	410.151(A)
Locations Not Permitted	410.151(C)
Bathtub and Shower Areas	410.10(D)
Support	410.151(D)
Locations	Art. 410 Part II
Aircraft Hangars Class I Locations	513.4
Not Within Class I Locations	513.7
Commercial Garages Class I Locations	511.4
Over Class I Locations	511.7(B)
Corrosive	410.10(B)
Hazardous (Classified)	Art. 500
Anesthetizing Locations (Above)	517.61(B)
Anesthetizing Locations (Within)	517.61(A)
Class I, Division 1	501.130(A)
Class I, Division 2	501.130(B)
Class I, Zone 0, 1, and 2 Locations	Art. 505
Class II, Division 1	502.130(A)(1)
Class II, Division 2	502.130(B)
Class III, Division 1 & 2	503.130
In Clothes Closets	410.16
In Ducts & Hoods, Commercial Cooking Hoods Only	410.10(C)
In Ducts for Dust, Loose Stock, or Vapor, Prohibited	300.22(A)
In Ducts or Plenums for Environmental Air	300.22(B)
Zone 20, 21, and 22 Locations	506.20
In Show Windows	410.14
In Trees	410.36(G)
Near Combustible Material	410.11
Other Space for Environmental Air	300.22(C)
Outdoors	225.7
Lampholders, Wet and Damp Locations	410.96
Location Below Energized Conductors	225.25
Support by Trees	410.36(G)
Wet or Damp Locations	410.10(A)
Over 1000 Volts — Electric Discharge Type	Art. 410 Part XIII
Over 1000 Volts — Lockable Disconnecting Means	410.141(B)
Over Bathtub	410.10(D)
Hydromassage Bathtub	680.72
Mobile Homes	550.14(D)
Park Trailers	552.54(B)
Recreational Vehicles	551.53(B)
Over Combustible Material	410.12
Spas & Hot Tubs	
Indoors	680.43(B)
Outdoors	680.22(C)
Outdoors (General)	680.22(C)
Swimming Pools	680.22(C)
Underwater	680.23
Therapeutic Pools & Tubs	
General	680.60
Permanently Installed	680.61
Tubs (Hydrotherapeutic Tanks)	680.62(F)
Underwater	680.23
Wet & Damp	410.10(A)

See HAZARDOUS (CLASSIFIED) LOCATIONSFerm's Finder

Low-Voltage Lighting	Art. 411
Medium Base Lampholder	
(Not Over 120 Volts Between Conductors)	210.6(B)
Mogul Base Lampholder	
(Not Over 277 Volts to Ground)	210.6(C)
Outdoors	225.7
Not Over 120 Volts (Dwellings)	210.6(A)
Not Over 600 Volts Between Conductors	210.6(D)
Outdoor Lighting	225.7
Outlet Box must be Accessible Under	
Surface Mounted Electric Discharge and LED (Fixture) Luminaires	410.24(B)
Suspended Ceiling, Securely Fastened	410.36(B)
Outline Lighting	Art. 600
1000 Volts or Less	Art. 410 Part XIII
More Than 1000 Volts	Art. 410 Part XIII
Overcurrent Protection	
For Conductors, (Fixture) Luminaire Wires	210.20(B)
For (Fixture) Luminaire Wires	240.5
For Lampholders	210.21(A)
For Rated Ampacity	Table 402.5

    Permissible Load 210.23
Pendants
    Aircraft Hangars . . . . 513.7(B)
    Commercial Garages . . . . 511.7(A)(2)
    Conductors for Incandescent Lamps 410.54
    Hazardous (Classified) Locations
        Class I, Division 1 Locations . . . . 501.130(A)(3)
        Class I, Division 2 Locations . . . . 501.130(B)(3)
        Class II, Division 1 Locations . . . . 502.130(A)(3)
        Class II, Division 2 Locations . . . . 502.130(B)(4)
        Class III, Division 1 & 2 Locations . . . . . . 503.130(C)
        Hospitals (Hazardous Anesthetizing Locations) . . . . . .
            . . . . 517.61(B)(3) Ex. 2
        Not in Clothes Closets . . . . 410.16(C)
        Not in Theater Dressing Rooms 520.71
        Not Over Bathtubs . . . . 410.10(D)
        Show Windows 210.62, 410.14
    Retrofit Kits
        Defined . . . . Art. 100
        Listing Requirements for Luminaires, Lampholders and Lamps
            . . . . .410.6
    Roof Decking
        Installed Under. . . . . 410.10(F)
    Supports . . . . Art. 410 Part IV
        Boxes for . . . . 314.27(A)
        Ceiling-Suspended (Paddle) Fans, Including Lights 314.27(C)
        Class I, Division 1 Locations . . . . 501.130(A)(4)
        Class II, Division 1 Locations . . . . 502.130(A)(4)
        Swimming Pool Areas . . . . 680.22(B)
    Taps
        See (FIXTURE) Luminaire WHIPS . . . . . . .Ferm's Finder
    Temperature Limits
        On Conductors in Outlet Boxes 410.21
        Special Provisions for Flush and Recessed 410.115
    Track Lighting
        Track Lighting (Defined). . . . . Art. 100
        See LIGHTING TRACK . . . . .Ferm's Finder
    Used as a Raceway 410.64
    Voltage Limitations
        Not Over 120 Volts Between Conductors (Dwellings) . . . . .
            . . . . .210.6(A) & (B)
        Not Over 150 Volts Between Conductors
            Swimming Pool Underwater Lights 680.23(A)(1), (4), and (5)
        Not Over 277 Volts to Ground. . . . . 210.6(C)
            Lighting Equipment Outdoors . . . . 225.7(C)

        Not Over 600 Volts Between Conductors. . . . . . . 210.6(D)
            Lighting Equipment Outdoors. . . . . 225.7(D)
    Whips, (Fixture) Luminaire
        See (FIXTURE) Luminaire WHIPS . . . . . . .Ferm's Finder
    Wire within 3" of a Ballast Must Be Rated Not Lower Than 90°C
        . . . . 410.68
    Wired Fixture (Luminaire) Sections. . . . . 410.137(C)
    Wires, (Fixture) Luminaire
        See (FIXTURE) Luminaire WIRES . . . . . . .Ferm's Finder
        Wiring of. Art. 410 Part VI

FLASH PROTECTION (Warning Labels) 110.16
See . . . . .Standard for Electrical Safety in the Workplace
See . . . . Product Safety Signs and Labels

FLASHERS, TIME SWITCHES 404.5
. . . . 600.6(B)

FLAT CABLE ASSEMBLIES (Type FC) . . . . .Art. 322
    Definition. 322.2
    Installation . . . . Art. 322 Part II
    Listing Requirements 322.6
    Protective Covers . . . . 322.10(3)
    Rating of Branch Circuit . . . . 322.10(1)
    Terminal Block Identification . . . . 322.120(C)
    Uses Not Permitted 322.12
    Uses Permitted 322.10
    See (GQKT) . . . . UL Product iQ
    See (GQRS) . . . . UL Product iQ

FLAT CONDUCTOR CABLE (Type FCC) . . . . . Art. 324
    Branch-Circuit Ratings . . . . 324.10(B)
    Conductor Identification . . . . .324.120(B)
    Definitions 324.2
    Installation .Art. 324 Part II
        Alterations . . . . 324.56(A)
        Anchoring 324.30
        Cable Connections and Insulating Ends . . . . . 324.40(A)
        Connections to Other Systems . . . . 324.40(D)
        Coverings for Floors 324.30
        Crossings 324.18
        Enclosure and Shield . . . . 324.40(C)
        Receptacles . . . . 324.42(A)
        Receptacles and Housings . . . . 324.42(B)
    Listing Requirements 324.6
    Shields . . . . 324.40(C)

System Height	324.10(G)
Transition Assemblies	324.56(B)
Uses Not Permitted	324.12
Uses Permitted	324.10
See (IKKT)	UL Product iQ
See (IKMW)	UL Product iQ

FLEXIBLE CORD & CABLE . . . . **Art. 400**
   See CORDS . . . . Ferm's Finder

FLEXIBLE METAL CONDUIT: Type FMC . . . . . **Art. 348**
   Angle Connectors (Not Concealed) 348.42
   Bends 348.24
   Bonding
      Equipment Bonding Jumper 348.60
      In Hazardous (Classified) Locations 250.100
      Required If Service Raceway . . . . 230.43(15)
      Signs . . . . 600.7(B)(4)
      Where Used as Equipment Grounding Conductor   250.118(5)
   Definition of 348.2
   Fittings Listed 348.6
   Grounding . . . . 250.118(5)
      Permitted as Equipment Grounding Conductor . . 250.118(5)
   Listing Requirements 348.6
   Maximum and Minimum Sizes 348.20
   Number of Conductors in, General 348.22
      Number of, in Metric Designator 12 (3/8") . . . Table 348.22
      Same Size in . . . . Informative Annex C
   . . . . Table C.3
      Same Size, Compact Conductors . . . . Informative Annex C
   . . . . Table C.3A
   Raceway for Service-Entrance Conductors . . . . . 230.43(15)
   Support of 348.30
   Uses Not Permitted 348.12
   Uses Permitted 348.10
      1.8 m (6 ft) Maximum Length . . . . 348.20(A)(2)
      Class I, Division 2 Locations for Flexible Connection . . . . .
   . . . . 501.10(B)(2)(2)
      Elevators, Dumbwaiters, Escalators, etc. . . . .
   . . . . 620.21(A)(1) through (A)(3)(a)
      Enclosing Motor Leads . . . . 430.245(B)
      (Fixture) Luminaire Taps . . . . 348.20(A)(2)(c)
      Intrinsically Safe Locations 504.20
      Metric Designator 12 (3/8") Flex (Uses Permitted)   348.20(A)
      Part of Listed Assembly . . . . 348.20(A)(5)
      Utilizing Equipment . . . . 348.20(A)(2)(a)
      Wired Luminaire (Fixture) Sections . . . . 410.137(C)
   See FLEXIBLE METAL CONDUIT . . . . Ferm's Finder
   See (DXUZ) . . . . UL Product iQ

FLEXIBLE METALLIC TUBING: TYPE FMT . . . . **Art. 360**
   Bends 360.24
      Radii for Fixed Bends . . . . Table 360.24(B)
      Radii for Flexing Use . . . . Table 360.24(A)
   Fittings Listed 360.6
   Grounding 360.60
      Permitted as Grounding Means . . . . 250.118(7)
   Listing Requirements 360.6
   Maximum and Minimum Size 360.20
   Number of Conductors 360.22
      Metric Designator 16 (1/2-in.) & 21 (3/4-in.) Trade Size . . .
   . . . . 360.22(A)
      Metric Designator 12 (3/8-in.) Trade Size . . . . . 360.22(B)
   Splices and Taps 360.56
   Uses Not Permitted 360.12
   Uses Permitted 360.10
      3/8" Permitted . . . . 368.20(A) Ex. 1 and Ex. 2
      Ducts or Plenums for Environmental Air . . . . . 300.22(B)
      For Approved Assemblies or Luminaires . 360.20(A) Exc. 2
      Other Space for Environmental Air . . . . 300.22(C)
      Tap Conductors to Flush or Recessed Luminaires . . . 410.117(C)
   See (ILJW) . . . . UL Product iQ

FLOATING BUILDINGS . . . . **Art. 555 Part III**
   Covered by Code . . . . 90.2(A)(1)
   Definition of 555.2
   Feeder Conductors 555.51
   Ground-Fault Protection 555.53
   Grounding 555.54
   Insulated Neutral Required 555.55
   Service Conductors 555.50
   Services and Feeders, Installation 555.52

FLOODPLAIN PROTECTION . . . . 708.10(C)(3)

FLOOR BOXES
   For Receptacle Spacing . . . . 210.52(A)(3)
   For Receptacle Use . . . . 314.27(B)
   Grounding . . . . 250.146(C)

FLOOR RECEPTACLES

Grounding of . . . . 250.146(C)
In Wet Locations (Protection). . . . . 406.9(D)
May Count as Required Wall Receptacles . . . . 210.52(A)(3)
Meeting Rooms. 210.65
Used with Listed Boxes . . . . 314.27(B)

FLUORESCENT LIGHTING LUMINAIRES (FIXTURES)
1000 Volts or Less . . . . Art. 410 Part XII
Auxiliary Equipment, Remote from . . . . 410.137(A)
Ballast Protection Required . . . . 410.130(E)
Circuit Breakers Used to Switch . . . . 240.83(D)
Clothes Closets, Permitted in . . . . 410.16(A)(2)
Mounting Location . . . . 410.16(C)
Connection of 410.24
Cord-Connected 410.62
Disconnecting Means Required . . . . 410.130(G)
Load Calculations, Value . . . . 220.14(D)
More than 1000 Volts . . . . Art. 410 Part XIII
Raceways 410.64
Snap Switches for . . . . 404.14(A)(1) and (B)(2)

FLUSH LIGHTS, See FIXTURES (LUMINAIRES) LIGHTING,
Flush Lights . . . . Ferm's Finder

FOREIGN MATERIALS (PAINT, PLASTER,CLEANERS, ABRASIVES, ETC). . . . . 110.12(B)

FORMAL INTERPRETATIONS 90.6

FORMING SHELLS, UNDERWATER POOL LIGHTS
Bonding of . . . . 680.26(B)
Definition. 680.2
No-Niche Luminaires (Fixtures) . . . . 680.23(D)
Underwater Audio Equipment . . . . 680.27(A)
Wet-Niche Luminaires (Fixtures) . . . . 680.23(B)

FOUNTAINS . . . . 680 Part V
Applicability of Article 680 . . . . 680.1
Self-Contained, Portable Less Than 1.5 m (5 ft) 680.50
Bonding (Equipotential Bonding) . . . . 680.26(B)
Of Piping Systems 680.53
Cord- and Plug-Connected 680.56
Ground-Fault Circuit-Interrupter Protection . . . . 680.51(A)
For Cord- and Plug-Connected . . . . 680.56(A)
Grounding, Equipment Required to be 680.54
Junction Boxes and Other Enclosures. 680.52
Lighting Luminaires (Fixtures), Submersible Pumps, etc. 680.51
Methods of Grounding 680.55
Signs, at Fountains 680.57
See SWIMMING POOLS . . . . Ferm's Finder

FOYERS, Dwelling Unit (Receptacles) . . . . 210.52(I)

FUEL CELL SYSTEMS . . . . Article 692
Connection to Grounded System . . . . 200.3 Ex.
Connection to Other Circuits. . . . . Art. 692 Part VII
Definition. 692.2
Electric Vehicle (draw current from) 625.2
Emergency Systems . . . . 700.12(E)
Health Care Facilities . . . . 517.30(B)
Installation 692.4
Listing Requirement 692.6
Grounding . . . . Art. 692 Part V
Installed by Qualified Persons. . . . . 692.4(C)
Modular Data Systems 646.17
Optional Standby System . . . . 701.12(F)
Output Characteristics 692.61
Outputs Over 1000 Volts . . . . Art. 692 Part VIII
Power Production Equipment. . . . . 705.2 Info Note
Service, Supply Side. . . . . 230.82(6)
Power Source Identification . . . . 692.4(B)
Unbalanced Connections 692.64
Utility-Interactive Point of Connection 692.65
Wiring Methods . . . . Art. 692 Part IV

FUEL DISPENSING- SEE GASOLINE . . . . . . Ferm's Finder

FULL-LOAD CURRENT, MOTORS
Alternating Current
   Single-Phase . . . . Table 430.248
   Three-Phase . . . . Table 430.250
   Two-Phase . . . . Table 430.249
Direct Current . . . . Table 430.247

FURNACES, See HEATING EQUIPMENT, SPACE
Ferm's Finder

FUNCTIONALLY GROUNDED PHOTOVOLTAIC SYSTEM
. . . . 690.2

FUSES
Ampere Ratings for Fuse and Inverse Time Circuit Breakers Table 240.6(A)
Arc Energy Reduction. 240.67

Cable Limiters, Supply Side at Service . . . . 230.82(1)
Cable Limiters, Supply of Circuits Over 150 V to Ground 240.40
Cartridge . . . . 240 Part VI
   Maximum Voltage, 300-Volt Type . . . . 240.60(A)
Classification of 240.61
Disconnect Ahead of 240.40
Electronically Actuated Fuse
   Feeders, Additional Requirements 240.101
   Over 1000 Volts (Defined) . . . . Art. 100, Part II
   Transformers, Overcurrent Protection . Table 450.3(A) Note 4
   Under 600 Volts (Defined) . . . . Art. 100, Part I
Ground-Fault Protection, Services If Fused Switch Used230.95(B)
Locked or Sealed 230.92
   Rooms or Enclosures . . . . 110.26(F)
Lower Voltage Permitted 240.61
Marking of
   Cartridge Fuses . . . . 240.60(C)
   Plug Fuses . . . . 240.50(B)
Mounting Height of Switches . . . . 404.8(A)
Next Higher Size Permitted (800 Amps or Less) . . . . 240.4(B)
Not in the Grounded Conductor, (Generally) . . . . 230.90(B)
   Fuses Used as Motor Overload Protection 430.36
   Provisions for Connecting in Grounded Conductor . . 240.22
Over 1000 Volts . . . . Art. 240 Part IX
Parallel 240.8
Plug . . . . Art. 240 Part V
   Not Over 125 Volts Between Conductors . . . . 240.50(A)(1)
   Not Over 150 Volts to Ground . . . . 240.50(A)(2)
   Not To Be Installed in Lampholders 410.90
   Replacement Only, Edison Base . . . . 240.51(B)
   Screw Shell Connected to Load Side . . . . 240.50(E)
   Type S Required for New Work 240.52
Protection from Tampering, Specific Circuits 230.93
Readily Accessible 240.24
Renewable Fuses . . . . 240.60(D)
   Access to Occupants, Service Equipment . . . . . 230.72(C)
   Fused Switches . . . . 404.8(A)
Semiconductor . . . . 430.52(C)(5)
Service Equipment Overcurrent Protection . . . . . . 230.90(A)
Standard Ratings 240.6
Transformers, Fuse Ratings 450.3
   1000 Volts and Less . . . . Table 450.3(B)
   Over 1000 Volts . . . . Table 450.3(A)
Voltage Ratings 240.61

Wet or Damp Locations 240.32
   Enclosures for 312.2
   Switch Enclosures for 404.4
See OVERCURRENT PROTECTION . . . . . . Ferm's Finder
See SHORT-CIRCUIT CURRENT AVAILABLE   Ferm's Finder
See (JCQR) . . . . UL Product iQ

G

GAGES (AWG), CONDUCTORS 110.6
Conductor Properties . . . . Chapter 9, Table 8

GALVANIC ACTION
Electrical Metallic Tubing 358.14
Intermediate Metal Conduit 342.14
Rigid Metal Conduit 344.14

GARAGES (COMMERCIAL) . . . . Art. 511
Battery Charging Equipment . . . . 511.10(A)
Class I Locations
   Defined in 511.3
   Group Classification (Atmosphere) . . . . 500.5(A)
Definitions
   Major Repair Garage . . . . Art. 100
   Minor Repair Garage . . . . Art. 100
Electric Vehicle Charging . . . . 511.10(B)
Elevators & Escalators In 620.38
Fuels, Heavier-Than-Air . . . . . . 511.3(C) and Table 511.3(C)
Fuels, Lighter-Than-Air . . . . . . 511.3(D) and Table 511.3(D)
Fuel Dispensing Equipment . . . . 511.4(B)(1)
Ground-Fault Circuit-Interrupter Protection, Receptacles 511.12
Grounding & Bonding . . . . Art. 250
   Additional Requirements Class I Locations 501.30
   At Service . . . . 250.92(B)
   Bonding Conductors and Jumpers 250.102
   General 511.16
   In Hazardous (Classified) Locations 250.100
Lightning Protection 250.106
   Surge Arresters . . . . Art. 242
   Surge Protection 501.35
   Surge Protective Devices [Transient Voltage (TVSS)] . Art. 242
   See Standard for the Installation of Lightning Protection Systems,
   NFPA 780 . . . . NFPA 780
Portable Lighting Equipment (Hand Lamps) . . . . 511.4(B)(2)

Sealing 501.15
    Horizontal As Well As Vertical Boundaries 511.9
Wiring Methods
    Above Class I Locations 511.7
        Equipment above . . . . 511.7(B)
    In Class I Locations (General) . . . . Art 501
        For Garages 511.4
    Rigid Nonmetallic Conduit Permitted . . . . 511.4(A)
    Underground Wiring. 511.8
See HAZARDOUS (CLASSIFIED) LOCATIONS . . Ferm's Finder
See Standard for Parking Structures, NFPA 88A . . . . NFPA 88A
See Code for Motor Fuel Dispensing Facilities and Repair Garages, NFPA 30A . . . . NFPA 30A

GARAGES (NONDWELLING)
GFCI Protection . . . . 210.8(B)(8)

GARAGES (RESIDENTIAL)
Floor Area Not Included for Lighting Load Calculations . 220.12
Garage Branch Circuits . . . . .210.11(C)(4)
Grade-Level Accessory Buildings . . . . 210.8(A)(2)
Ground-Fault Circuit-Interrupter Protection Requirements . . .
. . . . 210.8(A)(2)
Grounding Requirements If Remote 250.32
Lighting Outlets Required . . . . 210.70(A)(2)
Receptacles Required . . . . 210.52(G)(1)

GAS PIPE BONDED . . . . 250.104(B)
See National Fuel Gas Code, NFPA 54. . . . . .NFPA 54

GAS PIPE NOT AS GROUNDING ELECTRODE 250.52(B)(1)

GAS TUBE SIGN AND IGNITION CABLE
Signs and Outline Lighting . . . . 600.32(B)
See (ZJQX). UL Product iQ

GAS TUBE SIGN TRANSFORMERS
Signs and Outline Lighting 600.21, 600.23
See (PWIK) . . . . UL Product iQ

GASOLINE BULK STORAGE PLANTS . . . . Art. 515

See BULK STORAGE PLANTS . . . . Ferm's Finder

GASOLINE (MOTOR FUEL) DISPENSING . . . . . Art. 514
Aboveground Fuel Storage . . . . 514.3(A)(3)
Boatyard and Marinas
    Classification of Locations . . . . 514.3(C)
    Hazardous (Classified) Locations. 555.11
Circuit Disconnects (General) 514.11
    Attended Self-Service Stations . . . . 514.11(B)
    Unattended Self-Service Stations . . . . 514.11(C)
Class I Locations
    Group Classification (Atmosphere) . . . . 500.6(A)
    Extent of Locations. . . . . 514.3(B)(1) & (2)
Conductors (Gas & Oil Resistant) 501.20
See (ZLGR) . . . . UL Product iQ
Definition. . . . . Art. 100
Dispensing at Bulk Storage Plants. 515.10
Dispensing at Marinas and Boatyards. 555.11
Dispensing Equipment, Provisions to Remove All External
    Voltage Sources During Maintenance and Servicing . . 514.13
Emergency Disconnect and Controls . . . . 514.11(A)
Grounding & Bonding . . . . Art. 250
    Additional Requirements Class I Locations. 501.30
    At Service . . . . 250.92(B)
    Equipment Bonding Jumpers 250.102
    General. 514.16
    In Hazardous (Classified) Locations 250.100
Hand Lamps . . . . 511.4(B)(2)
Intrinsically Safe Systems . . . . Art. 504
Lightning Protection (Surge) 250.106
    Surge Arresters . . . . Art. 242
    Surge Protection 501.35
See . . Standard for the Installation of Lightning Protection Code
Marinas and Boatyards
Classification of Locations . . . . 514.3(C)
Hazardous (Classified) Locations 555.11
Sealing 501.15
    At Dispenser . . . . 514.9(A)
    Horizontal As Well As Vertical Boundaries . . . . 514.9(B)
Switch for Dispensing or Pumping Equipment . . . . 514.11(A)
    Handle Ties (Not Permitted) . . . . 514.11(A)
Wiring Methods
    Above Class I Locations 514.7
    Below Class I Locations (Underground) 514.8

Within Class I Locations 514.4

See Code for Motor Fuel Dispensing Facilities and Repair Garages

See Chapter 13, Farms and Remote Sites, NFPA 30A-2003 NFPA 30A-2003

See . . . Tables 514.3(B)(1) & (2) for Location Classifications

See Code for Motor Fuel Dispensing Facilities and Repair Garages

See GARAGES (COMMERCIAL) . . . . Ferm's Finder

See HAZARDOUS (CLASSIFIED) LOCATIONS Ferm's Finder

See (ZLGR) . . . . *UL Product iQ*

GENERAL CARE (CATEGORY 2) PATIENT SPACES . 517.18
Definitions 517.2

See HEALTH CARE FACILITIES . . . . Ferm's Finder

GENERAL REQUIREMENTS FOR WIRING METHODS . . . . .Art. 300
Boxes or Fittings, Where Required 300.15

Changing Raceway or Cable to Open Wiring 300.16

Conductors 300.3

   Different Systems . . . . 300.3(C)

      1000 Volts or Less . . . . 300.3(C)(1)

      Restrictions for Class 2 & 3 Circuit Conductors . . . .
. . . . .300.3(C)(1) Info. Note 1

   Installation with Other Systems (Nonelectrical) 300.8

   Insulating Bushings for Raceways . . . . 300.4(G)

   Number and Sizes of, in 300.17

   Of Same Circuit . . . . 300.3(B)

      Enclosures (Auxiliary Gutters) . . . . 300.3(B)(4)

      Grounding and Bonding Conductors . . . . . 300.3(B)(2)

      Nonferrous Wiring Methods . . . . 300.3(B)(3)

      Paralleled Installations . . . . 300.3(B)(1)

   Supporting of Conductors in Vertical Raceways 300.19

Exhaust and Ventilating Ducts 300.22

Exit Enclosures (Stair Towers) 300.25

Feed-Through Neutral Connections . . . . 300.13(B)

Fire Rated Cables and Conductors, Support of . . . . 300.19(B)

Free Length of Wire at Outlets 300.14

Induced Currents Ferrous Metal Enclosures or Ferrous Metal Raceways 300.20

Mechanical and Electrical Continuity

   Conductors . . . . 300.13(A)

   Raceways and Cables 300.12

   Raceways and Enclosures 300.10

Over 1000 Volts . . . . .Art. 300 Part II

   Aboveground Wiring Methods 300.37

   Aboveground Wiring Methods within raceways in Wet Locations 300.38

   Braid-Covered Insulated Conductors–Open Installation 300.39

   Conductor Bending Radius 300.34

   Conductors of Different Systems . . . . 300.3(C)(2)

   Covers Required 300.31

   Insulation Shielding 300.40

   Moisture or Mechanical Protection Metal-Sheathed Cables 300.42

   Protection Against Induction Heating 300.35

   Underground Installations . . . . 300.50(A)

      Minimum Cover Requirements . . . . Table 300.50

Protection Against

   Corrosion 300.6

   Physical Damage 300.4

Raceways as a Means of Support . . . . 300.11(B)

Raceways Exposed to Different Temperatures 300.7

   Expansion Joints . . . . 300.7(B)

Sealing Raceways, Change in Temperature . . . . . 300.7(A)

Securing and Supporting . . . . 300.11(A)

Spread of Fire or Products of Combustion 300.21

Through Metal Framing Members . . . . 300.4(B)

Through Wood Studs, Joists, Rafters . . . . 300.4(A)

Underground Installations 300.5

   Minimum Cover Requirements . . . . Table 300.5

Voltage Limitations 300.2

Wiring in Ducts, Plenums, Air-handling Spaces 300.22

GENERATORS . . . . Art. 445
Ampacity of Conductors 445.13

Connection Supply Side of Service Permitted . . . . . 230.82(5)

Critical Operations Power Systems

   Generator Set . . . . 708.20(F)

   Outdoor, Permanently Installed . . . . 708.20(F)(5)(a)

   Outdoor, Portable . . . . 708.20(F)(5)(b)

Disconnecting Means and Shutdown 445.18

   Disconnecting Means . . . . 445.18(A)

   Generators Installed in Parallel . . . . 445.18(C)

   Shutdown of Prime Mover . . . . 445.18(B)

Emergency Shutdown, One- and Two-Family Dwellings 445.18(D)

Emergency Shutdown, Remote . . . . 445.18(C)

Emergency Systems . . . . Art. 700

   See EMERGENCY SYSTEMS . . . . Ferm's Finder

Ground-Fault Circuit-Interrupter Protection

   Receptacles, 15kW or Smaller Portable Generators . . . 445.20

      Bonded Neutral Generators . . . . 445.20(B)

Unbonded (Floating Neutral) Generators . . . . 445.20(A)
Grounding
    AC Systems 250.20
        Conductor to be Grounded 250.26
        Separately Derived Systems 250.30
        As Source If Separately Derived System . . . . . . 250.20(C)
    Equipment (Fixed) 250.112
    Objectionable Currents 250.6
    Over 1000 Volts . . . . 250 Part X
        Impedance Grounded Neutral Systems 250.187
    Permanently Installed 250.35
    Portable and Vehicle-Mounted . . . . 250.34(A)&(B)
In Class I, Division 1 Locations . . . . 501.125(A)
In Class I, Division 2 Locations . . . . 501.125(B)
Interconnected Electric Power Production Sources . . . Art. 705
Legally Required Standby Systems . . . . Art. 701
Listing, Stationary Type 445.6
Nameplate, Stationary Generators Over 15kW 445.11
Neutral Bonded to Frame
    Field Modifications 445.11
    Marked by Manufacturer 445.11
Optional Standby Systems . . . . Art. 702
    Outdoor, Permanently Installed . . . . 702.12(A)
    Outdoor, Portable . . . . 702.12(B)
    Power Inlets Rated at 100 Amperes or Greater . . . 702.12(C)
Overcurrent Protection
    Shall Be Provided 445.12
    Where Generator Equipped . . . . 445.13(B)
Permanently Installed, Grounding 250.35
Portable Generators 250.34
    Separately Derived Systems 250.30
Premises Wiring System Derived From . . . . 250.20(B)
    DC Systems 250.162
        Point of Connection 250.164
        Size of DC Grounding Electrode Conductor . . . 250.166
Recreational Vehicles 551.30
    Pre-Wiring . . . . 551.47(R)
    Supply Conductors . . . . 551.30(E)
    See TRANSFER SWITCHES . . . . Ferm's Finder

GOOSENECKS, SERVICE CABLES 230.54

"GREENFIELD" . . . . Art. 348
    See FLEXIBLE METAL CONDUIT . . . . Ferm's Finder

GRAIN HANDLING & STORAGE AREAS
Bonding . . . . 502.30(A)
    At Service Equipment 250.92
    Equipment Bonding Jumpers 250.102
    In Hazardous (Classified) Locations 250.100
Class II Area Location Classification . . . . . 500.5(C)(1) & (2)
Class II Group Classifications . . . . 500.6(B)
Flexible Cords 502.140
General Requirements . . . . Art. 502
Grounding and Bonding . . . . Art. 250
    Additional Requirements 502.30
Lighting Luminaires (Fixtures) in 502.130
Lightning Protection (Surge) 250.106
    Surge Arrestors . . . . Art. 242
    Surge Protection 502.35
*See Standard for the Installation of Lightning Protection Systems,
NFPA 780* . . . . NFPA 780
Motors & Generators
    In Class II, Division 1 Locations . . . . .502.125(A)
    In Class II, Division 2 Locations . . . . .502.125(B)
Receptacles in 502.145
Sealing 502.15
Switches in 502.115
Wiring Methods
    Class II, Division 1 . . . . 502.10(A)
    Class II, Division 2 . . . . 502.10(B)
See Recommended Practice for the Classification of . . *Combustible Dusts and of Hazardous (Classified) Locations for Electrical Installations in Chemical Process Areas*
See HAZARDOUS (CLASSIFIED) LOCATIONS Ferm's Finder

GROMMETS REQUIRED . . . . 300.4(B)(1)
For Flexible Cords and Cables 400.14

GROOVES, SHALLOW, CABLES AND RACEWAYS INSTALLED
. . . . 300.4(F)

IN GROUND BAR, EQUIPMENT 408.40
Feeder Conductor Grounding Means 215.6
Isolated Ground Receptacles . . . . 250.146(D)

GROUND CLAMPS 250.70
Connection to Be Accessible . . . . 250.68(A)
Connection to Be Clean 250.12
Connection of Grounding and Bonding Equipment
    Permitted Methods 250.8
One Wire Only to a Ground Clamp 250.70

Protection of Clamp 250.10
See (KDER). . . . . *UL Product iQ*

GROUND DETECTORS
AC Systems 50 to 1000 Volts Not Grounded . . . . . 250.21(B)
    Required on All Ungrounded Systems 250.21
Electric Cranes in Class III Locations, Power Supply   503.155(A)
High-Impedance Grounded Neutral Systems 250.36
Isolated Power Systems in Health Care Facilities
    Line Isolation Monitor . . . . 517.160(B)
    Optional for Critical Care Spaces . . . . 517.19(E)
    Within Hazardous (Classified) Anesthetizing Locations . . . .
     . . . . 517.61(A)(2)
Portable or Mobile Equipment over 1000 Volts . . . 250.188(D)
See ISOLATING TRANSFORMERS . . . . Ferm's Finder

GROUND-FAULT CIRCUIT INTERRUPTERS (GFCI)
Accessory Buildings, Dwellings . . . . 210.8(A)(2)
Appliances
    Accessibility 422.5
    Dishwashers . . . . 422.5(A)(7)
    Drinking Water Coolers, Electric. . . . . 422.5(A)(2)
    High-Pressure Spray Washers . . . . 422.5(A)(3)
    Sump pumps . . . . 422.5(A)(6)
    Tire Inflation Machines (For Public Use) . . . . . 422.5(A)(4)
    Type and Location . . . . 422.5(B)
    Vacuum Machines (Automotive for Public Use) . 422.5(A)(1)
    Vending Machines . . . . 422.5(A)(5)
Basements, Unfinished, Dwelling Units . . . . . . 210.8(A)(5)
Bathrooms
    Dwellings . . . . 210.8(A)
    Other Than Dwelling Units . . . . 210.8(B)
    Patient Care, Health Care Facilities 517.21
Bathtubs, Hydromassage 680.71
Bathtubs and Shower Stalls . . . . 210.8(A)(9)
Boathouses . . . . 210.8(A)(8)
Boat Hoists . . . . 210.8(C)
Carnivals, Circuses, Fairs, etc. 525.23
Commercial Garages 511.12
Conductive Heated Floors of Bathrooms, Kitchens, and Hydromassage Bathtub Locations . . . . 424.44(E)
Construction Sites 590.6
Cord Sets (Temporary Installations) . . . . 590.6(A)(1)
Crawl Spaces, at or below Grade Level, Dwellings . 210.8(A)(4)
Crawl Space Lighting Outlets . . . . 210.8(C)
Damp Locations- Indoor (Dwelling Units) . . . . . 210.8(A)(11)
Definition . . . . Art. 100 Part I
Deicing and Snow-Melting Equipment, Impedance, Heating 426.32
Dishwasher Outlets (Dwelling Units) . . . . 422.5(A)(7)
Determining Distance (Receptacle) 210.8
Dwellings . . . . 210.8(A)
Elevators, etc. 620.6
Feeders 215.9
Fountains
    Cord- and Plug-Connected Equipment 680.56
    Signs . . . . 680.57(B)
Garages, Commercial 511.12
Garages, Dwellings . . . . 210.8(A)(2)
Garages, Nondwelling . . . . 210.8(B)(8)
Generators (Portable), Receptacles, 15kW or Smaller . . . 445.20
Health Care Facilities, Wet Locations . . . . 517.20(A)
    Pool and Tubs for Therapeutic Use . . . . Art. 680 Part VI
    Therapeutic Tubs (Hydrotherapeutic Tanks) . . . 680.62(A)
        Receptacles within 1.5 m (5 ft.) . . . . 680.62(E)
High Pressure Spray Washers . . . . 422.5(A)(3)
HVAC Equipment . . . . 210.8(E)
HVAC Outdoor Outlets (Dwelling Units) . . . . 210.8(F)
Hydromassage Bathtub 680.71
Indoor Wet Locations, Other than Dwelling Units . 210.8(B)(6)
Kitchens, Dwelling Units . . . . 210.8(A)(6)
Kitchens, Other Than Dwelling Units . . . . 210.8(B)(2)
Laundry Areas . . . . 210.8(A)(10)
Locker Rooms . . . . 210.8(B)(7)
Marinas & Boatyards . . . . 210.8(A)(8), 555.35(B)
Mobile Home Service Equipment, Additional Receptacles . . 550.32(E)
Mobile Homes . . . . 550.13(B)
    Pipe Heating Cable Outlet . . . . 550.13(E)
Other Than Dwelling Units . . . . 210.8(B)
    Basements (Unfinished/Non-Habitable) . . . . . 210.8(B)(10)
    Bathrooms . . . . 210.8(B)(1)
    Crawl Spaces . . . . 210.8(B)(9)
    Garages and Service Bays . . . . 210.8(B)(8)
    Kitchens . . . . 210.8(B)(2)
    Locker Rooms . . . . 210.8(B)(7)
    Outdoors . . . . 210.8(B)(4)
    Rooftops . . . . 210.8(B)(3)
    Showering Facilities . . . . 210.8(B)(7)
    Sinks . . . . 210.8(B)(5)
    Wet Locations (Indoor) . . . . 210.8(B)(6)

Outdoors, Dwelling Units . . . . 210.8(A)(3)
Outdoors, Nondwelling . . . . 210.8(B)(4)
Park Trailers, Where Required. . . . . 552.41(C)
    Pipe Heating Cable Outlet . . . . 552.41(D)
Pipeline and Vessel Heating, Impedance, Heating. . . . 427.27
Readily Accessible Location 210.8
Receptacle Replacement . . . . 406.4(D)

    Note: Where replacements are made at receptacle locations that are required to be GFCI types, the replacements must be GFCI protected.

Receptacles
    Dwellings . . . . 210.8(A)
    Other Than Dwellings . . . . 210.8(B)
Replacement of . . . . 406.4(D)
Recreational Vehicle Parks. 551.71
Recreational Vehicles
    Receptacle Locations, Required . . . . 551.41(C)
    Shower Luminaires (Fixtures) . . . . 551.53(B)
    With One Branch Circuit . . . . 551.40(C)
Residential Occupancies . . . . 210.8(A)
Rooftop Receptacles, Nondwelling . . . . 210.8(B)(3)
Signs (Outdoor Portable) . . . . 600.10(C)(2)
Sink
    Dwelling Units. . . . . 210.8(A)(7)
    Nondwelling . . . . 210.8(B)(5)
Spas & Hot Tubs
    Indoor Installations
        Lighting Luminaires (Fixtures), Outlets, and Paddle Fans . . . . 680.43(B)(2)
        Receptacles Providing Power to . . . . 680.43(A)(3)
        Receptacles within 3.0 m (10 ft.) . . . . 680.43(A)(2)
        Outdoor . . . . 680.42(A)(2)
        Outlet Supplying Self-Contained or Package Units. . . 680.44
Sump Pumps . . . . 422.5(A)(6)
Swimming Pools
    Electrically Operated Pool Covers . . . . 680.27(B)(2)
    Motors, Other Than Dwelling Units . . . . 680.21(A)(1)
    Protection of Wiring in Enclosures. . . . . 680.23(F)(3)
    Requirement for Luminaires, Paddle Fans . . . . 680.22(B)(4)
    Requirement for Receptacles. . . . . 680.22(A)(4)
    Storable Pools 680.32
    Types Allowed 680.5
    Underwater Luminaires, Relamping . . . . 680.23(A)(3)
    Therapeutic Pools & Tubs . . . . 680.62(A)
    Receptacles within 1.5 m (6 ft) . . . . 680.62(E)

Tire Inflation Machines (For Public Use) . . . . . . 422.5(A)(4)
Unfinished Basements . . . . 210.8(A)(5)
Vacuum Machines (Automotive For Public Use) . . 422.5(A)(1)
Vending Machines . . . . 422.5(A)(5)
Wet Locations
    Indoor Locations, Dwelling . . . . 210.8(A)(11)
    Indoor Locations, Nondwelling . . . . 210.8(B)(6)
See (KCXS) . . . . UL Product iQ

GROUND FAULT CURRENT PATH . . . . ART. 100
    Ground-Fault Current Paths, Examples . . Art. 100, Info. Note

GROUND-FAULT CURRENT PATH, EFFECTIVE . ART. 100
    Metal Enclosures 250.109

GROUND-FAULT DETECTION, DIRECT-CURRENT SYSTEMS
    Grounded. . . . . 250.167(B)
    Marking. . . . . 250.167(C)
    Ungrounded . . . . 250.167(A)

GROUND-FAULT PROTECTION OF EQUIPMENT
    Boatyards and Marinas 555.35
    Building or Structure Main Disconnecting Means . . . 240.13
    Definition of . . . . Art. 100 Part I
    Emergency Systems, Signals . . . . 700.6(D)
    Feeders 215.10
    Feeders- Temporary (No GFP) . . . . 210.15, Ex. No. 3
    Fixed Outdoor Deicing and Snow-Melting Equipment
        Impedance Heating 426.32
        Resistance Heating. 426.28
    Fuses . . . . 230.95(B)
    Health Care Facilities 517.17
    Legally Required Standby Systems, Not Required. . . . 701.26
    Marinas and Boatyards 555.35
    Orderly Shutdown (Branch Circuit) . . . . 210.13 Ex. 1
    Orderly Shutdown (Feeder) . . . . 215.10 Ex. 1
    Orderly Shutdown (Service) . . . . 230.95 Ex.
    Performance Testing . . . . 230.95(C)
    Pipeline and Vessel Heating
        Impedance Heating 427.27
        Resistance Heating. 427.22
    Settings. . . . . 230.95(A)
    Service Equipment 230.95
    Signs and Outline Lighting, Transformers and Electronic
        Power Supplies (Secondary-Circuit Ground Fault Protection) .

. . . . 600.23(B)
See (KDAX) . . . . *UL Product iQ*

GROUND-FAULT SENSING AND RELAYING EQUIPMENT
See (KDAX) . . . . *UL Product iQ*

GROUND RING, ELECTRODE . . . . 250.52(A)(4)
Size of Conductor to . . . . 250.66(C)

GROUND RODS . . . . **Art. 250 Part III**
25 Ohms or Less Resistance to Ground . . . 250.53(A)(2)EX.
CATV and Radio Distribution Systems. 800.100
    At Mobile Homes — Grounding and Bonding . . . 800.106
Communication Circuits (Primary Protector) 805.93
    At Mobile Homes 800.106
Intersystem Bonding Termination . . 250.94(A), 805.93(B), 820.100(B)
Lightning Protection Systems, Required Bonding. . . . 250.106
    Use of Strike Termination Devices (Air Terminals) . . . 250.60
Network-Powered Broadband Systems 800.100
    At Mobile Homes 800.106
Maximum No. 6 Ground Wire to a Ground Rod . . . 250.66(A)
Multiple Rods Distance Apart . . . . . 250.53(A)(3) Info. Note
Radio and Television Equipment . . . . 810.21(F)
Size, Length & Depth to Be Driven . . . .
    Auxiliary CATV and Radio Distribution Systems . . 820.100
    Communication Circuits. . . . . .805.100(B)
    Network-Powered Broadband Systems . . . . . .800.100(B)
    Radio and Television Equipment . . . . 810.21(F)
Grounding Electrodes. 250.54
    Permitted. 250.54
See (KDER) . . . . *UL Product iQ*
See (KDSH) . . . . *UL Product iQ*

GROUNDED
Circuit and System, AC. . . . . .Art. 250 Part II
Circuit and System, DC . . . . Art. 250 Part VIII
Conductor to Service Equipment . . . . 250.24(C)
Conductors
    Continuity of Grounded Conductor . . . . 200.2(B)
    Delta Connected Service. . . . . .250.24(C)(3)
    Feeder Identification 215.12
    Grouping for Multiple Circuits. . . . . 210.4(D)
    Grouping with Same Circuit . . . . 200.4(B)
    Identification 200.6
    Installation . . . . 200.4(A)
    Multiple Circuits . . . . 200.4(B)
    Not Used for Grounding Equipment. . . . . . . .250.142(B)
    Not Used for Grounding. . . . . .250.24(A)(5)
    Sizing of . . . . Table 250.102(C)(1)
    Switches Controlling Lighting Loads. . . . . 404.2(C)
    Theaters, Extra-Hard Usage Cords . . . . 520.44(C)
    To Be Grounded, AC Systems 250.26
    To Be Grounded, DC Systems 250.162
    Use for Grounding Permitted
        At Separate Buildings or Structures . . 250.32(B)(1) Ex. (2)
        Load-Side Equipment . . . . 250.142(B), Ex. 1–4
    Separately Derived Systems. . . . . 250.30(A)
    Supply-Side Equipment . . . . .250.142(A)
Definition of . . . . Art. 100 Part I
Direct-Current Systems, Ground-Fault Detection . .250.167(B)
Effectively
    Effective Ground-Fault Current Path. . . . . . . 250.4(A)(5)
    Metal In-Ground Support Structure . . . . . .250.52(A)(2)
Equipment . . . . 250 Part VI
Lighting Systems 30 Volts or Less
    Secondary Circuits Not Permitted to Be Grounded . 411.6(A)
    Secondary Circuits Not Permitted to Be Grounded 250.22(4)
Location of System Connections, AC . . . . 250.24(A)
Other Conductor and Enclosures 250.86
Overcurrent Device Permitted in Grounded Conductor . .230.90(B)
Point of Connection, DC 250.164
Power Systems in Anesthetizing Locations 517.63
Separately Derived Systems 250.30
Service Enclosures and Raceways 250.80
Sizing, Grounded Conductor . . . . Table 250.102(C)(1)
Use & Identification of Grounded Conductors . . . . . Art. 200
    For Branch Circuits . . . . 210.5(A)
    Identification of . . . . 310.6(A)
    Means of Identification. 200.6
See NEUTRAL OR GROUNDED CONDUCTOR   *Ferm's Finder*

GROUNDING . . . . **Art. 250**
Agricultural Buildings . . . . 547.5(F)
Air-Conditioning (outdoor roof) 440.9
Air-Conditioning, Room 440.61
Alternating Current Systems 250.20
Anesthetizing Locations 517.62
Antenna. 810.21
Appliances (General) . . . . Art. 250 Part VI

2020 Ferm's Fast Finder | 97

Clothes Washers 250.114
Dishwashers 250.110
   Cord- and Plug-Connected (If Permitted) 250.114
Existing Counter Mounted Cooking Units. 250.140
Freezers 250.114
In Mobile Homes . . . . 550.16(B)(3)
In Park Trailers 552.56
In Recreational Vehicles 551.55
Ranges 250.140
Wall Mounted Ovens 250.140
Audio Signal Processing, Amplification & Reproduction . . 640.7
Bars (In Panelboards) 408.40
   Feeder Conductor Grounding Means 215.6
Boxes 250.86
   Attachment of Conductors to Boxes 250.148
   Equipment Grounding. . . . . Art. 250 Part VI
   Metal Boxes 314.4
   Methods of Equipment Grounding . . . . . Art. 250 Part VII
   Other Conductor Enclosures 250.86
CATV Systems, To Other Systems 250.94
   Cable Grounding. 820.100
   Conductive Shield of Coaxial Cable 820.93
   Equipment Grounding. 820.103
   Intersystem Bonding Termination . . . . 820.100(B)
   Separately Derived Systems. 250.30
Circuits (AC) . . . . Art. 250 Part II
Circuits (DC). . . . . Art. 250 Part VIII
Clothes Dryers 250.140
Clothes Washers 250.114
Communications Systems (Methods) . . . . . . Art. 800 Part IV
   Bonding for Communication Systems 250.94
   Cable and Primary Protector . . . . 805.93(A)
   Cable and Primary Protector at Mobile Homes. . 800.106(A)
   Devices, Listing Required 800.180
   Intersystem Bonding Termination . . . . . . 800.100(B)(1)
Conductors
   Aluminum or Copper-Clad Aluminum Not Permitted Near
   the Earth or on Masonry Walls (Equipment Grounding). . . .
   . . . . 250.120(B)
      Grounding Electrode Conductors . . . . 250.64(A)
      Physical Damage. . . . . 250.64(B)
   Definition (GROUNDED or GROUNDING) Art. 100 Part I
   Equipment Grounding Conductors . . . . . Art. 250 Part VI
      Continuity. 250.124

      Identification of 250.119
      Installation 250.120
      Size of 250.122
      Types 250.118
   Line-Side of Service Equipment 250.66
      Minimum . . . . Table 250.66
      Sizing . . . . 250.102(C)
   Load-Side of Service Equipment 250.122
      Adjustment for Increase in Phase Conductor Size 250.122(B)
      Flexible Cord and (Fixture) Luminaire Wires . . 250.122(E)
      Minimum . . . . Table 250.122
      Motor Circuits . . . . 250.122(D)
      Multiple Circuits, Same Raceway . . . . . . 250.122(C)
      Parallel Conductors . . . . 250.122(F)
      Sizing 250.122
   Material for Grounding Electrode Conductor 250.62
   Objectionable Current Over 250.6
Conductor Connections
   Boxes, to and in 250.148
   Clean Surfaces 250.12
   Conductors and Equipment 250.8
   Grounding Electrode, to 250.68
   Grounding Electrode, Means of Connection 250.70
   Protection of Attachment 250.10
   Solder Connections Must be Mechanically Secure 110.14(B)
   Wiring Device Terminals, Identification of 250.126
Conductor to Be Grounded. 250.26
Conductor Required 250.64
   General. 250.4
   Swimming Pool Circuits (Specific) . . . . Art. 680
   Note: Installing supplementary equipment grounding
   conductors in metal raceways is no substitute for effectively
   bonding those raceways . . . . 250.96(B) Info. Note
Connections . . . . 250.8(A) & (B)
Cranes & Hoists 610.61
   Class III, Division 1 & 2 Locations 503.30
Direct Current Systems 250.162
   Point of Connection 250.164
   Separately Run from Circuit Conductors . . . 250.134 Ex. 2
   Size 250.166
   Use of Grounded Conductor (Load-side Equipment) 250.142(B) Ex. 3
Dishwashers, Permanently Connected 250.110
Dishwashers, Cord- and Plug-connected 250.114

Dryers (Clothes) 250.140
   Mobile Homes 550.16
Earth Return Not Permitted . . . . 250.4(A)(5)
Earth Return Not Permitted . . . . 250.4(B)(4)
   (Supplementary) Auxiliary Electrodes 250.54

      Note: Supplementary grounding electrodes shall be permitted to augment the equipment

      grounding conductors specified in Section 250.118. The earth shall not be used as an effective ground-fault current path.

Electrode Conductor
   Connection of, to Electrode 250.68
      Clean Surfaces 250.12
      Methods 250.70
   Definition of . . . . Art. 100
   Enclosures and Raceways for . . . . 250.64(E)

      Note: Metal enclosures protecting the grounding electrode conductors must be bonded at both ends to the grounding electrode conductor.

   Installation of 250.64
      DC Systems 250.164
      Separately Derived AC Systems (Grounded)
          . . . . 250.30(A)(3) and (4)
      Separately Derived AC Systems (Ungrounded) . 250.30(B)
      Service-supplied AC Systems 250.24
   Protection of Connection Accessibility . . . . . . 250.68(A)
   Size of Required, AC Systems 250.66
   Size of Required, DC Systems 250.166
   Splice of Not Generally Permitted . . . . 250.64(C)
      Busbar Splice Permitted . . . . 250.64(C)(2)
      Irreversible Compression Connector or Exothermic
      Welding or Riveting Permitted . . . . 250.64(C)
      Multiple Enclosure Services, Permitted for. . . . 250.64(D)
   Taps Permitted . . . . 250.64(D)
Electrodes . . . . Art. 250 Part III
   Aluminum Electrodes Not Permitted . . . . . 250.52(B)(2)
   Auxiliary Grounding Electrodes 250.54
   Bonding Together
      Air Terminals 250.60
      All Electrodes Bonded Together to Form System . . 250.50
      CATV Systems . . . . 800.100(B) & (D)
      Common Electrode 250.58
      Conductor Taps . . . . 250.64(D)(1)
      For Communications Circuits . . . . . 800.100(B) & (D)
      For Radio and Television Equipment 810.58
      In Agricultural Buildings . . . . 547.9(A)(5)
      Lightning Protection Systems 250.106
   Made (Installed) Rod, Pipe, and Plate and Other Electrodes 250.52
      Metal Well Casing Permitted as Electrode . . . 250.52(A)(8)
      Network-Powered Broadband Systems. . 800.100(B) & (D)
      Bonding for Communication Systems 250.94
      Two or More Buildings from Common Service . 250.32(A)
      Underground Gas Line Not Permitted . . . . . 250.52(B)(1)
   Burial Depth, Ground Rings . . . . 250.53(F)
   Burial Depth, Plate Electrodes . . . . 250.53(A)(5)
   Burial Depth, Rod and Pipe . . . . 250.53(A)(4)
   Types of, Permitted to Be Used for System . . . . 250.52(A)
   Types of, Not Permitted to Be Used for System . . 250.52(B)
Definition of Grounding Electrode . . . . Art. 100
Enclosures and Raceways, Existing Installations . . 250.86 Ex. 1
   Branch-Circuit Extensions . . . . 250.130(C)
Equipment . . . . Art. 250 Part VI
   Cord- and Plug-Connected 250.114
   Fastened in Place (Fixed) 250.110
   Fixed (Specific) 250.112
   Nonelectric 250.116
Faceplates, See FACEPLATES, Grounding Required Ferm's Finder
Fixed Equipment . . . . Art. 250 Part VI
   Effectively Grounded 250.136
   Methods of Grounding . . . . Art. 250 Part VII
      Fastened in Place 250.134
      Swimming Pool 680.6
      Wet-Niche, Dry-Niche, and No-Niche . . . . 680.23(F)(2)
Flexible Metal Conduit
   See FLEXIBLE METAL CONDUIT . . . . . . Ferm's Finder
Freezers 250.114
Generators
   Fastened in Place 250.110
   Portable and Vehicle-Mounted 250.34
   Separately Derived Systems . . . . 250.30(A)(5) and (6)
Hazardous (Classified) Locations 250.100
   See HAZARDOUS (CLASSIFIED) LOCATIONS Ferm's Finder
Health Care Facilities
   See HEALTH CARE FACILITIES . . . . Ferm's Finder
High Impedance Grounded Neutral 250.36
   Separately Derived Systems . . . . 250.30(A)
High Voltage System . . . . Art. 250 Part X
   See OVER 1000 VOLTS, NOMINAL . . . . . . Ferm's Finder
Information Technology Equipment

Equipment Grounding and Bonding. 645.15

System Grounding 645.14

Insulated Ground Wire

Identification 250.119

Insulation Integrity. 110.7

Marinas and Boatyards, Branch Circuits and Feeders 555.37(B)

Panelboards, In Patient Care Spaces 517.14

Patient Bed Location Receptacles [Critical Care (Category 1)]
. . . . 517.19(B)

Patient Bed Location Receptacles [General Care (Category 2)]. . . . . 517.18(B)

Patient Care Spaces. . . . . 517.13(B)

Pool Luminaires (Fixtures) and Related Equipment . . . . . .
. . . . 680.23(F)(2)

Pool Motors . . . . 680.21(A)(1)

Wet-Niche Luminaires (Fixtures) with NM Conduit. . . . .
. . . . 680.23(B)(2)(b)

Insulated (Isolated) Grounds

Counted in Conductor Fill for Boxes. . . . . . . 314.16(B)(5)

For Receptacles (IG) . . . . 250.146(D)

Identification of Receptacles and Cover Requirements 406.3(D)

Receptacles for Audio Signal Processing, etc. . . . . . 640.7(C)

Receptacles in Patient Care Vicinity (not allowed) . . . 517.16(A)

Receptacles outside Patient Care Vicinity. . . . . . 517.16(B)

Where Used . . . . 250.96(B) & Info. Note

Liquidtight Flexible Metal Conduit

See LIQUIDTIGHT FLEXIBLE METAL CONDUIT . . .
. . . . Ferm's Finder

Luminaires (Lighting Fixtures) 410.40

Metal Boxes

Continuity and Attachment to 250.148

Electrical Continuity. 300.10

For Service Conductors 250.80

Other Than for Service Conductors 250.86

Outlet, Device, Junction Boxes, etc. 314.4

Over 1000 Volts . . . . .Art. 250 Part X

Metal Elbows, Underground in Nonmetallic Conduit Run

Other Than Service Connected to EGC . . . . . 250.86 Ex. 3

Service Raceways Connected to Grounded Service Conductor. . . . . 250.80 Ex.

Metal Enclosures . . . . Art. 250 Part IV

1000 Volts and Higher 250.190

See CONTINUOUS DUTY. . . . . .Ferm's Finder

Metal Frames of Buildings

As Grounding Electrode . . . . .250.52(A)(2)

For Separately Derived System Grounding Electrode . . . .
. . . . 250.30(A)(6)(2)

Required Bonding Where Not Used As Electrode .250.104(C)

Metal Gas Piping, Aboveground, Upstream of Shutoff 250.104(B)

Underground Not Permitted As Grounding Electrode . . . . .
. . . . 250.52(B)(1)

Metal Piping

Gas Piping. .250.104(B)

Other Piping. . . . . .250.104(B)

Water Piping

For Separately Derived Systems . . . . .250.30(A)(6)

Required Bonding Where Not Used As Electrode 250.104(D)

Use As Grounding Electrode . . . . .250.52(A)(1)

Note: Interior metal water piping located more than 1.52 m (5 ft) from the point of entrance to the building cannot be used as a part of the grounding electrode system or as a conductor to interconnect electrodes that are part of the grounding electrode system . . . . .250.68(C)(1)

Metal Well Casings

As Part of Grounding Electrode System. . . . . . .250.52(A)(8)

Required Grounding With Submersible Pump . . . 250.112(M)

Metallic Entrance Conduits Containing

Communication Circuits. 800.49

Community Antenna Television and Radio Distribution Systems
. . . . 800.49

Network-Powered Broadband Communications Systems. . . .
. . . . 800.49

Optical Fiber Cables 770.49

Premises-Powered Broadband Communications Systems. . . .
. . . . 800.49

Meters

Continuity around Water Meters. . . . . .250.53(D)(1)

Effective Grounding Path . . . . 250.68(B)

Instruments, Relays and . . . . Art. 250 Part IX

In Class I Locations 501.105

On Load-Side of Service . . . . .250.142(B)

Methods

Common Grounding Electrode 250.58

Conductor Connections, Grounded System . . . 250.130(A)

Conductor Connections, Ungrounded System . .250.130(B)

Cord- and Plug-Connected Equipment 250.138

Effective Ground-Fault Current Path, Definition. . . Art. 100

For Grounding Electrode Conductor . . . . . 250.68(B)

Equipment Considered Effectively Grounded 250.136

Fixed Equipment. 250.134

Note: Fixed equipment grounding conductors must be run in

the raceway with the circuit conductors or in the case of direct burial conductors right with them in the trench.

 Frames of Ranges and Clothes Dryers 250.140

 Main Bonding Jumper 250.28

 Path at Services. 250.24

 Raceways, Short Sections of 250.132

 System Bonding Jumper 250.28

 Underground Service Cable 250.84

 Use of Grounded Circuit Conductor

  Load-Side Equipment . . . . 250.142(B)

  Supply-Side Equipment . . . . 250.142(A)

 Note: Connections that depend on solder not permitted.

Mobile Homes 550.16

Motors, Grounding All Voltages . . . . Art. 430 Part XIII

 At Swimming Pools . . . . 680.6(3)

 Connection at Terminal Housings . . . . 430.12(E)

 Equipment Grounding. . . . . Art. 250 Part VI

Multiple Circuit Connections 250.144

 Equipment Bonding Jumper In Parallel . . . . . 250.102(C)

 Single Equipment Grounding Conductor . . . . 250.122(A)

Neutral (Grounded Conductor) to Main Service Equipment . . .
. . . . 250.24(C)

 Connection to Equipment Grounding Conductor . . 250.130

Objectionable Current 250.6

Optical Fiber Cables 770.100

 Length of Conductor. . . . . 770.100(A)(4)

 Listing Requirements. 770.180

Other Systems

 See GROUNDING, Electrodes, Bonding Together Ferm's Finder

Over 250 Volts to Ground 250.97

Over 1000 Volts . . . . Art. 250 Part X

Panelboards

 General. 408.40

 In Critical Care (Category 1) Spaces . . . . 517.19(E)

 In Patient Care Spaces 517.14

 Method of, for Swimming Pools . . . . 680.25(A)

 Required Grounding for Swimming Pools . . . . . 680.6(7)

 Used as Service Equipment . . . . 408.3(C)

Ranges 250.140

 Mobile Homes . . . . 550.16(B)

Receptacles 406.3

 Circuit Extensions . . . . 250.130(C)

 Connection to Boxes 250.146

 General, Identification, Use, Grounding Poles . . . 406.10(B)

 Isolated Ground . . . . .406.3(D)

  For Audio Signal Processing Equipment. . . . . . 640.7(C)

 Replacements. . . . . .406.4(D)

Recreational Vehicles

 Bonding of Non-Current-Carrying Parts. 551.56

 General Requirements 551.54

 Insulated Neutral. 551.54(C)

 Interior Equipment. 551.55

 Receptacles to Be Grounding Type 551.52

Recreational Vehicle Parks. . . . . 551.76(A)

 Supply Configurations (Receptacles) 551.81

 Supply Equipment 551.76

Refrigerating Equipment (outdoor rooftops) 440.9

Sealtight Flexible Metal Conduit

See LIQUIDTIGHT FLEXIBLE METAL CONDUIT Ferm's Finder

Separate Buildings 250.32

 Agricultural Buildings . . . . 547.9(B)(3)

 Class I Locations . . . . 501.30(A) Ex.

 Class II Locations . . . . 502.30(A) Ex.

 Class III Locations . . . . 503.30(A) Ex.

 For Swimming Pool Equipment 680.6

 Panelboards as Service Equipment . . . . 408.3(C)

 Panelboards 408.40

Separately Derived Systems 250.30

 Means of. 250.30

 With 60/120 Volts to Ground . . . . 647.6(A)

Service Equipment

 Alternating Current Systems, Service-Supplied 250.24

 Bonding of Services 250.92

  Methods of . . . . 250.92(B)

 Break the Grounded Conductor in the Main Disconnect. . . 230.75

 Common Grounding Electrode 250.58

 Concentric and Eccentric Knockouts. . . . . 250.92(B)

 Continuity of Raceways and Enclosures 300.10

 Enclosures 250.80

  All Service Enclosures . . . . 250.92(A)

  Continuity of 300.10

  Methods. . . . . 250.92(B)

 Exposed Structural Steel, Bonding of. . . . . . . 250.104(C)

 Connection to Equipment Grounding Conductor . . 250.130

 Main Bonding Jumper and System Bonding Jumper . . 250.28

 Main Disconnect . . . . 250.24(A)(1)

 Grounding Electrode Conductor . . . . . Art. 250 Part III

 See GROUNDING, Electrode Conductor . . . Ferm's Finder

Grounding Electrodes
    See GROUNDING, Electrodes . . . . Ferm's Finder
Solar Photovoltaic Systems 690.41
Symbol . . . . 250.126 Info. Note Figure
Unnecessary Bends and Loops Avoided   .250.4(A)(1) Info. Note
Wind Electric Systems 694.40

GROUNDING CONNECTIONS
Location of System Grounding Connections
    Arrange to Prevent Objectionable Current . . . . . . 250.6(A)
    Equipment Bonding Jumper Line-Side . . . . 250.102(C)
    Equipment Bonding Jumper Load-Side . . . . 250.102(D)
        Point of Connection for DC Systems . . . . 250.164(A)
        Service Supplied AC Systems, Grounded . . . . . 250.24(A)
        Service Supplied AC Systems, Ungrounded . . . . 250.24(E)
    Main and System Bonding Jumper(s) 250.28
        Grounded Conductor (Neutral) Must Be Bonded to Service Equipment . . . . 250.24(C)
        Installation . . . . .250.24(A)(4)
        Material . . . . 250.28(A)
    Metallic Piping and Exposed Structural Steel 250.104(B) & (C)
    Minimum Size Load Side, Equipment Bonding Jumper . . . . Table 250.122
        Size of, on Supply Side of Service . . . . . . .250.102(C)
    Minimum Size, Supply Side Equipment Bonding Table 250.122
        Jumper (12.5% rule) . . . . Table 250.102(C)(1)
    Multiple Raceways, Load-Side, Equip. Bonding Jumper . . . . 250.102(D)
    Multiple Raceways, Supply-Side Bonding Jumper .250.102(C)
    Size . . . . 250.28(D)
    Supplemental Ground Required . . . . .250.53(D)(2)
    Use of Grounded Circuit Conductor . . . . .250.142(A)
    Use of Grounded Circuit Conductor on
        Electrode-Type Boilers Over 1000 Volts . . . . . 490.72(E)
        Grounding for Electrode-Type Boilers 490.74
        Load Side of Mains . . . . .250.142(B)
    Water Line, Bond Around Unions in . . . . . .250.53(D)(1)
        For Effective Grounding Path . . . . 250.68(B)
    Water Meters & Filters, Bond Around . . . . . .250.53(D)(1)
        For Effective Grounding Path . . . . 250.68(B)
    Water Pipe Used to Ground a System Shall Be Supplemented . . . . .250.53(D)(2)
    Signs 600.7
    Solar Photovoltaic Systems . . . . .Art. 690 Part V
    Specific Equipment 250.112

Supplementary Grounding, Auxiliary Grounding Electrode 250.54
    Permitted to Connect to Equipment Grounding Conductor . . 250.54
Swimming Pools
    Equipment to Be Grounded 680.6
    Methods of Grounding 680.6
    Wet-Niche Luminaires (Fixtures) . . . . 680.23(B)
    See SWIMMING POOLS, Bonding & Grounding Ferm's Finder
Terminal Bar 408.40
Transformers 450.10
    Grounding Autotransformers 450.5

GROUNDING ELECTRODE CONDUCTORS
Burial Depth .250.64(B)(4)
Definition . . . . Art. 100
Installation 250.64
Location, Common Location . . . . 250.64(D)(3)
Securing and Protection from Physical Damage . . . . 250.64(B)
See GROUNDING, Electrode Conductors . . . . . Ferm's Finder
Taps . . . . .250.64(D)(1)

GROUNDING ELECTRODES
Definition . . . . Art. 100
Direct-Current Sizing 250.166
Not Permitted As . . . . 250.52(B)
Types
    Concrete-Encased Electrode . . . . 250.52(A)(3)
    Ground Ring . . . . 250.52(A)(4)
    Local Metal Underground Systems or Structures 250.52(A)(8)
    Metal In-Ground Support Structure . . . . 250.52(A)(2)
    Metal Underground Water Pipe . . . . 250.52(A)(1)
    Other Listed Electrodes . . . . 250.52(A)(6)
    Plate Electrodes . . . . 250.52(A)(7)
    Rods and Pipes Electrodes . . . . 250.52(A)(5)
See GROUNDING, Electrodes . . . . Ferm's Finder

GROUPING OF CONDUCTORS
MC Cable – Single Cables Grouped . . . . 330.80(B)
Multiwire Branch Circuits in Panelboards . . . . 210.4(B)

GROUPING OF DISCONNECTS
Outside Feeders and Branch Circuits 225.34
Services 230.72

GUARDING, GUARDS
Circuit-Breaker Handles . . . . 240.41(B)

Circuit Breakers and Fuses During Operation .... 240.41(A)
Construction Sites, Over 600 Volts 590.7
Elevators, Dumbwaiters, Escalators, etc., Machine Room   620.71
Generators 445.15
Live Parts, General 110.27
    Over 1000 Volts 110.34
Motors and Motor Controllers 430.232
    For Attendants 430.233
    Portable Motors Over 150 Volts to Ground 430.243
    Solar Photovoltaic Systems . . . . 690.33(B)
Storage Batteries 480.9
Transformers 450.8
Wind Electric Systems . . . . . 694.10(C) Informational Note
X-Ray Installations 517.78

GUEST ROOMS and GUEST SUITES
Arc-Fault Circuit-Interrupter Protection . . . . 210.12(C)
Bathrooms (overcurrent devices not located) . . . . . 240.24(E)
Cooking (permanent provisions) 210.17
Fixed Multioutlet Assemblies . . . . 220.14(H)
Lighting Outlets . . . . 210.70(B)
Overcurrent Device (location) . . . . 240.24(B)(1)(2)
Receptacle Placement 210.60
Tamper-Resistant Receptacles . . . . 406.12(2)
Voltage Limitations (Branch-Circuit) . . . . 210.6(A)

GUTTERS, AUXILIARY . . . . Art. 366
Ampacity of Conductors
    In Nonmetallic . . . . 366.23(B)
    In Sheet Metal . . . . 366.23(A)
Extension Beyond Equipment . . . . 366.12(2)
    No Restriction on Length for Elevators. 620.35
    Permitted Length for Elevators . . . . 366.12 Ex.
Grounding, Connected to an ECG 366.60
Indoor and Outdoor Use
    Nonmetallic . . . . 366.10(B)(1) and (2)
    Sheet Metal . . . . .366.10(A)(1) and (2)
Number of Conductors in
    For Elevators 620.33
    Nonmetallic . . . . 366.22(B)
    Sheet Metal. . . . . 366.22(A)
Supports
    Nonmetallic . . . . 366.30(B)
    Sheet Metal. . . . . 366.30(A)
Wire Bending Space . . . . 366.58(A)

H

HALLWAY
AFCI Protection . . . . 210.12(A)
Lighting Outlet . . . . .210.70(A)(2)
Mobile Home . . . . 550.13(D)
Outlets (Dwelling Units) . . . . 210.52(H)
Switches (Grounded Conductor) . . . . 404.2(C)

HAND LAMPS/PORTABLE LIGHTING EQUIPMENT
Commercial Garages (Unswitched & Insulated) . . 511.4(B)(2)
General, Portable Luminaires (Formerly Hand Lamps) . . 410.82
Grounding Not Required with Isolating Transformer
 . . . . 250.114(4)(g) Ex.
Grounding Required . . . . 250.114(4)(e)

HANDHOLE ENCLOSURES . . . .
Size, Over 1000 Volts 314.71
Systems 1000 Volts and Less 314.30
Systems over 1000 Volts . . . . 314.70(C)

HANDLE TIES
Feeder Disconnecting Means . . . . 225.33(B)
Multiwire Branch Circuits . . . . .240.15(B)(1)
Service Disconnecting Means . . . . 230.71(B)

HANGARS, AIRCRAFT . . . . Art. 513
See AIRCRAFT HANGARS . . . . .Ferm's Finder

HARMONICS
Cablebus 368.258
Design Consideration for Neutral Currents (Branch Circuit) . . .
. . . . 210.4(A) Info. Note 1
Electrical Duct Banks . . . . .Informative Annex B, Figure Notes
Existing Neutral Overheating . . . . 310.10(G)Ex.2 Info. Note
Feeder or Service Neutral, Consideration of 220.61(C)(2) Info. Note
Neutral Considered Current-Carrying, Due to . . . 310.15(E)(3)
Non-Linear Loads in Excess of 50 percent 400.5
Non-Linear Loads Major Portion   Table 520.44(C)(3)(a) Note
Temperature Limitations internally due to 310.14(A)(3) Info. Note 1
See NONLINEAR LOAD . . . . .Ferm's Finder

HAZARDOUS (CLASSIFIED) LOCATIONS . . . . Art. 500
Approval for Class and Properties (Equipment) . . . . . 500.8(A)
Bonding and Grounding 250.100
    Class I Locations 501.30

Class II Locations 502.30

Class III Locations 503.30

Cable Seals

    Class I, Division 1 . . . . 501.15(D)

    Class I, Division 2 . . . . 501.15(E)

    Zone 0 . . . . .505.16(A)(2)

    Zone 1 . . . . .505.16(B)(7)

    Zone 2 . . . . .505.16(C)(2)

Classification of Group (Atmosphere)

    Class I, Groups A, B, C, and D . . . . 500.6(A)

    Class I, Groups IIA, IIB, and IIC (in Zone Locations) . .505.6

    Class II, Groups E, F, and G . . . . 500.6(B)

Classification of Locations, Rooms, Sections and Areas . . .500.5

Note: Each room, section, or area of an occupancy hall should be considered individually in determining its classification.

    Class I Locations . . . . 500.5(B)

        Division 1 . . . . 500.5(B)(1)

        Division 2 . . . . 500.5(B)(2)

        Zone 0 . . . . 505.5(B)(1)

        Zone 1 . . . . 505.5(B)(2)

        Zone 2 . . . . 505.5(B)(3)

    See . . Recommend Practice for the Classification of Flammable Liquids, Gases, or Vapors and of Hazardous (Classified) Locations for Electrical Installations in Chemical Process Areas

    Class II Locations . . . . 500.5(C)

        Division 1 . . . . 500.5(C)(1)

        Division 2 . . . . 500.8(C)(2)

    See . . . . Recommended Practice for the Classification of Combustible Dusts and of Hazardous (Classified) Locations for Electrical Installations in Chemical Process Areas

    Class III Locations . . . . 500.5(D)

        Division 1 . . . . 500.5(D)(1)

        Division 2 . . . . 500.5(D)(2)

Class I Locations (Zones) . . . . Art. 505

Class I Locations . . . . Art. 501

    Bonding in . . . . 501.30(A)

        At Service 250.92

        General 250.100

    Chemicals by Groups . . . . 500.6(A)

    For Zone 0, 1, & 2 . . . . .505.6

    See Recommend Practice for the Classification of Recommended Practice for the Classification of Flammable Liquids, Gases, or Vapors and of Hazardous (Classified) Locations for Electrical Installations in Chemical Process Areas

    Conductor Insulation 501.20

        See (ZLGR) . . . . UL Product iQ

Control Transformers & Resistors 501.120

Definitions (as applied to Hazardous Locations)Art. 100, Part III

Documentation Required 500.4

Dual Classifications of Areas . . . . 505.7(B)

Dual Classifications of Areas . . . . 506.7(B)

Equipment Temperature . . . . 500.8(C)(4)

Flexible Connections

    Division 1 . . . . 501.10(A)(2)

    Division 2 . . . . 501.10(B)(2)

Flexible Cords

    For Flexibility Division 2 Locations . . . . . . 501.10(B)(2)

    For Process Control Instruments (Division 2) . . 501.105(B)(6)

    Use of 501.140

Grounding 501.30

    At Service 250.92

    General 250.100

    Zone 0, 1, & 2 Locations 505.25

    Zone 20, 21, & 22 Locations 506.25

Health Care Facilities, In . . . . Art. 517 Part IV

Live Parts Not Exposed 501.25

Luminaires (Lighting Fixtures) 501.130

Meters, Instruments, and Relays 501.105

Motors and Generators 501.125

Multiwire Branch Circuits, Prohibited . . . . 210.4(B)

Optical Fiber Cable

    Division 1 . . . . 501.10(A)(1)(e)

    Division 2 . . . . 501.10(B)(1)(7)

Protection Techniques 500.7

Receptacles and Attachment Plugs 501.145

Reciprocating Engine-Driven Generators and Compressors . . 501.125(B) IN

Reclassification of Areas, Zone 0, 1, and 2 Locations 505.7(C)

Reclassification of Areas, Zone 20, 21, and 22 Locations506.7(C)

Sealing & Drainage 501.15

Signaling, Alarm, Remote Control & Communication. 501.150

Special Precautions 505.7

Specific Locations . . . . Art. 510

Spray Application, Dipping, Coating, and Printing Processes Using Flammable Combustible Materials . . . . . Article 516

Supervision of Work, Qualified Person,

    Class I, Zone 0, 1, and 2 Locations . . . . 505.7(A)

        Special Precautions 506.7

Surge Arresters, Over 1 kV . . . . Art. 242

See Standard for the Installation of Lightning Protection Systems, NFPA 780 . . . . NFPA 780

Types and Installation	501.20
Switches, Circuit Breakers, Motor Controllers & Fuses	501.115
Transformers & Capacitors	501.100
Utilization Equipment	501.135
Wiring Methods	501.10
Aircraft Hangars	513.4
Bulk Storage Plants	515.4
Underground Wiring	515.8
Class I, Division 1	501.10(A)
Class I, Division 2	501.10(B)
Class I, Zone 0	505.15(A)
Class I, Zone 1	505.15(B)
Class I, Zone 2	505.15(C)
Commercial Garages	511.4
Definitions	
Major Repair Garage	511.3(D)
Major and Minor Repair Garage	511.3(C)
Gasoline (Motor Fuel) Dispensing and Service Stations	514.4
Underground Wiring	514.8
Health Care, within Hazardous Anesthetizing Locations	517.61
Major Repair Garage Classification	511.3(D)
Minor Repair Garage Classification	511.3(C)
Spray Application, Dipping, and Coating Processes	516.6
Class II Locations	Art. 502
Bonding in	502.30(A)
At Service	250.92(B)
General	250.100
Control Transformers & Resistors	502.120
Documentation Required	500.4
Dusts by Group	500.6(B)
See Recommended Practice for the Classification of Combustible Dusts and of Hazardous (Classified) Locations for Electrical Installations in Chemical Process Areas	
Flexible Connections	
Division 1	502.10(A)(2)
Flexible Cords	
Division 1 Locations	502.10(A)(2)(5)
Division 2 Locations	502.10(B)(2)
Installation	502.140(B)
Luminaries (Lighting Fixtures)	502.130
Permitted Uses	502.140(A)
Grounding	502.30
At Service	250.92
General	250.100
Live Parts, Not Exposed	502.25
Luminaires (Lighting Fixtures)	502.130
Motors and Generators	502.125
Optical Fiber Cable	
Division 1	502.10(A)(1)(4)
Division 2	502.10(B)(1)(8)
Receptacles and Attachment Plugs	502.145
Sealing	502.15
Signaling, Alarm, Remote Control & Communication and Meters, Instruments, and Relays	502.150
Special Precautions	505.7
Special Precautions	506.7
Surge Arresters	Art. 242
Types and Installation	502.35
See Standard for the Installation of Lightning Protection Systems, NFPA 780	NFPA 780
Switches, Circuit Breakers, Motor Controllers & Fuses	502.115
Transformers & Capacitors	502.100
Utilization Equipment	502.135
Ventilation Piping	502.128
Wiring Methods	
Class II, Division 1	502.10(A)
Class II, Division 2	502.10(B)
Class III Locations	Art. 503
Bonding in	503.30(A)
At Service	250.92
General	250.100
Control Transformers & Resistors	503.120
Cranes & Hoists	503.155
Documentation Required	500.4
Special Requirements	610.3
See CRANES AND HOISTS	Ferm's Finder
Fibers or Flyings, Easily Ignitable	500.5(D)
Flexible Connections	503.10(A)(3)
Flexible Cords	503.140
Grounding	503.30
At Service	250.92
General	250.100
Live Parts Not Exposed	503.25
Luminaires (Lighting Fixtures)	503.130
Motors and Generators	503.125
Operating Temperature (Dust)	503.5
Receptacles	503.145
Signaling, Alarm, Remote Control & Communication	503.150

Storage Battery Charging Equipment 503.160
Switches, Circuit Breakers, Motor Controllers & Fuses 503.115
Transformers & Capacitors 503.100
Utilization Equipment 503.135
Ventilating Piping 503.128
Wiring Methods
    Class III, Division 1 . . . . 503.10(A)
    Class III, Division 2 . . . . 503.10(B)
Classified (Specific Locations) . . . . Art. 510
    Aircraft Hangars . . . . Art. 513
        See AIRCRAFT HANGARS . . . . Ferm's Finder
    Bulk Storage Plants. . . . . Art. 515
        See BULK STORAGE PLANTS . . . . . . Ferm's Finder
    Commercial Garages . . . . Art. 511
        See COMMERCIAL GARAGES . . . . . . Ferm's Finder
    Finishing Processes. . . . . Art. 516
        See FINISHING PROCESSES . . . . Ferm's Finder
    Health Care Facilities. . . . . Art. 517
        See HEALTH CARE FACILITIES . . . . . Ferm's Finder
    Motor Fuel Dispensing Facilities . . . . Art. 514
        See GASOLINE (MOTOR FUEL) DISPENSING . Ferm's Finder
    Spray Application, Dipping & Coating . . . . . . Art. 516
        See SPRAY APPLICATION, DIPPING & COATING, ETC. Ferm's Finder
Luminaires (Fixtures) & Fittings
    Class I, Division 1 . . . . 501.130(A)
    Class I, Division 2 . . . . 501.130(B)
    Class I, Zones 0, 1, and 2 . . . . 505.20
    Class I, Zones 20, 21, 22 . . . . 506.20
    Class II, Division 1 . . . . 502.130(A)
    Class II, Division 2 . . . . 502.130(B)
    Class III, Divisions 1 and 2. . . . . 503.130
    Fixture (Luminaires) Fittings *See* (IGIV) . . . . UL Product iQ
    Fixtures (Luminaires) Paint Spray Booth
        See (IFYJ) . . . . UL Product iQ
    Fixtures, (Luminaires) Recessed Type
        See (IGBW) . . . . UL Product iQ
    Fixtures, (Luminaires), see (IFUX). . . . . . . UL Product iQ
    Spray Applications, Booths, etc. . . . . 516 Part III
    Note: Equipment subject to accumulation of residue or dusts is not permitted in spray areas unless specifically listed for the application. . . . . 516.6(C)4
Ignition Temperature Information . . . . 500.8(C)
NFPA 325, *Guide to Fire Protection Hazard Properties of Flammable Liquids, Gases, and Volatile Solids contained in the NFPA "Fire Protection Guide to Hazardous Materials"–2001 edition"*
See Recommended Practice for the Classification of Flammable Liquids, Gases, or Vapors and of Hazardous (Classified) Locations for Electrical Installations in Chemical Process Areas . . . .
NFPA 497. *See Recommended Practice for the Classification of Combustible Dusts and of Hazardous (Classified) Locations for Electrical Installations in Chemical Process Areas* . . . . . . NFPA 499
Intrinsically Safe Systems . . . . Art. 504
    As Protection Technique . . . . 500.7(E)
    See INTRINSICALLY SAFE SYSTEMS . . . . Ferm's Finder
    Identification of Occupancies 500.9
    Protection Techniques, General 500.7
    Class I, Zone 0, 1, and 2 Locations 505.8
    Class I, Zone 20, 21, and 22 Locations 506.8
Reference Standards
    Class I, II, and III, Divisions 1 and 2 Locations . . . 500.4(B)
    Class I, Zone 0, 1, and 2 Locations 505.4
    Class I, Zone 20, 21, and 22 Locations 506.4
    Suitability — Equipment. . . . . 500.8(A)
Wiring under Hazardous Areas
    Aircraft Hangars 513.8
    Bulk Storage Tanks 515.8
    Commercial Garages 511.4
    Motor Fuel Dispensing Facilities 514.8
See ZONE 0, CLASS 1 HAZARDOUS (CLASSIFIED) LOCATIONS . . . . Ferm's Finder
See ZONE 20, 21, 22 CLASS 1 HAZARDOUS (CLASSIFIED) LOCATIONS . . . . Ferm's Finder

HDPE TYPE CONDUIT. . . . . Art. 353
Class I, Division 1 . . . . 501.10(A)(1)(a) Ex.
Commercial Garages . . . . 511.8 Ex.
Conduit and Tubing Fill Tables . . . . Informative Annex C
Joints, Made by Heat Fusion, Electrofusion or Mechanical Fittings . . . . 353.48 Info. Note
Motor Fuel Dispensing . . . . 514.8 Ex. No. 2
Services
Recreational Vehicles . . . . 551.80(B)
Service-Entrance . . . . 230.43(17)
Underground Wiring Method . . . . 230.30(B)(4)

HEADERS
Cellular Concrete Floor Raceways . . . . 372.18(A)
    Connections to Cabinets and Other Enclosures . . . 372.18(B)
    Inserts . . . . 372.18(D)

Maximum Number of Conductors. 372.22
Splices and Taps 372.56
Cellular Metal Floor Raceways 374.1
Maximum Number of Conductors. 374.22
Splices and Taps 374.56
Definition of (Headers)
For Cellular Concrete Floor Raceways 372.2
For Cellular Metal Floor Raceways 374.2

HEADROOM & WORKING CLEARANCE AT EQUIPMENT
1000 Volts, Nominal, or Less 110.26
Clear Spaces . . . . 110.26(B)
Depth of Working Space . . . . 110.26(A)(1)
Entrance to Working Space. . . . . 110.26(C)
Height of Working Space. . . . . 110.26(A)(3)
Width of Working Space . . . . 110.26(A)(2)
Over 1000 Volts 110.32
Depth of Working Space . . . . 110.34(A)
Entrance and Access to Work Space 110.33
Work Space about Equipment 110.34

HEALTH CARE FACILITIES (HCF) . . . . **Art. 517**
Ambulatory Health Care Facility
Communications . . . . Art. 517 Part VI
Definition of. 517.2
Essential Electrical System, HCF's 517.25
Wiring and Protection . . . . .Art. 517 Part II
Essential Electrical System, Other HCF's. 517.45
Wiring and Protection . . . . .Art. 517 Part II
See Standard for Health Care Facilities, NFPA 99 . . .NFPA 99
Applicability . . . . 517.10(A)
Application to Other Article- Essential Electrical System . 517.26
Areas Not Covered . . . . 517.10(B)
Battery Systems. . . . . .517.30(B)(3)
Definition of 517.2
Essential Electrical System- Power Sources 517.30
Alternate Power Sources
Battery Systems. . . . . .517.30(B)(3)
Fuel Cell Systems . . . . .517.30(B)(2)
Generating Units . . . . .517.30(B)(1)
Capacity of Systems . . . . 517.31(D)
Coordination. . . . . 517.31(G)
Feeders from Alternate Power Sources . . . . . . 517.31(F)
Optional Loads . . . . .517.31(B)(1)
Separate Branches . . . . 517.31(A)

Sources of Power 517.30
Transfer Switches . . . . 517.31(B)
Fire Alarm Systems . . . . Art. 517 Part VI
Fuel Cells . . . . 517.30(B)(2)
General . . . . Art. 760
See FIRE ALARM SYSTEMS . . . . .Ferm's Finder
See Standard for Health Care Facilities, NFPA 99 . . .NFPA 99
See Life Safety Code, NFPA 101. . . . . NFPA 101
See National Fire Alarm Code, NFPA 72 . . . . . . .NFPA 72
Generating Units . . . . .517.30(B)(1)
Governing Body, Definition 517.2
Ground-Fault Protection of Equipment. . . . .
517.17(D)
Grounding
Certain Spaces, Nonapplicability. . . . . .517.10(B)(1)
Definitions of Patient Equipment and Reference Point. .517.2
Feeders, Critical Care Spaces . . . . 517.19(E)
Hospitals. . . . . .Art. 517 Part II
Anesthetizing Locations 517.62
Critical Care (Category 1) Spaces . . . . . .Art. 517 Part II
Specific Requirements 517.19
General Care (Category 2) Spaces . . . . . .Art. 517 Part II
Specific Requirements 517.18
Patient Care Spaces . . . . .Art. 517 Part II
Critical Care (Category 1) 517.19
General Care (Category 2) 517.18
Panelboards 517.14
Receptacles and Fixed Equipment 517.13
Therapeutic Pools & Tubs . . . . Art. 680 Part VI
X-Ray Area 517.78
Nursing Homes and Limited Care Facilities . .Art. 517 Part II
Nonapplicability, Certain Spaces. . . . . .517.10(B)(2)
Panelboards . . . . 517.19(E)
Ground-Fault Circuit-Interrupter Protection
Not Required Certain Critical Care (Category 1) Spaces . . 517.21
Therapeutic Equipment . . . . 680.62(A)
Therapeutic Pools & Tubs . . . . Art. 680 Part VI
Receptacles within 6 ft. . . . . 680.62(E)
Therapeutic Tubs (Hydrotherapeutic Tanks) . . . 680.62(A)
Wet Procedure Locations, Receptacles and Fixed Equipment. .
517.20
Wet Procedure Locations Defined (under Patient Care Space)
517.2
Ground-Fault Protection of Equipment
Feeders 215.10

Additional Level Required (Applicability) . . . . 517.17(A)

Where Protection Provided . . . . 517.17(B)

Not Required for Emergency Systems 700.31

Selectivity . . . . 517.17(C)

Services 230.95

Signal Devices, Emergency Systems . . . . 700.6(D)

Testing . . . . 517.17(D)

Hazardous (Classified) Locations . . . . Art. 517 Part IV

See HAZARDOUS (CLASSIFIED) LOCATIONS . . Ferm's Finder

See Standard for Health Care Facilities, NFPA 99 . . . NFPA 99

Hospitals

Critical Branch, Selected Equipment 517.33

Definition of 517.2

Essential Electrical System 517.30

Branches of (Life Safety Branch and Critical Branch) 517.31

Capacity of Systems . . . . 517.31(D)

Coordination . . . . 517.31(G)

Equipment System Branch

Equipment System Connection to Alternate Power . 517.35

Separation of Circuits . . . . 517.31(C)(1)

Feeders from Alternate Power Sources . . . . . . 517.31(F)

Mechanical Protection . . . . .517.31(C)(3)

Optional Loads . . . . 517.31(B)(1)

Sources of Power 517.30

Transfer Switches . . . . 517.31(B)

Wiring Requirements . . . . 517.31(C)

Legally Required Standby . . . . Art. 701

Life Safety Branch, Selected Equipment 517.33

See Standard for Health Care Facilities, NFPA 99 . . . NFPA 99

Inhalation Anesthetizing Location . . . . Art. 517 Part IV

See Standard for Health Care Facilities, NFPA 99 . . . NFPA 99

Isolated Grounding Receptacles, Not Used In Patient Care Vicinity 517.16(B)

Isolating Transformers

Critical Care (Category 1) Spaces (Optional) . . .517.19(F) & (G)

Isolated Power Systems 517.160

Low-Voltage Circuits . . . . 517.64(C)

Wet Procedure Locations 517.20

See Standard for Health Care Facilities, NFPA 99 . . . NFPA 99

Medical Office (Dental Office), Definition 517.2

Not Covered- Areas . . . . 517.10(B)

Nursing Homes and Limited Care Facilities

Definitions, Both .517.2

Essential Electrical Systems 517.25

Capacity of System . . . . 517.42(C)

Connection to Equipment Branch 517.44

Connection to Life Safety Branch 517.43

Emergency Systems, General . . . . Art. 700

Legally Required Standby . . . . Art. 701

Separation from Other Circuits . . . . 517.42(D)

Sources of Power 517.41

See Standard for Health Care Facilities, NFPA 99 . . . NFPA 99

Panelboards (Bonding & Grounding) 517.14

Critical Care (Category 1) Spaces . . . . 517.19(E)

Patient Care Space

Critical Care (Category 1) Space

Grounding of Receptacles and Fixed Equipment . . 517.13

Requirements for 517.19

Definitions 517.2

General Care (Category 2) Space

Grounding of Receptacles and Fixed Equipment . . 517.13

Requirements for 517.18

Grounding of Receptacles and Fixed Equipment . . 517.13

See Standard for Health Care Facilities, NFPA 99 . . . NFPA 99

Ground-Fault Circuit-Interrupter Protection 517.21

Wet Procedure Locations 517.20

Receptacle(s)

Above Hazardous Locations . . . . 517.61(B)

Covers That Limit Pediatric Access . . . . 517.18(C)

Exemption from, Certain Critical Care (Category 1) Spaces 517.21

Explosionproof . . . . 517.61(A)(3) & (5)

Ground-Fault Circuit-Interrupter 517.20

Grounding

Connecting Grounding Terminal to Box 250.146

Critical Care (Category 1) Spaces . . . . 517.19(B)

General Care (Category 2) Spaces . . . . 517.18(B)

General 517.13

Hospital Grade, Required

Above Hazardous Locations . . . . .517.61(B)(5)

Critical Care (Category 1) Spaces . . . . 517.19(B)

General Care (Category 2) Spaces . . . . 517.18(B)

In Other Than Hazardous Locations . . . . . .517.61(C)(2)

Note: Non-hospital grade receptacles must be replaced with hospital-grade receptacles upon any modifications of use, renovation, or as existing receptacles need replacement.

Insulated Equipment Grounds 517.13

Covers . . . . 406.3(D)(2)

Critical Care (Category 1) Patient Bed Location . . 517.19(B)

General Care (Category 2) Patient Bed Location . . 517.18(B)
Isolated Ground . . . . 250.146(D)
    Use and Identification of . . . . 406.3(D)(1)
        Isolated Power Systems 517.160
Isolated Grounding Receptacles, Not Used Inside Patient Care Vicinity . . . . 517.16(A)
Low Voltage . . . . 517.64(F)
Number of (Hospitals)
    Operating Rooms . . . . 517.19(C)
    Patient Bed Locations, Critical Care (Category 1) 517.19(B)
    Patient Bed Locations, General Care (Category 2) 517.18(B)
Receptacles, 50- and 60-ampere, 250 Volts
    Above Hazardous Locations . . . . 517.61(B)(6)
    In Other Than Hazardous Locations. . . . . . 517.61(C)(3)
Sealing Fittings . . . . 517.61(B)(4)
Sources of Power 517.30
Tamper Resistant . . . . 517.18(C)
Therapeutic Pools & Tubs . . . . 680 Part VI
    Bonding . . . . 680.62(B)
    Methods of Bonding . . . . 680.62(C)
    Grounding . . . . 680.62(D)
    Methods of . . . . 680.62(D)
    Ground-Fault Circuit-Protection, for Equipment . 680.62(A)
    For Receptacles . . . . 680.62(E)
    Luminaires (Lighting Fixtures)
        Indoor Areas . . . . 680.22(B)
        Permanently Installed Pools . . . . 680.61 Ex.
        Tub Areas . . . . 680.62(F)
        Underwater 680.23
Wiring Methods . . . . Art. 517 Part II
    Insulated Equipment Grounding Conductor 517.13
        Critical Care (Category 1) Space Bed Location Receptacles . . . . 517.19(B)
        General Care (Category 2) Space Bed Location Receptacles . . . . 517.18(B)
    Mechanical Protection of Essential Electrical System 517.31(C)(2)
X-RAY Equipment . . . . Art. 517 Part V

Note: Radiation safety and performance requirements of several classes of x-ray equipment are regulated under Public Law 90-602 and are enforced by the Department of Health and Human Services 517.70 Info. Note 1

See Standard for Health Care Facilities, NFPA 99 . . . . NFPA 99

HEAT DETECTING CIRCUIT INTERRUPTER. . . 440.65(3)

HEAT PUMPS . . . . Art. 440

Branch-Circuit Conductors . . . . Art. 440 Part IV
Branch-Circuit Short-Circuit and Ground-Fault Protection Art. 440 Part III
Calculations
    See ELECTRIC HEAT (SPACE), Calculations of Load Ferm's Finder
Controllers for . . . . Art. 440 Part V
Definitions 440.2
Disconnecting Means . . . . Art. 440 Part II
Motor-Compressor & Branch-Circuit Overload Protection . . . . Art. 440 Part VI

HEATING APPLIANCES . . . . Art. 422
Central Heating Equipment (Other Than Electric) 422.12
Flexible Cords for 422.16
Infrared Lamp Type 425.14
    Lampholder Types and Ratings 422.48
Signals for 422.42
Subdivision of Load . . . . 422.11(F)

HEATING CABLES . . . . Art. 424 Part V
See ELECTRIC HEAT (SPACE), Cables . . . . . Ferm's Finder

HEATING EQUIPMENT, SPACE (Fixed) . . . . . . Art. 424
Mobile Home Outside, Provisions for . . . . 550.20(B)
See DEICING AND SNOW-MELTING . . . . . Ferm's Finder
See DUCT HEATERS . . . . Ferm's Finder
See ELECTRIC HEAT (SPACE) . . . . Ferm's Finder
See FIXED ELECTRIC HEATING EQUIPMENT FOR PIPELINES AND VESSELS . . . . Ferm's Finder
See HEAT PUMPS . . . . Ferm's Finder
See INDUCTION AND DIELECTRIC HEATING Ferm's Finder
See INFRARED LAMP HEATING . . . . Ferm's Finder

HEATING PANELS AND HEATING PANEL SETS . . . . Art. 424 Part IX
For Pipeline and Vessel Heating . . . . 427.23(B)
See ELECTRIC HEAT (SPACE), Panels and Sets . Ferm's Finder

HEIGHT OF WORKING SPACE . . . . 110.26(A)(3)

HERMETIC REFRIGERANT MOTOR COMPRESSORS . . . . Art. 440
Ampacity and Rating 440.6
Branch-Circuit Conductors . . . . Art. 440 Part IV
Branch-Circuit Short-Circuit and Ground-Fault Protection . . . . Art. 440 Part III
Controllers for Motor-Compressors . . . . Art. 440 Part V

Compressor and Branch-Circuit Overload Protection. . . . . . .
. . . . Art. 440 Part VI
Defined. . . . . Art. 100
Disconnecting Means . . . . Art. 440 Part II
Room Air Conditioners . . . . Art. 440 Part VII
   Protection Devices 440.65
      Arc-fault circuit interrupter (AFCI) . . . . 440.65(2)
      Heat Detecting Circuit Interrupter (HDCI). . . 440.65(3)
      Leakage-Current Detector-Interrupter (LCDI) . 440.65(1)

H.I.D. LAMP TYPE LUMINAIRES (FIXTURES)
See (IEWX). . . . . UL Product iQ
See (IEXT) . . . . UL Product iQ

HIGH IMPEDANCE GROUNDED NEUTRAL 250.36
AC Systems 50 to 1000 Volts Not Required To Be Grounded 250.21
   Equipment Bonding Jumper . . . . 250.36(E)
   Grounding Electrode Conductor Connection Location 250.36(F)
   Grounding Impedance Location . . . . 250.36(A)
   Grounded System Conductor . . . . 250.36(B)
   Marked Legibly and Permanently. . . . . 408.3(F)(3)
   Neutral Point to Grounding Impedance Conductor Routing . . .
   . . . . 250.36(D)
   Separately Derived Systems . . . . 250.30(A) Ex.
   System Grounding Connection . . . . 250.36(C)
   1kV and Over 250.187
      Equipment Grounding Conductors . . . . 250.187(D)
      Grounding Impedance Location . . . . 250.187(A)
      Neutral Identified and Insulated . . . . 250.187(B)
      System Neutral Connection . . . . 250.187(C)

HIGH-LEG (STINGER or WILD LEG)
Caution Signs . . . . 408.3(F)(1)
Feeder Identification 110.15
Marking of in Switchboards and Panelboards . . . . . 408.3(F)
Marking of, General 110.15
   No Delta Breakers in Single-Phase Panels . . . . 408.36(C)
Phase Arrangement, Switchboards and Panelboards. . . 408.3(E)
Plug Fuses Permitted . . . . 240.50(A)(2)
Service Conductor Identification 230.56
Signage Requirements, Caution. . . . . 408.3(F)(1)
Switchboard or Panelboard Permanently Field Marked . 408.3(F)

HIGH VOLTAGE, OVER 1000 VOLTS NOMINAL
Busways . . . . Art. 368 Part IV
Capacitors .Art. 460 Part II
Definitions (Over 1000 Volts, Nominal) . . . . .Art. 100 Part II
Equipment . . . . .Art. 490 Part II
Grounding . . . . .Art. 250 Part X
Medium Voltage Cable: Type MV . . . . Art. 311
Motors . . . . Art. 430 Part XI
Outside Branch Circuits and Feeders . . . . . . Art. 225 Part III
Overcurrent Protection . . . . Art. 240 Part IX
Portable Cables . . . . Art. 400 Part III
Pull and Junction Boxes. . . . . Art. 314 Part IV
Requirements for Electrical Installations . . . . Art. 110 Part III
Resistors and Reactors . . . . Art. 470 Part II
Services . . . . Art. 230 Part VIII
Solar Photovoltaic Systems . . . . Art. 690 Part IX
Solar Photovoltaic Systems (Large Scale) . . . . Art. 691
Transformers, see TRANSFORMERS . . . . Ferm's Finder
Tunnel Installations. . . . . Art. 110 Part IV
Wind Electric Systems . . . . Art. 694 Part VIII
Wiring Methods . . . . .Art. 300 Part II
See OVER 1000 VOLTS, NOMINAL . . . . Ferm's Finder

HOIST, STAGE LIGHTING
Defined. 520.2
Wiring Information. 520.40

HOISTS AND CRANES . . . . Art. 610
See CRANES & HOISTS . . . . Ferm's Finder

HOISTWAY (ELEVATOR) CABLE
See (MSZR) .UL Product iQ

HOISTWAYS
Definition of . . . . Art. 100 Part I
Door Interlock Wiring . . . . 620.11(A)
Wiring in
   Class 2, or 3 Remote Control and Signaling . . . 725.136(H)
   Lightning Protection Conductors . . . . 620.37(B)
   Location of and Protection for Cables 620.43
   Main Feeders . . . . 620.37(C)
   Power-Limited (PLFA) Circuits in Hoistways . . .760.136(F)
   Suspension of Traveling Cables. 620.41
   Uses Permitted . . . . 620.37(A)

Wiring Methods, General. 620.21
   Elevators, Between Risers, Limit Switches, etc. 620.21(A)(1)(a)
   For Class 2 Power-Limited Circuits . . . . . 620.21(A)(1)(a)
   Wiring Methods . . . . 620.21(A)(1)(c)

HOODS & DUCTS
　See DUCTS & HOODS . . . . Ferm's Finder

HORTICULTURAL LIGHTING EQUIPMENT Art. 410, Part XVI

HOSPITALS
　See HEALTH CARE FACILITIES . . . . Ferm's Finder

HOT TUBS & SPAS . . . . **Art. 680 Part IV**
　Definitions 680.2
　Disconnecting Means for Equipment — Simultaneous . . 680.13
　Electrically Powered Pool Lift . . . . Art. 680 Part VIII
　　Bonding 680.83
　　Defined 680.2
　　Listed 680.81
　Emergency Switch Required, User Access to 680.41
　Ground-Fault Circuit-Interrupter Protection Required . . . . . .
　. . . . . 680.42(A)(2)
　Indoor Installations 680.43
　　Bonding . . . . 680.43(D)
　　　Methods of . . . . 680.43(E)
　　Disconnecting Means, Equipment 680.13
　　Disconnecting Means, Motors . . . . Art. 430 Part IX
　　Electric Heaters 680.10
　　Grounding . . . . 680.43(F)
　　　Methods of 680.6
　　Lighting Luminaires (Fixtures) and Ceiling Fans . . 680.43(B)
　　　Ground-Fault Circuit-Protection . . . . 680.43(B)(1)
　　Receptacles . . . . 680.43(A)(2)
　　　Ground-Fault Circuit-Protection, within 3.0 m (10 ft.) . . .
　　　. . . . . 680.43(A)
　　　Location . . . . 680.43(A)(1)
　　　Outlet Box Hood . . . . 406.9(B)(1)
　　　Tamper-Resistant 406.12
　　　That Provide Power to Spas and Hot Tubs . . 680.43(A)(3)
　　　Weather-Resistant . . . . 406.9(A) and (B)
　　Switches [1.5 m (5 ft.) away] . . . . 680.43(C)
　Outdoor Installations 680.42
　　Bonding . . . . 680.42(B)
　　Flexible Connections . . . . 680.42(A)
　　Wiring Per Article 680 Parts I, II, and IV . . . . . . Art. 680
　Storable Hot Tub (defined) 680.2
　See SPAS AND HOT TUBS . . . . Ferm's Finder
　See SWIMMING POOLS . . . . Ferm's Finder
　See (WBYQ) . . . . UL Product iQ

HOTELS & MOTELS
　Arc-Fault Circuit-Interrupter Protection . . . . 210.12(C)
　Assembly Occupancies (finish rating) . . . . 518.4(C)
　Branch-Circuit Limitations . . . . 210.6(A)
　General Lighting Loads . . . . Table 220.12
　　Demand Factors . . . . Table 220.42
　GFCI for Bathrooms, Kitchens . . . . 210.8(A), 210.18, 210.60
　Lighting Outlets Required . . . . 210.70(B)
　Meeting Rooms 210.65
　Receptacle
　　Loads 220.44
　　Receptacle Placement . . . . 210.60(B)
　　Receptacles Required 210.52
　　Tamper-Resistant 406.12
　See Life Safety Code, NFPA 101 . . . . NFPA 101

HOUSEBOATS . . . . **Art. 555**
　See FLOATING BUILDINGS . . . . Ferm's Finder
　HVAC Equipment (Air Conditioning and Refrigerator) Art. 440

HVAC EQUIPMENT
　GFCI Protection (Outdoor), Dwelling Unit . . . . . . 210.8(F)
　Receptacle . . . . 210.8(E)

HYBRID SYSTEM
　Example, Photovoltaic Systems . . . . Figure 690.1(b)
　Unbalanced Interconnections, Unbalanced Voltage . . . 705.100
　Utility-Interactive Inverters, Interconnection 705.82

HYDROMASSAGE BATHTUBS . . . . **680 Part VII**
　Accessibility 680.73
　Bonding . . . . 680.74(B)
　GFCI Protection Required for Circuit 680.71
　Other Electrical Equipment 680.72
　See (NCHX) . . . . UL Product iQ

I

IDENTIFICATION OF
　Branch Circuits on Panel Doors . . . . 408.4(A)
　Conductors
　　Branch Circuit, General 210.5
　　For General Wiring 310.10

High (Delta) Leg
- For Conductors or Busbar 110.15
- For Services 230.56
- General 110.15
- Intrinsically Safe Systems . . . . 504.80(C)
- Motor Control Centers . . . . 430.97(B)
- Phase Arrangement, Switchboards and Panelboards 408.3(E)
- Sensitive Electronic Equipment . . . . 647.4(C)
- Ungrounded, Multiwire Branch Circuits 110.15

Disconnecting Means 110.22
- Panelboard Circuits and Modifications 408.4

Emergency System Wiring . . . . 700.10(A)

Equipment Grounding Conductor 250.119

Feeders
- Direct-Current Systems, Ungrounded Conductors 215.12(C)(2)
- Equipment Grounding Conductor . . . . 215.12(B)
- Grounded Conductor . . . . 215.12(A)
- Supplied From More Than One Nominal Voltage System . . . . 215.12(C)(1)

(Fixture) Luminaire Wires, Grounded Conductor 402.8

Flexible Cords
- Equipment Grounding Conductor 400.23
- Grounded Conductor 400.22
- Luminaire (Fixture) Wires, Grounded Conductor 402.8

Grounded Conductor . . . . Art. 200
- Conductors of Different Systems . . . . 200.6(D)
- Conductors for General Wiring . . . . 310.6(A)
- For Branch Circuits . . . . 210.5(A)
- In Flexible Cord . . . . 200.6(C)
- In Multiconductor Cables . . . . 200.6(E)
- Sizes Larger Than 4 AWG . . . . 200.6(B)
- Sizes 6 AWG or Smaller . . . . 200.6(A)
- Terminals 200.10
- Use of Color, General . . . . 200.7(A)
  - Circuits 50 Volts or More . . . . 200.7(C)
  - Circuits Less Than 50 Volts . . . . 200.7(B)

Identification of Terminals
- Aluminum, for . . . . 110.14(A)
- Devices Connected to Grounded Conductor, for 200.10
- Grounding-type Receptacles, Adapters, etc. . . . . 406.10(B)
- Intrinsically Safe Systems . . . . 504.80(A)
- More Than One Conductor, for . . . . 110.14(A)

Intrinsically Safe System Wiring . . . . 504.80(B)

Isolated Ground Receptacles . . . . 406.3(D)

Isolated Power Systems, Health Care Facilities . . 517.160(A)(5)

Panelboard Circuits 408.4

Receptacles Connected to Critical Branch, Hospitals . 517.34(A)

Service Disconnecting Means . . . . 230.70(B)

Ungrounded Conductors . . . . 310.6(C)

See Also CONDUCTORS, Identification of . . . . Ferm's Finder

ILLUMINATION ABOUT ELECTRICAL EQUIPMENT . . . 110.26(D)

ILLUMINATION FOR EMERGENCY SYSTEMS . . . 700.16

IMMERSION DETECTION OF APPLIANCES 422.41

IMMERSION HEATERS 422.44

IMPEDANCE GROUNDING
- Grounded Neutral System, 1 kV and Over 250.187
- Grounded Neutral System, High-Impedance 250.36
- See HIGH IMPEDANCE GROUNDED NEUTRAL Ferm's Finder

IN SIGHT FROM
Definition of . . . . Art. 100

Disconnecting Means For
- Air-Conditioning Equipment 440.14
- Air-Conditioning Unit, Single-Phase Room, Attachment Plug 440.63
- Appliances, Motor Operated Rated over 1/8 HP . . 422.31(C)
- Appliances, Permanently Connected Over 300 VA 422.31(B)
- Duct Heaters 424.65
- Electric Discharge Lighting, Over 1000 Volts . . . 410.141(B)
- Electrically Driven or Controlled Irrigation Machines . . . . 675.8(B)
- Electric Signs . . . . 600.6(A)
- Fixed Electric Space Heating Equipment 424.19
- Induction and Dielectric Heating Equipment 665.12
- Mobile Home Service Equipment . . . . 550.32(A)
- Motors . . . . Art. 430 Part IX
  - Control Circuits 430.74
  - Controller . . . . 430.102(A)
  - Motor and Driven Machinery . . . . 430.102(B)
- Phase Converters . . . . 455.8(A)
- Shore Power Connections . . . . 555.36(B)
- Swimming Pool, Spa, and Hot Tub Equipment 680.13
  - Emergency Switch for Spas and Hot Tubs 680.41

INCANDESCENT LAMPS . . . . Art. 410

Access to Other Boxes. 410.118

Aircraft Hangers . . . . 513.7(C)

Clearance of Luminaires (Fixtures) 410.116

Clothes Closets, In 410.16

Dressing Rooms of Theaters. 520.72

Garages . . . . 511.7(B)

Lamp Wattage Marking 410.120

Medium and Mogul Bases 410.103

Recessed 410.115

INCIDENTAL METAL PARTS

Not Likely to Become Energized . . . . 422.15(C)

INDEPENDENT

Emergency Lighting Branch Circuits 700.17

Means of Support for Wiring above Ceilings . . . . . 300.11(A)

    Fire-Rated Assemblies . . . . .300.11(A)(1)

    Non-Fire-Rated Assemblies. . . . . .300.11(A)(2)

Supports for Service Drops 230.29

Wiring, Emergency Circuits . . . . 700.10(B)

Note: *See* specific occupancy or system for further information on independent wiring, sources of supply, etc.

INDIVIDUAL BRANCH CIRCUIT

Appliances . . . . 422.10(A)

Definition of . . . . Art. 100 Part I

Electric Sign or Outline Lighting . . . . 600.5(A)

Electrode-Type Boiler Over 1000 Volts . . . . 490.72(A)

Elevator Car Lighting, HVAC 620.22

Fixed Electric Space-Heating Equipment . . . . . . 424.4(A)

Hydromassage Bathtubs. 680.71

Marinas and Boatyards . . . . 555.33(A)(3)

Motor Circuits 430.52

Office Furnishings . . . . 605.9(B)

Permissible Loads 210.22

Receptacles on . . . . 210.21(B)(1)

    Permissible Load 210.23

Sign 600.5

INDUCTION (ARRANGE WIRING TO PREVENT HEATING BY)

Busbar Arrangements . . . . 408.3(B)

For Three-Way and Four-Way Switching . . . . . . 404.2(A)

General . . . . 300.3(B)

Impedance Heating Systems 426.33

In Metal Enclosures or Raceways 300.20

Over 1000 Volts 300.35

Single Conductor Type UF Cable . . . . 340.10(2)

Single Conductors in Cable Trays . . . . 392.20(D)

Underground Installations . . . . 300.5(I)

INDUCTION & DIELECTRIC HEATING EQUIPMENT . . . . Art. 665

Ampacity of Supply Conductors 665.10

Definitions 665.2

Disconnecting Means 665.12

Grounding, Guarding & Labeling . . . . Art. 665 Part II

Industrial & Scientific Application . . . . Art 665 Part 1

Output Circuit 665.5

Overcurrent Protection 665.11

Remote Control 665.7

INDUSTRIAL CONTROL EQUIPMENT

See (NIMX) . . . . *UL Product iQ*

INDUSTRIAL CONTROL PANELS. . . . . ART. 409

Arc-Flash Hazard Warning, Marking 110.16

Busbars and Conductors 409.102

Conductor, Minimum Size and Ampacity 409.20

Defined . . . . Article 100

Disconnecting Means 409.30

Enclosures 409.100

Enclosure Type, Marking 110.28

Grounding 409.60

Installation . . . . Art. 409, Part II

Marking. 409.110

More Than One Power Source, Marking 409.110

Overcurrent Protection 409.21

Service Equipment 409.108

Short-Circuit Current Rating 409.22

    Documentation . . . . 409.22(B)

    Installation . . . . 409.22(A)

Spacing between Uninsulated Parts 409.106

Wire Space 409.104

See (NITW) . . . . *UL Product iQ*

INDUSTRIAL MACHINERY . . . . Art. 670

Clearance, Working Space . . . . 670.1 Info. Note 2

Conductor Size . . . . 670.4(A)

Definition of 670.2

Exceeds Available Fault Current 670.5

Disconnecting Means . . . . 670.4(B)

Fixed Resistance and Electrode Industrial Process Heating Equipment . . . . Art. 425
    Boilers . . . . Art. 425 Part VI and VII
    Branch Circuits 425.4
    Control and Protection. . . . . Art. 425 Part III
    Disconnecting Means 425.19
    Listed Equipment .425.6
    Locations. 425.12
    Infrared Lamp Industrial Heating Equipment 425.14
    Marking of Heating Equipment . . . . Art. 425 Part IV
    Process Duct Heaters. . . . . .Art. 425 Part V
    Process Electrode-Type Boilers . . . . . . .Art. 425 Part VII
    Process Resistance-Type Boilers. . . . . . . Art. 425 Part VI
        Overcurrent Protection 425.72
        Overpressure Limit Control 425.74
        Overtemperature Limit Control 425.73
Machine Nameplate Data. 670.3
Overcurrent Protection . . . . 670.4(C)
See Electrical Standard for Industrial Machinery . . . . NFPA 79
Short-Circuit Current Rating 670.5
Surge Protection. 670.6

INFORMATION TECHNOLOGY EQUIPMENT . . Art. 645
    Abandoned Cables . . . . 645.5(G)
    Ampacity of Branch-Circuit Conductors . . . . 645.5(A)
    Cable Routing Assemblies. . . . . 645.3(F)
    Cables under Raised Floors . . . . 645.5(E)
        Air-handling Spaces . . . . 300.22(D)
    Connecting Cables to Branch Circuit . . . . 645.5(B)
    Critical Operations Data System 645.2
    Dedicated Zones for Disconnecting Means 645.10
    Definition. 645.2
    Disconnecting Means 645.10
    Engineering Supervision 645.25
    Grounding, Equipment 645.15
    Grounding, System 645.14
    Interconnecting Cables . . . . 645.5(C)
    Marking of Equipment 645.16
    Neutral Load
        Conductor Considered Current-Carrying . . . . . 310.15(E)
        Feeder or Service 220.61
        Multiwire Branch Circuits . . . . . . 210.4(A) Info. Note 2
    Optical Fiber Cables . . . . 645.3(H)
    Penetrations of Fire-Resistant Room Boundary . . . . 645.3(A)
    Physical Protection . . . . 645.5(D)
    Plenums. . . . . 645.3(B)
    Power Distribution Unit 645.17
    Remote Disconnect Control 645.2
    Room, Definition 645.2
    Securing in Place . . . . 645.5(F)
    Selective Coordination 645.27
    Special Requirements for Room 645.4
    Surge Protection 645.18
    Under Raised Floors . . . . 645.5(E)
        Branch Circuits . . . . .645.5(E)(1)
        Cords, Cables, Grounding Conductors. . . . . .645.5(E)(2)
        Optical Fiber Cables . . . . .645.5(E)(3)
        Table- Cables Installed Under Raised Floor. . .Table 645.10(B)(5)
    Uninterruptible Power Supplies (UPS) 645.11
    See Definition of Nonlinear Load . . . . Art. 100 Part I
    Zone 645.2
    Zones, Identified for Disconnect 645.10
        Remote Disconnect Controls . . . . 645.10(A)
    Critical Operations Data Systems. . . . . 645.10(B)

INFORMATIONAL NOTES . . . . 90.5(C)

INFORMATIVE ANNEXES . . . . .90.5(D)
    Informative ANNEX A, Product Safety Standards . . . Annex A
    Informative ANNEX B, Application Information for Ampacity Calculation . . . . Annex B
    Informative ANNEX C, Conduit and Tubing Fill Tables
        For Conductors and Fixture Wires of the Same Size   Annex C
    Informative ANNEX D, Examples . . . . Annex D
    Informative ANNEX E, Types of Construction (see 334.10) . . . Annex E
    Informative ANNEX F, Availability and Reliability for Critical Operations Power Systems;
    and Development and Implementation of Functional Performance Tests (FPTs) for
    Critical Operations Power Systems . . . . Annex F
    Informative ANNEX G, Supervisory Control and Data Acquisition (SCADA) . . . . Annex G
    Informative ANNEX H, Administration and Enforcement Annex H
    Informative ANNEX I, Recommended Tightening Torque Tables Annex I
    Informative ANNEX J, ADA Standards for Accessible Design . . Annex J

INFRARED LAMP HEATING APPLIANCES
    Branch Circuits
        Fixed Electric Space-Heating . . . . 424.4(A)

Industrial Heating Appliances 425.14
Permissible Load . . . . 210.23(C)
Industrial Heating 422.48
Overcurrent Protection . . . . 422.11(C)

INHIBITOR REQUIRED ON
Aluminum Connections If Required by Listing or Manufacturer .
. . . . 110.3(B)
Not to Adversely Affect 110.14
See (DVYW) . . . . *UL Product iQ*

INNERDUCT
Coaxial Cables . . . . 800.110(A)(3)
Communication Wires and Cables . . . . 800.110(A)(3)
Definition. . . . . Art. 100
Installed Within Listed Metal Raceway . . . . 770.113(E)
Network-Powered Broadband Communication Cables . . . . . .
800.110(A)(3)
Optical Fiber Cables . . . . 770.110(A)(3)
Risers in Metal Raceways . . . . 800.113(E)

INSERTS
Cellular Concrete Floor Raceways . . . . 372.18(D)
Cellular Metal Floor Raceways . . . . 374.18(C)
Underfloor Raceways 390.75

INSPECTIONS AND TESTS
Equipment Over 1000 Volts, Nominal
Outside Branch Circuits and Feeders. 225.56
Pre-energization and Operating Tests. . . . . 110.41(A)
Test Reports . . . . 110.41(B)

INSTALLATION AND USE INSTRUCTIONS . . . . 110.3(B)
See (AALZ) . . . . *UL Product iQ*
INSTITUTIONS, EMERGENCY LIGHTING . . . Art. 700
See *Life Safety Code, NFPA 101*. . . . . NFPA 101

INSULATED (ISOLATED) EQUIPMENT GROUNDING CONDUCTOR
See GROUNDING, *Insulated Ground Wire* . . . Ferm's Finder

INSULATED FITTINGS & BUSHINGS
Cabinets, Cutout Boxes and Meter Socket Enclosures . 312.6(C)
Required for Conductors 4 AWG and Larger . . . . . 300.4(G)
Underground Installations . . . . 300.5(H)

INSULATED (ISOLATED) GROUND RECEPTACLES
Audio Signal Processing, Amplification, and Reproduction 640.7(C)

Conductor Fill in Outlet, Device & Junction Boxes 314.16(B)(5)
Connecting Receptacle Grounding Conductor to Box 250.146(D)
Continuity and Attachment of Conductor to Boxes 250.148 Ex.
Faceplates for . . . . 406.3(D)(2)
Identification of Receptacles . . . . 406.3(D)
Insulated from Mounting Means, Reduction of Noise 250.146(D)
In Health Care Facilities . . . . 517.16(A)
Inside Patient Care Vicinity . . . . 517.16(A)
Outside Patient Care Vicinity . . . . 517.16(B)
Special-Purpose Receptacles . . . . 517.19(H)
Patient Care Vicinity . . . . 517.16(A)
Inside . . . . 517.16(A)
Outside. . . . . 517.16(B)
Reduction of Electrical Noise . . . . 250.96(B)
Sensitive Electronic Equipment . . . . 647.7(B)

INSULATING DEVICES & MATERIALS
Insulating Tape
See (OANZ) . . . . *UL Product iQ*
Insulation, Equivalent to Conductors. . . . . 110.14(B)

INSULATION
Conductors
Construction and Application 310.4
Corrosive Conditions, Suitable for . . . . 310.10(G)
Dry Locations . . . . 310.10(A)
Dry and Damp Locations . . . . 310.10(B)
Exposed to Direct Sunlight. . . . . 310.10(D)
General Requirement for 310.1
Identification 310.6
Temperature Limitations . . . . 310.14(A)(3)
Wet Locations . . . . 310.10(C)
Equipment . . . . 110.3(A)(4)
Luminaire (Fixture) Wire Construction and Application . . 402.3
Flexible Cord Construction Specifications . . . . . . 400 Part II
Type and Usage 400.4
Integrity of 110.7
Service Conductors
Overhead Service Conductors 230.22
Service-Entrance Conductors 230.41
Underground Service Conductors 230.30
Shielding — Over 1000 Volt Cables, Grounding 300.40
Splices and Joints . . . . 110.14(B)
In Flexible Cord to Retain Insulation Properties 400.13
Thermal

Branch-Circuit Wiring above for Space-Heating Cables 424.36

Within 75 mm (3 in.) of Recessed Luminaires (Fixtures). . . . 410.116(B)

INTERCONNECTED ELECTRIC POWER PRODUCTION SOURCES . . . . **Art. 705**
    Definitions 705.2
    Directory 705.10
    Disconnecting Means
        Device . . . . 705.20(1)(d)
        Equipment . . . . 705.20(1)(d)
        Sources 705.20
        Equipment Approval 705.6
    Generators . . . . 705.30(D)
    Ground-Fault Protection 705.32
    Interrupting and Short-Circuit Current Rating 705.16
    Loss of Primary Source 705.40
    Loss of 3-Phase Primary Source 705.40
    Microgrid Systems . . . . .Art. 705 Part II
    Overcurrent Protection 705.30
    Scope 705.1
    System Installation 705.8
    Unbalanced Interconnections 705.45
    Warning Labels (load side) 705.12

INTEGRATED ELECTRICAL SYSTEMS . . . .**Art. 685**

INTEGRATED GAS SPACER CABLE . . . . **Art. 326**
    Ampacity of 326.80
    Bends
        Number of 326.26
        Radius 326.24
    Conductors 326.104
    Conduit 326.116
        Dimensions . . . . .Table 326.116
    Definition 326.2
    Insulation 326.112
    Uses Not Permitted 326.12
    Uses Permitted 326.10

INTERCOMMUNICATIONS SYSTEMS . . . . **Art. 805**
    Class III Locations 503.150

INTERCONNECTED ELECTRIC POWER PRODUCTION SOURCES . . . .**Art. 705**
    Circuit Sizing and Current 705.28
    Definitions 705.2
    Equipment Approval 705.6
    Interrupting and Short-Circuit Current Rating 705.16
    Load-Side Source Connections 705.12
    Microgrid Systems . . . . Art. 705, Part II

Overcurrent Protection **705.30**

Output Characteristics **705.14**

Power Control Systems **705.13**

Scope **705.1**

Supply-Side Source Connections **705.11**

System Installation **705.8**

Wiring Methods **705.28**

INTERCONNECTORS

Nonmetallic-Sheathed Cable . . . . **334.40(B)**

INTERMEDIATE METAL CONDUIT . . . . **Art. 342**
  Bends
    How Made 342.24
    Number of 342.26
  Bushings 342.46
  Construction - Made of 342.100
  Couplings and Connectors
    Running Threads Not Permitted . . . . 342.42(B)
    Threadless . . . . 342.42(A)
  Definition 342.2
  Dimensions and Percent Area of . . . .Chapter 9 Table 4 (IMC)
  Dissimilar Metals 342.14
  Marking [Note 1.5 m (5 ft.) Intervals] 342.120
  Minimum and Maximum Sizes 342.20
  Not To Be Used as a Means of Support for Cables or
    Class 2 and 3 Circuit Conductors 725.143
    Communications Circuits . . . . 805.133(B)
    Community Antenna TV & Radio Distribution Systems . . . . .820.133(B)
    Fire Alarm Circuit Conductors 760.143
    Network-powered Broadband Systems . . . . . . 830.133(B)
    Nonelectrical Equipment (General) . . . . 300.11(B)
  Number of Conductors in 342.22
    Combinations of Conductors (General) Chapter 9, Table 1, Notes

to Tables
- Dimension of Conductors in . . . . . . Chapter 9 Table 5
- Compact Stranded. . . . . Chapter 9 Table 5A
- Same Size . . . . Table C4 and C4(A)

Physical Damage- Severe . . . . 342.10(E)
Reaming and Threading 342.28
Securing and Supports 342.30
Severe Physical Damage (use permitted) . . . . 342.10(E)
Splices and Taps 342.56
Uses Permitted 342.10
Wet Locations . . . . 342.10(D)
See RIGID METAL CONDUIT . . . . Ferm's Finder
See (DYBY) . . . . UL Product iQ

INTERRUPTING CAPACITY OF BREAKERS AND FUSES
Circuit Breakers, Series Ratings 240.86
Circuit Impedance, Short-Circuit Current Ratings and Other Characteristics 110.10
Contribution from Interconnected Power Sources 705.16
Individual Pole Consideration, Circuit Breakers 240.85 Info. Note
Interrupting Ratings, Sufficient for 110.9
Marking
- Cartridge Fuses and Fuseholders . . . . 240.60(C)
- Circuit Breakers . . . . 240.83(C)

INTERSYSTEM BONDING TERMINATION
Definition of . . . . Art. 100
Communication Circuits . . . . 800.100(B)
CATV Systems . . . . 800.100(B)
Installation 250.94
Network Powered Broadband Communication Systems. . 800.100(B)
Radio and Television Equipment . . . . 810.21(F)

INTRINSICALLY SAFE SYSTEMS . . . . **Art. 504**
Bonding 504.60
Conductors, Separation of 504.30
Definitions . . . . Art. 100
Enclosures . . . . 504.10(C)
Equipment Listed 504.4
Equipment Installation 504.10
Grounding 504.50
Hazardous (Classified) Locations Protection Technique 500.7(E)
- Class I, Zone 0, 1, and 2 Locations. . . . . 505.8(C)
Identification 504.80
- Color Coding . . . . 504.80(C)
- Terminals. . . . . 504.80(A)
- Wiring . . . . 504.80(B)
Sealing 504.70
Separation 504.30
- From Different Intrinsically Safe Circuit Conductors 504.30(B)
- From Grounded Metal . . . . 504.30(C)
- From Nonintrinsically Safe Circuit Conductors . . 504.30(A)
Simple Apparatus
- Control Drawing. . . . . 504.10(A)
- Definition . . . . Art. 100
- Enclosures . . . . 504.10(C)
- Installation . . . . 504.10(D)
- Location . . . . 504.10(B)
Wiring Methods 504.20

INVERTER
Connected to Grounded System . . . . 200.3 Ex
Definition of 690.2
Definition of Multimode Inverter. 705.2
Recreation Vehicles 551.32
Used with Fuel Cell Systems . . . . Art. 692
Used with Interconnected Electric Power Production Sources. . . . . Art. 705
Used with Solar Photovoltaic Systems. . . . . Art. 690
Used with Wind Electric Systems . . . . Art. 694
Utility-Interactive, Definition. . . . . Art. 100

IRRIGATION MACHINES (Electrically Driven of Controlled)
. . . . **Art. 675**
Bonding 675.14
Branch-Circuit Conductors. 675.9
Center-Pivot Irrigation Machines . . . . Art. 675 Part II
- Continuous-Current Rating . . . . 675.22(A)
- Equivalent Current Rating 675.22
- Locked-Rotor Current . . . . 675.22(B)
Collector Rings 675.11
Conductors, More Than Three in Raceway or Cable 675.5
Connectors 675.17
Control Panel Marking 675.6
Definitions 675.2
Disconnecting Means
- For Individual Motors and Controllers . . . . 675.8(C)
- Main Controller . . . . 675.8(A)
- Main Disconnecting Means . . . . 675.8(B)
Energy from More Than One Source 675.16

Grounding 675.12
    Methods of 675.13
Irrigation Cable .675.4
    See (OFFY) . . . . UL Product iQ
Irrigation Machines
    Continuous Current Rating . . . . 675.7(A)
    Equivalent Current Ratings 675.7
    Locked-Rotor Current . . . . 675.7(B)
Lightning Protection 675.15
    Grounding Electrode System . . . . Art. 250 Part III
    See Standard for the Installation of Lightning Protection Systems, NFPA 780 . . . . NFPA 780
Lockable Disconnect . . . . 675.8(B)
Several Motors on One Branch Circuit 675.10

ISLAND COUNTERTOP RECEPTACLE SPACES . . . . 210.52(C)(2)

ISLANDING, LOSS OF INTERCONNECTED POWER . . . . 705.40 Info. Note 1

ISOLATED
Arcing or Suddenly Moving Parts 240.41
Capacitors, Isolating Means, Over 1000 Volts . . . . 460.24(B)
Conductor Enclosures and Raceways . . . . 250.86 Ex. 3
Definition of (as applied to location) . . . . Art. 100 Part I
Electric Snow-Melting and Deicing 426.12
Elevators and Similar Equipment . . . . 620.5(B)
Equipment Ground, Technical Power System . . . . 640.2 (Def.)
Ground Receptacle, Reduction of Electrical Noise . 250.146(D)
Impedance Heating 426.30
Isolating Means, Equipment Over 1000 Volts 490.22
Live Parts Guarded Against Accidental Contact . . . . 110.27(A)
Motors and Controllers Over 50 Volts . . . . 430.232(3)
Other Than Service Raceways . . . . 250.86 Ex. 3
Phase Conductors, in Close Proximity . . . . . . 300.5(I) Ex. 2
Resistors and Reactors . . . . 470.18(B)
Service Raceways Underground — Metal Elbows . . 250.80 Ex.
Short Sections of Raceway 250.132
Sign and Outline Lighting Parts, Grounding 600.7
Metal Fittings at Swimming Pools, Exempt from Bonding . . . . 680.26(B)(5)
Water Pipe, Multi-occupancy Buildings . . . . . . 250.104(A)(2)

ISOLATED (INSULATED) GROUND RECEPTACLES
Inside Patient Care Vicinity . . . . 517.16(A)
Outside Patient Care Vicinity . . . . 517.16(B)
See INSULATED (ISOLATED) GROUND RECEPTACLES . . . . Ferm's Finder

ISOLATED POWER SYSTEMS
Cranes and Hoists Over Class III Locations . . . . . .503.155(A)
Critical Care (Category 1) Patient Space (Optional) . 517.19(F)
    Individual Circuits for Certain Critical Branch Equipment . . . . .517.31(C)(2)
    System Grounding . . . . 517.19(G)
    Wet Locations, Permitted in Lieu of GFCI Protection . .517.20(A)
Electrolytic Cells, Portable Equipment . . . . 668.21(A)
Health Care Facilities 517.160
Impedance Heating Systems 426.31
Induction and Dielectric Heating Equipment 665.5
Swimming Pool Transformers and Power Supplies . 680.23(A)(2)

ISOLATING TRANSFORMERS
AC Systems 50 to 1000 Volts Not Required to Be Grounded 250.21
Crane and Hoist, Using Track As Circuit Conductor 610.21(F)(2)
Electrolytic Cells, Portable Equipment . . . . 668.20(B)
    Power Supply and Receptacles . . . . 668.21(A)
Health Care Facility Installations . . . . 517.160(A)(1)
    Restriction on Rooms Served . . . . 517.160(A)(4)
Impedance Heating Systems 426.31
Lighting Systems 30 Volts or Less . . . . 411.6(B)
Swimming Pool Use . . . . .680.23(A)(2)

J

JOINTS
Expansion
    Busways, Over 1000 Volts, Nominal 368.244
    Earth Movement 1000 Volts and Less . . . . 300.5(J)
    Earth Movement Over 1000 Volts, Nominal . . . . 300.50(C)
    Electric Space Heating Cables in Concrete or Poured Masonry Floors 424.44
    Fixed Electric Heating for Pipelines and Vessels 427.16
    Fixed Outdoor Electric Deicing and Snow-Melting . . . . Art. 426 Part III
    Heating Panels & Heating Panel Sets . . . . 424.98(C)
    Metal Raceways 250.98
    Nonmetallic Auxiliary Gutters 366.44
    Nonmetallic Wireways 378.44

Raceways Exposed to Different Temperatures . . . . 300.7(B)
   Rigid PVC Conduit
      Expansion Characteristics Tables. . . . . Table 352.44
      Securing and Supporting, Provisions for 352.30
      To Be Provided 352.44
   RTRC Conduit . . . . Table 355.44
Grounding Electrode Conductor . . . . 250.64(C)
Insulating, at Luminaires (Fixtures) . . . . 410.36(D)
Insulation of . . . . 110.14(B)
Strain at, Flexible Cords 400.14

JOISTS (WOOD or METAL)
Air-Handling Space . . . . 300.22(C) Ex.
Armored Cable
   Exposed 320.15
   In Accessible Attics. 320.23
   Through or Parallel to Framing Members 320.17
Boxes on 314.23
Cables and Raceways Parallel to . . . . 300.4(D)
Concealed Knob-and-Tube Wiring 394.23
   Through or Parallel to Framing Members 394.17
Electric Space-Heating Cables . . . . 424.41(I) & (J)
   In Concrete or Poured Masonry Floors. . . . . . . 424.44(C)
Holes Through or Notches in 300.4
Nonmetallic-Sheathed Cable . . . . 334.15(C)
Open Wiring
   Crossing . . . . 398.15(C)
   In Attics 398.23
   Through 398.17

JUMPERS, BONDING
Definition of . . . . Art. 100 Part I
Equipment
   Attachment. . . . . 250.102(B)
   Definition of . . . . Art. 100
   Installation . . . . 250.102(E)
   Load Side of Service . . . . 250.102(D)
   Material . . . . 250.102(A)
   Supply Side of Service . . . . 250.102(C)
Expansion Joints, Raceways 250.98
Grounding Electrode System . . . . 250.53(C)
Grounding Electrodes 250.50
Grounding-Type Receptacles 250.146
Hazardous (Classified) Locations
   Class I Locations 501.30
   Class II Locations 502.30
   Class III Locations 503.30
   Intrinsically Safe Systems 504.60
   Zone 0, 1, and 2 Locations. 505.25
   Zone 20, 21, and 22 Locations. 506.25
Health Care Facilities (Patient Care Vicinity) . . 517.19(D) & (E)
High-Impedance Grounded Neutral Systems . . 250.36(E) & (G)
Intersystem Bonding Termination. 250.94
Main
   AC Systems (Grounded) 250.28
   DC Systems 250.168
   Definition . . . . Art. 100
Other Systems 250.94
Piping Systems
   Metal Water Piping. . . . . 250.104(A)
   Other Metal Piping . . . . 250.104(B)
Separately Derived Systems . . . . 250.30(A)(1) & (2)
Service Equipment . . . . 250.92(B)
Service Supplied AC Systems, Wire or Busbar . . 250.24(A)(4)
Structural Metal, Exposed. . . . . 250.104(C)
System Bonding Jumper 250.28
System Bonding Jumper . . . . 250.30(A)(1)
Definition . . . . Art. 100, Part I

JUNCTION AND PULL BOXES . . . . Art. 314
Accessible 314.29
Accessible after Electric-Discharge and LED Luminaires (Fixtures) Installed. . . . . 410.24(B)
Boxless Devices
   See BOXLESS DEVICES . . . . Ferm's Finder
Ceiling-Suspended (Paddle) Fan Support . . . . 314.27(C)
Conductors
   Entering Boxes (General). . . . . 300.4(G)
   Entering Boxes and Conduit Bodies 314.17
      From Electrical Nonmetallic Tubing. 362.46
      From Intermediate Metal Conduit. 342.46
      From High Density Polyethylene Conduit. 353.46
      From Nonmetallic Underground Conduit with Conductors 354.46
      From Reinforced Thermosetting Resin Conduit . . 355.46
      From Rigid Metal Conduit 344.46
      From Rigid PVC Conduit. 352.46
Conduit Bodies
   Cross-sectional Area of . . . . 314.16(C)(1)
   Dimensions of for Pulls 314.28

Over 1000 Volts	314.71
General	314.1
Marked Capacity of	314.16(C)(2)
Number of Conductors in	314.16(C)
Support of	314.16(C)(2)
Where Required	300.15

Continuity of Metal Enclosures

At Services	250.92(A)
Boxes	314.4
Electrical Continuity	300.10
Hazardous (Classified) Locations, General	500.8
Class I Locations	501.30
Class I, Zone 0, 1, and 2 Locations	505.25
Class I, Zone 20, 21, and 22 Locations	506.25
Class II Locations	502.30
Class III Locations	503.30
Intrinsically Safe Systems	504.50 & .60
In Hazardous (Classified) Locations	250.100
Mechanical Continuity	300.12
Other Enclosures	250.96
Over 250 Volts	250.97

Covers

Completed Installations	314.25
Compatible with Construction	314.28(C)
Extensions from	314.22 Ex.
Manholes	110.75(D)
Marking of	110.75(E)
Material	314.41
Over1000 Volts	314.72(E)
Marking	314.72(E)
Required	300.31
Suitable for Expected Handling	314.72(F)
Emergency Systems, Separation of Wiring	700.10(B)

Essential Electrical Systems. Separation of Wiring

Hospitals	517.31(C)
Nursing Homes	517.42(D)

Extension Rings

Box Volume Calculations	314.16(A)
Exposed Extensions	314.22
Floor Boxes	314.27(B)
For Cellular Concrete Floor Raceways	372.18(C)
Splices and Taps in	372.56
For Cellular Metal Floor Raceways	374.18(B)
Splices and Taps in	374.56
For Underfloor Raceways	390.74
Splices and Taps in	390.56
Free Length of Conductors In (General)	300.14
Deicing and Snow-Melting Equipment (Embedded)	426.22(E)
Deicing and Snow-Melting Equipment (Exposed)	426.23(A)
For Electric Space-Heating Cables	424.43(B)
Pipeline and Vessel Heating	427.18(A)

Grounding

See GROUNDING, Boxes	Ferm's Finder
High Voltage Systems, Over 1000 Volts	314 Part IV
See OVER 1000 VOLTS, NOMINAL	Ferm's Finder

Identification of

For Emergency Systems	700.10(A)
For Intrinsically Safe Systems	504.80(C)
For Underground Locations	314.29
Luminaire (Fixture) Support	410.30
Weight Limits for Support of Luminaires (Lighting Fixtures)	314.27(A) & (B)

Metallic Boxes

Conductors Entering	314.17(B)
Grounding of	
See GROUNDING, Boxes	Ferm's Finder
Sealing Unused Openings	110.12(A)
Thickness of Metal	314.40
Mounting	110.13(A)
In Concrete, Tile or Other Noncombustible	314.20
Repairing Noncombustible Surfaces	314.21
Supports	314.23
From Cable Tray Systems	392.18(G)
Nonmetallic, Permitted	314.3
Grounding Conductors	250.148(B)
Sealing Unused Openings	110.12(A)
Support	314.23
Provisions for	314.43
Number of Conductors in	314.16
Required	300.15
Round Boxes (Not Permitted)	314.2
Sizing of Pull & Junction Boxes (Standard)	314.16
Dimensions	314.28
Over 1000 Volts	314.71
Unused Openings (Closed)	110.12(A)
Volume Required Per Conductor	Table 314.16(B)
Warning Sign "Danger - High Voltage - Keep Out"	314.72(E)
Wet Locations	

General. .300.6
Prevent Moisture Entering or Accumulating 314.15
See SWIMMING POOLS . . . . .Ferm's Finder
See (BGUZ) . . . . UL Product iQ

K

KITCHENS
Appliance Load 220.53
Arc-Fault Circuit-Interrupter Protection . . . . 210.12(A)
Branch Circuits Required in Dwelling Units . . . . 210.11(C)(1)
   Mobile and Manufactured Homes . . . . 550.12(B)
   Recreational Vehicles. 551.42
   Park Trailers 552.46
Connection of
   Dishwasher and Trash Compactor . . . . 422.16(B)(2)
   Waste Disposal . . . . 422.16(B)(1)
Definition of . . . . Art. 100, Part I
Disconnection of Appliances in . . . . Art. 422 Part III
GFCI Protection of Receptacle Outlets in Dwelling Units . 210.8(A)
   Dishwasher Outlet . . . . 422.5(A)(7)
   Mobile and Manufactured Homes . . . . 550.13(B)
   Park Trailers . . . . 552.41(C)
   Recreational Vehicles . . . . 551.41(C)
GFCI Protection of Receptacle Outlets Other Than
   Dwelling Units . . . . 210.8(B)(2)
Grounding of Appliances in Cord- and Plug-Connected
   Fastened in Place or Connected by Permanent Wiring 250.114
      Methods. 250.110
      Ranges 250.140
      Specific Equipment. 250.112
Installation of Appliances . . . . Art. 422 Part II
Kitchen Equipment Load, Other Than Dwelling Units . . 220.56
Lighting Outlet, Wall Switch-Controlled in . . . . 210.70(A)(1)
Receptacle Outlets Required for
   Countertops . . . . 210.52(C)
      Position . . . . 406.5(E)
   General Provisions . . . . 210.52(A)
   Small Appliances . . . . 210.52(B)
      Mobile and Manufactured Homes . . . . 550.12(B)
      Park Trailers 552.41

Recreational Vehicles 551.41
Range Load Household 220.55
Sinks . . . . 210.8(A)(7)
Small Appliance Branch-Circuit Load, Dwelling Unit 220.52(A)
Mobile and Manufactured Homes . . . . 550.18(A)(2)

KNIFE SWITCHES
600 to 1000 Volts 404.13
Damp or Wet Locations 404.4
Enclosures . . . . 404.3(A)
Mounting Height 404.8
Position and Connection of 404.6
Ratings
   General-Use . . . . 404.13(C)
   Isolating Switches . . . . 404.13(A)
   Motor-Circuit . . . . 404.13(D)
   To Interrupt Current . . . . 404.13(B)
See SWITCHES . . . . Ferm's Finder

KNOB-&-TUBE WIRING (CONCEALED) . . . . . Art. 394
Boxes, Wiring Entering
   Metal Boxes . . . . 314.17(B)
   Nonmetallic Boxes . . . . 314.17(C)
Clearance
   General . . . . 394.19(A)
   Limited Conductor Space . . . . 394.19(B)
   Piping, Exposed Conductors . . . . 394.19(C)
Conductor Supports . . . . 394.30(A)
Conductors 394.104
Definition of 394.2
In Unfinished Attic and Roof Spaces 394.23
Insulation in Area of . . . . 394.12(5)
Splices 394.56
Through Walls, Floors, etc. 394.17
Tie Wires . . . . 394.30(B)
Uses Not Permitted 394.12
Uses Permitted 394.10

KNOCKOUTS
Concentric, Eccentric, or Oversized, Bonding Around
   Bonding at Service . . . . 250.92(B)
   Over 250 Volts . . . . 250.97 Ex.
   Grommets Required At Metal Studs . . . . 300.4(B)(1)
Openings to Be Closed
   General . . . . 110.12(A)

In Cabinet, Cutout Boxes and Meter Socket Enclosure. . . . .
. . . . 312.5(A)

Through Which Conductors Enter . . . . 314.17(A)

L

LABELED (Definition of) . . . . **Art. 100 Part I**
Manufactured and Mobile Homes
Manufactured Home Service Equipment . . . . . . 550.32(B)(7)
Mobile Home Service Equipment, 125/250-Volt Receptacle . . . 550.32(G)
Outside Heating/Air-Conditioning Equipment. . . . 550.20(B)
Park Trailers
    Outside Heating/Air-Conditioning Equipment . . 552.59(B)
    Prewired for Air-Conditioning . . . . 552.48(P)(3)
    Service Equipment . . . . 552.44(D)
Recreational Vehicles
    Prewired for Air-Conditioning . . . . .551.47(Q)(3)
    Prewired for Generator. . . . . .551.47(R)(4)
    Service Equipment . . . . 551.46(D)

LABELS
Cable Trays Containing Conductors Over 600 Volts . 392.18(H)
    Industrial Establishments. . . . . 392.18(H) Ex.
Cable Trays, Service Conductors 230.44
Circuit Directories 110.22
Direct-Current Ground-Fault Detection 250.167
Elevators, Etc.. . . . . 620.3(A)
Equipment Grounding Conductors. 250.119
Field-Applied Hazard . . . . 110.21(B)
Fixed Electric Space-Heating Equipment . . . . . . 424.92(C)
Flexible Cords. 400.20
In Panelboards . . . . 408.4(A)
    Source of Supply . . . . 408.4(B)
Induction and Dielectric Heating Equipment 665.23
Intrinsically Safe Systems, Identification 504.80
Manufacturer's Markings . . . . 110.21(A)
Recreational Vehicles
    Air-Conditioning, Pre-Wring . . . . 551.47(Q)
    Branch Circuits, Pre-Wiring . . . . 551.47(S)
    Designed . . . . 551.4(C)
    Electrical Entrance . . . . 551.46(D)

    Generator, Pre-Wiring . . . . 551.47(R)
Service, Supply Side. 230.82
Signs, Markings and Listings . . . . 600.4(C) and (D)
Solar Photovoltaic Systems
    Direct Current on or Inside a Building. . . . . . . 690.31(G)
    Rapid Shutdown Systems . . . . 690.56(C)
See ANSI Z535.4-2011, Product Safety Signs and Labels . . . . .
. . . . ANSI Z535.5

LABEL ON CABLE TRAYS
Containing Conductors Rated Over 600 Volts . . . . 392.18(H)
Where Service and Non-Service Conductors are Present 230.44 Ex.

LACQUERS AND PAINTS
Application of . . . . Art. 516
Classification of Atmospheres. . . . . 500.6(B)
Class I Locations, General . . . . 500.5(B)
    Requirements in . . . . Art. 501
    Zone 0, 1, and 2 Locations. . . . . Art. 505
To Be Removed for Grounding Continuity 250.12
Not to Contaminate Equipment, Integrity of Equipment 110.12(B)

LAMPHOLDERS . . . . 410 Parts VIII & XI
Circuits and Equipment, at Less Than 50 Volts 720.5
Combustible Materials, near 410.97
Cord-Connected . . . . 410.62(A)
Damp or Wet Locations 410.96
Double-Pole Switched 410.93
Heavy Duty (Rating of). . . . . 210.21(A)
Infrared Lamp Industrial Heating Appliances. 425.14
Mogul Base 410.103
Outdoor 225.24
Outlet Boxes . . . . 314.27(A)
Over Combustible Material, Unswitched Type 410.12
Pendant
    Bathtub and Shower Areas . . . . 410.10(D)
    Not Permitted in Clothes Closets . . . . 410.16(B)
    Not Permitted in Theater Dressing Rooms 520.71
Permissible on 30-ampere Branch Circuits . . . . . . 210.23(B)
Permissible on 40- and 50-ampere Branch Circuits . . 210.23(C)
Raceway Supported Enclosures for . . . . 314.23(F)
Screw Shell Type 410.90
Voltage Limitations . . . . 210.6(B)(C)(D)

LARGE-SCALE PHOTOVOLTAIC (PV) ELECTRIC POWER

PRODUCTION FACILITY . . . . Art. 691
   Applicable PV Systems (no less than 5000 kW) 691.1
   Arc-Fault Mitigation 691.10
   Conformance of Construction to Engineered Design . . .691.7
   Definitions. 691.2
   Direct Current Operating Voltage 691.8
   Disconnecting Means for Isolating Photovoltaic Equipment 691.9
   Engineered Design. .691.6
   Equipment Approval. 691.5
   Fence Bonding and Grounding. 691.11
   Special Requirements. 691.4

LAUNDRY RECEPTACLE OUTLETS, DWELLINGS
   Arc-Fault Circuit-Interrupter Protection . . . . 210.12(A)
   Branch-Circuit Load . . . . 220.52(B)
      Mobile and Manufactured Homes . . . . 550.18(A)(3)
   Laundry Area, Definition of 550.2
   Receptacles, GFCI Protected . . . . 210.8(A)(10)
   Required Branch Circuit . . . . .210.11(C)(2)
      Mobile and Manufactured Homes . . . . 550.12(C)
   Requirement for Receptacle . . . . 210.52(F)
      Mobile and Manufactured Homes . . . . . . .550.13(D)(7)
   Within 6 ft of Appliance . . . . 210.50(C)

LEAKAGE-CURRENT DETECTOR-INTERRUPTER (LCDI)
. . . . **440.65**
   Definition of 440.2
   Room Air Conditioners. 440.65

LEAKAGE CURRENT MEASUREMENT DEVICE (MARINAS AND BOATYARDS) . . . . **555.35(B)**

LED SIGN ILLUMINATION SYSTEM **600.33**
   Definition of 600.2

LEGALLY REQUIRED STANDBY SYSTEMS . . . . **Art. 701**
   Batteries. . . . . 701.12(C)
   Capacity and Rating 701.4
   Circuit Wiring . . . . .Art. 701 Part II
   Definition of 701.2
   Ground-Fault Protection of Equipment (Not Required). . .701.31
   Maintenance . . . . 701.3(C)
   Overcurrent Protection . . . . Art. 701 Part IV
      Accessible to Authorized Persons Only 701.30
      Selective Coordination 701.32
   Signals 701.6
      Ground-Fault Indication . . . . .701.6(D)
   Signs 701.7
   Sources of Power . . . . Art. 701 Part III
   Tests and Maintenance 701.3
   Transfer Equipment
      Electrically Operated, Mechanically Held 701.5
      Listing . . . . 701.5(A)
      Short-Circuit Current Rating (documented) . . . . .701.5(D)

LENGTHS
   Busways
      Maximum of Cord or Cable from Plug-In Device 368.56(B)(2)
      Maximum, Reduction without Overcurrent Device 368.17(B)
         Conduit Bodies . . . . 314.28(A)
         Conduit Bodies and Handhole Enclosures, Over 1000 Volts
. . . . 314.71
   Conduit and Tubing
      Note: *See* article for the specific type of conduit or tubing
   Flexible Cord, Specific Appliances . . . . 422.16(B)
   Free Conductor at Outlets, Junctions, and Switches . . . 300.14
      Deicing and Snow-Melting, Nonheating Leads. . . 426.23(A)
      Nonheating Leads of Resistance Elements, Pipelines and Vessels
. . . . 427.18(A)
      Nonheating Leads of Space-Heating Cable . . 424.43(B) & (C)
   Pull and Junction Boxes. 314.28
      Over 1000 Volts, Nominal 314.71
   Space-Heating Cable, Nonheating Leads 424.34
   Taps
      Branch Circuit . . . . 210.19(A)(4) Ex. 1
      Feeder . . . . 240.21(B)
      Motor Feeders 430.28
      Single Motor . . . . 430.53(D)(2)&(3)
      Supervised Industrial Installations . . . . . Art. 240 Part VIII
      Transformer Secondary Conductors . . . . 240.21(C)

LIFE SAFETY BRANCH
   Definition of 517.2
   Essential Electrical Systems
      Hospitals 517.33
         Separation from Other Circuits . . . . .517.31(C)(1)
      Nursing Homes and Limited Care Facilities 517.42
         Separation from Other Circuits . . . . 517.42(D)

LIFE SUPPORT EQUIPMENT, ELECTRICAL. . . **517.45(B)**
   Defined. 517.2

Essential Electrical Systems
   Hospitals and Health Care Facilities 517.29
   Nursing Homes and Limited Care Facilities 517.45
Ground-Fault Protection 517.17

LIGHTING
Airfield Lighting Cable . . . . 310.10(F)Ex. 2
   Branch Circuits, Calculations of Loads. 220.12
   Inductive and LED Loads . . . . 220.18(B)
   See CALCULATIONS. . . . . .Ferm's Finder
Cove 410.18
Crawl Spaces-GFCI Protection . . . . 210.8(C)
Decorative 410.160
Demand Factors 220.42
   See DEMAND FACTORS . . . . .Ferm's Finder
Electric Discharge Lighting
   Connection of Luminaires 410.24
   Cord-Connected . . . . 410.62(C)
   Definition . . . . Art. 100
   Hazardous (Classified) Locations
      Class I, Div. 2 . . . . 501.130(B)(6)
      Class II, Div. 2 . . . . 502.130(B)(5)
   Lamp Auxiliary Equipment 410.104
Energy Code, Designed and Constructed. . . . . . . 220.12(B)
Feeders, Calculation of Loads 220.42
   Show Window and Track Lighting 220.43
Festoon Lighting
   Conductor Size . . . . 225.6(B)
   Definition . . . . Art. 100 Part I
   Portable Stage Equipment 520.65
Fixtures See LUMINAIRES. . . . . .Ferm's Finder
LED Lighting
   Connection of Luminaires 410.24
   Cord Connected . . . . 410.62(C)
Lighting Assembly (Swimming Pools)
   Cord-and-Plug Connected . . . . 680.33(A)(B)
      Definition 680.2
   Through-Wall . . . . 680.23(E)
      Definition 680.2
Lighting Outlets
   All Occupancies . . . . 210.70(C)
   Crawl Spaces . . . . 210.8(C)
   Definition . . . . Art. 100 Part I
   Dwelling-Type Occupancies . . . . 210.70(A)
   Emergency Systems . . . . Art. 700
   Equipment Over 1000 Volts, Nominal. . . . . . . 110.34(D)
   Guest Rooms or Guest Suites. . . . . 210.70(B)
   Heating, Air-Conditioning and Refrigeration Equipment . . . 210.70(C)
   Motor Control Centers . . . . 110.26(D)
   Permanently Installed Pool Locations. . . . . 680.22(B)
   Service Equipment . . . . 110.26(D)
   Spa and Hot Tub Locations . . . . 680.43(B)
   Switchboards & Panelboards . . . . 110.26(D)
   See Life Safety Code, NFPA 101. . . . . NFPA 101
Loads for Non-Dwelling Occupancies 220.12

LIGHTING SYSTEMS, LOW VOLTAGE . . . . Art. 411
Branch Circuit .411.7
Class 2 Power Source (Connected to) 411.1
Definition. 411.3
Hazardous, (Classified) Locations 411.8
Listing Required 411.4
Location Requirements 411.5
Rating. 411.3
Scope. 411.1
Secondary Circuits
   Bare Conductors . . . . 411.6(C)
   Isolating Transformer Required. . . . . 411.6(B)
   Not to Be Grounded . . . . 411.6(A)
   Pools, Spas, and Fountains — Horizontal Clearance . 411.5(B)
   Walls, Floors, and Ceilings (Conductors Installed In) 411.5(A)

LIGHTING TRACK . . . . 410 Part XIV
Calculated Load for . . . . 220.43(B)
Connected Load . . . . .410.151(B)
Construction Requirements . . . . .410.155(A)
Definition. . . . . Art. 100
Fastening 410.154
Grounding .410.155(B)
Heavy-Duty Track 410.153
Installation . . . . 410.151(A)
Locations Not Permitted . . . . 410.151(C)
Support . . . . 410.151(D)
See (IFFR) . . . . UL Product iQ

LIGHTNING PROTECTION SYSTEM (GROUND TERMINALS)
Bonding to Grounding Electrode System 250.106

Irrigation Machines 675.15
Separation of Communications Wires 800.53
Use of Strike Termination Devices 250.60
See Standard for the Installation of Lightning Protection Systems. . . .
. . . . NFPA 780

LIGHTNING, SURGE ARRESTERS . . . . Art. 242
Antenna Discharge Units- Receiving Stations 810.20
Antenna Discharge Units- Transmitting Stations 810.57
Circuits Requiring Primary Protectors
    Communication Circuits . . . . 805.90(A)
        General Requirements for Protective DevicesArt. 805 Part III
        Installation of Conductors 805.50
Network-Powered Broadband Systems 830.90
    Grounding Methods . . . . Art. 830 Part IV
    Premises-Powered Broadband Communications Systems840.90
    Grounding Methods . . . . Art. 840 Part IV
Class I, Division 1 and 2 . . . . 501.35
Class II, Division 1 and 2 . . . . 502.35
Community Antenna TV . . . . Art. 820 Part III
Definition . . . . Art. 100 Part 1
Grounding Electrode Conductor in Metal Enclosures . . . 242.32
Installation . . . . 242.12(A)
Number Required 242.20
Selection of . . . . Art. 242
Services Over 1000 Volts 230.209
See Standard for the Installation of Lightning Protection Systems, NFPA 780 . . . . NFPA 780
See (OVGR) . . . . *UL Product iQ*
See . . . . *Soares Book on Grounding and Bonding*

LIMITED ACCESS WORKING SPACE . . . . . . 110.26(A)(4)

LIMITED CARE FACILITY
Definition of 517.2
Essential Electrical System 517.40
Wiring and Protection . . . . Art. 517 Part II

LIMITED POWER (LP) CABLES . . . . 725.179(I)
Transmission of Power and Data . . . . 725.144(B)

LINE ISOLATION MONITOR . . . . 517.160(B)
Definition 517.2
See(OWLS) . . . . *UL Product iQ*

LIQUIDTIGHT FLEXIBLE METAL CONDUIT: TYPE LFMC
. . . . Art. 350
Angle Connectors (Not Concealed) 350.42
Bends 350.24
    Number of 350.26
Bonding
    Equipment Bonding Jumper 350.60
    In Hazardous (Classified) Locations 250.100
    Required If Service Raceway . . . . 230.43(15)
    Where Used As Equipment Grounding Conductor . . 250.118(6)
Definition of 350.2
Grounding . . . . 250.118(6)
    Permitted As Grounding Means 350.60
Listing Requirements 350.6
Maximum and Minimum Sizes 350.20
Number of Wires in, General 350.22
    Number of Wires in 3/8" Size . . . . 350.22(B)
    Same Size Conductors, CompactInformative Annex C, Table C7A
    Same Size, in . . . . Informative Annex C, Table C7
Trimming (Rough Edges) 350.28
    6 ft. Maximum Length . . . . 350.30(A) Ex. No. 3 and 4
    Elevators, Dumbwaiters, Escalators, etc. . . . . . 620.21(A)(1)
    Enclosing Motor Leads . . . . 430.245(B)
Uses Not Permitted 350.12
Uses Permitted 350.10
    As Service and Feeders for Floating Buildings 555.52
    As Service Entrance . . . . 230.43(15)
See (DXHR) . . . . *UL Product iQ*

LIQUIDTIGHT FLEXIBLE NONMETALLIC CONDUIT . .
. . . . Art. 356
Angle Connectors (Not Concealed) 356.42
Bends 356.24
    Number of 356.26
Definitions
    Type LFNC-A . . . . 356.2(1)
    Type LFNC-B . . . . 356.2(2)
    Type LFNC-C . . . . 356.2(3)
Encasement in Concrete . . . . 356.10(7)
Equipment Grounding 356.60
Hazardous Location Installation . . . . 356.12(4)
Listing 356.6
Maximum and Minimum Size 356.20
Number of Wires in, General 356.22
    Type LFNC-A
        Same Size Conductors, Compact
        . . . . Informative Annex C, Table C6(A)

    Same Size, in . . . . . . Informative Annex C, Table C6

  Type LFNC-B

    Same Size Conductors, Compact . . . .
. . . . .Informative Annex C, Table C5(A)

    Same Size, in . . . . . . Informative Annex C, Table C5

  Type LFNC-C

    Manufacturer's information on internal area will have to be used to apply percent of cross-sectional fill area using Chapter 9 Table 1

Securing and Support. 356.30

Uses Not Permitted 356.12

Uses Permitted 356.10

  As Service Entrance . . . . 230.43(16)

  As Services and Feeders for Floating Buildings 555.52

  Class I, Division 2 Locations . . . . 501.10(B)(2)(4)

  Class II, Division 1 & 2 Locations .502.10(A)(2)(3) and (B)(2)

  Class III, Division 1 & 2 Locations . . . . . 503.10(A)(3)(3)

  Elevators, Dumbwaiters, Escalators, etc. 620.21

  Lengths Over 1.8 m (6 ft.) for Type LFNC-B . . . 356.10(5)

  Neon Secondary Conductors Over 1000 Volts . . 600.32(A)(1)

See (DXOQ) . . . . UL Product iQ

LISTED

Armored Cable: Type AC. 320.6

Assembly Type Cord . . . . 400.10(A)(11)

Audio System Equipment Near Bodies of Water . . . 640.10(B)

Circuit Breakers for SWD Duty . . . . 240.83(D)

Circuit Breakers for HID Duty . . . . 240.83(D)

Definition of . . . . Art. 100

Electrical Metallic Tubing. 358.6

Electrical Nonmetallic Tubing 362.6

Electric Discharge Lighting Systems of More Than 1000 Volts . . 410.140(A)

Electric Signs and Outline Lighting. 600.3

Electric Vehicle Charging Systems 625.5

Electrified Truck Parking Spaces, Various Equipment . . Art. 626

Electrodes. . . . . 250.52(A)(5)(b)

Enforcement of Code. 90.4

  Approval 110.2

  Examination . . . . 110.3(A)

  Installation and Use . . . . 110.3(B)

Examination of Equipment for Safety. 90.7

Extension Cord Sets. . . . . 240.5(B)(3)

Field Assembled Extension Cord Sets . . . . 240.5(B)(4)

Fixed Electric Space-Heating Equipment 424.6

Flat Conductor Cable. 324.6

Flexible Metal Conduit 348.6

  For Grounding . . . . .250.118(5)

  Flexible Metallic Tubing 360.6

  Fittings for Grounding . . . . .250.118(7)

  Fuel Cell Systems. 692.6

High Density Polyethylene Conduit 353.6

Intermediate Metal Conduit 342.6

Lighting Systems Operating at 30 Volts or Less 411.3

Liquidtight Flexible Metal Conduit. 350.6

  For Grounding. . . . . .250.118(6)

    Liquidtight Flexible Nonmetallic Conduit. 356.6

    Luminaires as Raceways . . . . 410.64(A)

Luminaires and Lampholders 410.6

Manufactured Wiring Systems 604.6

Mobile Homes . . . . 550.4(C)

Nonmetallic Auxiliary Gutters 366.6

Nonmetallic Extensions 382.6

Nonmetallic-Sheathed Cable 334.6

Nonmetallic Underground Conduit with Conductors . . .354.6

Nonmetallic Wireways 378.6

Optical Fiber Cables 770.179

Other Listed Electrodes — Chemical, Enhanced . . 250.52(A)(6)

Panelboards, Maximum Number of Overcurrent Devices 408.54

Reinforced Thermosetting Resin Conduit 355.6

Rigid Metal Conduit 344.6

Rigid Polyvinyl Chloride Conduit 352.6

Single Conductor Cables in Cable Trays . . . . 392.10(B)(1)(a)

Solar Photovoltaic Systems

  General Requirements . . . . 690.4(B)

  Over 1000 Volts 691.5

Storage Batteries (and Equipment) 480.3

Strut-Type Channel Raceway 384.6

Surface Metal Raceways 386.6

Surface Nonmetallic Raceways 388.6

Swimming Pool Double Insulated Pumps. . . . . . . 680.21(B)

Underground Feeder and Branch-Circuit Cable. 340.6

Wind Electric Systems . . . . 694.7(B)

LISTED & MARKED — ZONE LOCATIONS

Class I Locations .501.5

Class I, Zone 0, 1, and 2, General 505.9

Class I, Zone 0 . . . . 505.20(A)

Class I, Zone 1 . . . . 505.20(B)

Class I, Zone 2 . . . . 505.20(C)
Class II Locations 502.6
Class II, Zone 20, 21, and 22, General 506.9
    Zone 20 . . . . 506.20(A)
    Zone 21 . . . . 506.20(B)
    Zone 22 . . . . 506.20(C)
Class III Locations .503.6

LIVE PARTS
Appliances 422.4
Capacitors . . . . Art. 460 Part I
Definition of . . . . Art. 100 Part I
Equipment, Specific, Over 1000 Volts . . . . . Art. 490 Part III
Exposed 110.26
Generators 445.14
Guarding 110.27
Hazardous (Classified) Locations
    Class I, Division 1 & 2 . . . . 501.25
    Class II, Division 1 & 2 . . . . 502.25
    Class III, Division 1 & 2 . . . . 503.25
Luminaires, Lampholders, and Lamps 410.5
Lighting Systems, Electric Discharge
    1000 Volts or Less . . . . Art. 410 Part XII
    More than 1000 Volts . . . . Art. 410 Part XIII
Motion Picture and Television Studios, in. . . . . . . 530.15(A)
Motors and Controllers. . . . . Art. 430 Part XII
Over 1000 Volts, Work Space and Guarding . . 110.34(A)(D)(E)
    Outdoor Installations . . . . 110.31(C)
Theaters, in 520.7
Transformers, Guarding. . . . . .450.8(C)

LOADS
Appliances, Household Cooking 220.55
Branch Circuits
    Calculations . . . . Art. 220
    Continuous Loads on . . . . 210.20(A)
    Maximum 220.18
    Permissible 210.23
    Summary of . . . . Table 210.24
Continuous (Definition of). . . . . Art. 100 Part I
    Branch Circuits, on . . . . 210.20(A)
    Feeders 215.3
    Service-Entrance Conductors. . . . . 230.42(A)
    See CONTINUOUS LOAD . . . . Ferm's Finder
Demand

Appliances, 4 or More, Dwelling Units. 220.53
Clothes Dryers 220.54
Electric Heating 220.51
General Lighting 220.42
Household Cooking Appliances 220.55
Marinas and Boatyards 555.6
Mobile Homes 550.31
Motors 430.26
Receptacles, Other Than Dwellings . . . . .220.14(I)
Recreational Vehicles. 551.73
Stage Set Lighting 530.19
Use of Optional Calculations . . . . Art. 220 Part IV
See DEMAND FACTORS . . . . Ferm's Finder
Electric Cooking Appliances in Dwellings and Household Cooking Appliances Used in Instructional Programs 220.55
Electric Vehicle . . . . Art. 625
    Other Applicable Articles . . . . Table 220.3
Farm . . . . Art. 220 Part V
Feeder. Art. 220 Part III
Household Cooking Appliances Used in Institutional Programs .
. . . . 220.14(B)
Inductive
    Lighting . . . . 220.18(B)
    Signs, Disconnects . . . . 600.6(B)
    Switches, Snap Switch Ratings 404.14
    See NONLINEAR LOADS . . . . Ferm's Finder
    Kitchen Equipment, Other Than Dwelling Units . . 220.56
Laundry, Dwellings . . . . 220.52(B)
Lighting, Energy Code . . . . 220.12(B)
Marinas and Boatyards 555.6
Mobile Home Parks 550.31
Mobile Homes 550.18
Motors, Conductors . . . . .Art. 430 Part II
Motor Outlets . . . . 220.14(C)
Nonlinear (Definition of). . . . . Art. 100 Part I
See NONLINEAR LOADS . . . . Ferm's Finder
Other Loads, All Occupancies 220.14
    Banks and Office Buildings. . . . . 220.14(K)
    Dwelling Occupancies . . . . .220.14(J)
    Electric Dryers & Electric Cooking Appliances in Dwellings and Household Cooking Appliances Used in Instructional Programs
. . . . 220.14(B)
    Fixed Multioutlet Assemblies . . . . 220.14(H)
    Heavy-Duty Lampholders . . . . 220.14(E)
    Hotels and Motel Occupancies . . . . 220.14(M)

    Luminaires . . . . 220.14(D)

    Motor Outlets . . . . 220.14(C)

    Other Outlets . . . . 220.14(L)

    Receptacle Outlets . . . . .220.14(I)

    Show Windows . . . . 220.14(G)

    Sign and Outline Lighting . . . . 220.14(F)

    Specific Appliances or Loads . . . . 220.14(A)

Park Trailers 552.47

Permissible

    Branch Circuits 210.22

        Multiple-Outlet Branch Circuits 210.23

Recreational Vehicle Parks 551.73

Small Appliances, Dwellings . . . . 220.52(A)

Specific Appliances or Loads . . . . 220.14(A)

Stage Equipment 520.41

LOAD FACTOR . . . . *See* **Informative Annex B**

(For ampacities calculated under engineering supervision)

Application of Tables . . . . . Informative Annex B, 310.15(B)

Criteria Modifications . . . . . Informative Annex B, 310.15(C)

Examples Showing Use of Figure B-310.15(B)(2)(1) . . . .

. . . . Informative Annex B, 310.12

Interpolation ChartInformative Annex B, Fig. B., 310.15(B)(2)(1)

See Load Factor . . . . *Ferm's Charts and Formulas*

LOCKABLE DISCONNECTING MEANS 110.25

Air-Conditioning and Refrigerating Equipment 440.14

Appliances . . . . 422.31(B)

Carnivals, Circuses, Fairs and Similar Events

    Rides, Tents, Concessions 525.21

    Services 525.10

Cranes and Hoists . . . . 610.31 and 610.32

Electric-Discharge Lighting, More than 1000 Volts . 410.141(B)

Electric Vehicle Charging System 625.43

Electrically Driven or Controlled Irrigation Machines . 675.8(B)

Electrified Truck Parking Spaces

    Parking Space . . . . 626.24(C)

    Supply Wiring . . . . 626.22(D)

    Transport Refrigerated Units 626.31

Elevators, etc.

    Car Light, Receptacle(s), and Ventilation Disconnecting Means .

. . . . 620.53

    Disconnects, General 620.51

    Heating and Air-Conditioning Disconnecting Means . . . 620.54

    Utilization Equipment Disconnecting Means 620.55

Equipment Over 1000 Volts, Circuit Breakers 490.46

Equipment Over 1000 Volts, Interrupter Switches 490.44

Feeder Disconnecting Means . . . . 225.52(C)

Fire Pumps 695.4

Fixed Electric Space-Heating Equipment . . . 424.19(A) and (B)

Fixed Outdoor Electric Deicing and Snow-Melting Equipment . .

. . . . 426.51(A) and (D)

Generators 445.18

Induction and Dielectric Heating Equipment 665.12

Motors . . . . 430.102(A) and (B)

Motors, Over 1000 Volts 430.227

Motors with More Than One Source of Power 430.113

Outdoor Lamps 225.25

Sensitive Electronic Equipment, Lighting Equipment . . 647.8(A)

Signs . . . . 600.6(A)

Transfer Equipment .702.5

Transformers 450.14

LOCKER ROOMS, GFCI PROTECTION . . . . 210.8(B)(7)

LOCKNUTS (DOUBLE)

Bonding At Service, Not Permitted for . . . . 250.92(B)

Bonding Over 250 Volts to Ground 250.97

Continuity of Metal Raceways 300.10

Hazardous Areas (Bonding), Not Permitted for 250.100

    Class I Locations . . . . 501.30(A)

    Class II Locations . . . . 502.30(A)

    Class III Locations . . . . 503.30(A)

    Intrinsically Safe Systems . . . . 504.60(A)

    Zone 0, 1, and 2 Locations . . . . 505.25(A)

    Zone 20, 21, and 22 Locations . . . . 506.25(A)

Mobile Homes . . . . 550.15(F)

Park Trailers . . . . 552.48(B)

Recreational Vehicles . . . . 551.47(B)

Under Insulating Bushings . . . . 300.4(G)

LOW-VOLTAGE FIXED ELECTRIC SPACE-HEATING EQUIPMENT

Branch Circuits 424.104

Energy Sources 424.101

Installation 424.103

Listed Equipment 424.102

Scope 424.100

LOW VOLTAGE LIGHTING SYSTEMS . . . . Art. 411

(Note: operating at no more than 30 volts ac or 60 volts dc- see 411.1 Scope)

Branch Circuit	411.7
Class 2 Power Source (Connected to)	411.1
Hazardous, (Classified) Locations	411.8
Listing Required	411.4
Location Requirements	411.5
Pools, Spas, and Fountains — Horizontal Clearance	411.5(B)
Rating	411.3
Scope	411.1
Secondary Circuits	
Bare Conductors	411.6(C)
Isolating Transformer Required	411.6(B)
Not to Be Grounded	411.6(A)
Pools, Spas, and Fountains — Horizontal Clearance	411.5(B)
Walls, Floors, and Ceilings (Conductors Installed In)	411.5(A)

LOW-VOLTAGE SUSPENDED CEILING POWER DISTRIBUTION SYSTEMS

Circuits not to be Grounded	250.22(6)
Conductor Sizes and Types	393.104
Connections	393.57
Connectors	393.40(A)
Definitions	393.2
Disconnecting Means	393.21
Enclosures	393.40(B)
Grounding	393.60
Installation	393.14
Interconnection of Power Sources	393.45(B)
Listing Requirements	393.6
Overcurrent Protection	393.45(A)
Reverse Polarity	393.45(C)
Scope	393.1
Securing and Supporting	393.30
Splices	393.56
Uses Not Permitted	393.12
Uses Permitted	393.10

LOW-VOLTAGE SYSTEMS

Audio Signal Processing, Amplification & Reproducing	Art. 640
Circuits and Equipment Operating at Less Than 50 Volts	Art. 720
Class 1, 2 & 3 Systems	Art. 725
See CLASS 1, 2 & 3 REMOTE CONTROL CIRCUITS Ferm's Finder	
Communications Circuits	Art. 805
See COMMUNICATIONS CIRCUITS	Ferm's Finder
Community Antenna TV & Radio Distribution Systems	Art. 820
See COMMUNITY ANTENNA TV & RADIO DISTRIBUTION SYSTEMS	Ferm's Finder
Different Systems in Same Enclosure	
See CONDUCTORS, Different Systems in Same Enclosure	Ferm's Finder
Fire Alarm Systems	Art. 760
See FIRE ALARM SYSTEMS	Ferm's Finder
Grounding, AC Systems Less than 50 Volts	250.20(A)
Health Care Facilities, Equipment and Instruments	517.64
Lighting	Art. 411
Motor Control Circuits	Art. 430 Part VI
Network-Powered Broadband Communications Systems	Art. 830
Park Trailers	Art. 552 Part II
Premises-Powered Broadband Communications Systems	Art. 840
Radio & TV Equipment	Art. 810
Recreational Vehicles	Art. 551 Part II
Definition	551.2

LUGS & TERMINAL CONNECTIONS

Aluminum to Copper (Permitted if Listed)	110.14
Connection of Grounding and Bonding Equipment	250.8
Fine Stranded Conductors	110.14
Identification of Wiring Device Terminals	
Equipment Grounding Terminals	250.126
Receptacles, Adapters, Connectors, Plugs	406.10(B)
Grounded Conductor Connections	200.10
No Conductor Larger Than 10 AWG Under a Binding Screw	110.14(A)
Not More Than One Conductor Under a Binding Screw	110.14(A)
Solar Photovoltaic Wiring Systems	690.31(C)(4) and (C)(5)
Splices	110.14(B)
Suitable for the Purpose (Listed)	110.14

    Note: Use oxide inhibitor on aluminum conductor terminations where required.

Temperature Limitations	110.14(C)
See Aluminum Conductor Terminations	Ferm's Charts and Formulas
See Tightening Torque Information	Ferm's Charts and Formulas
See (DVYW)	UL Product iQ
See (ZMVV)	UL Product iQ

LUMINAIRES . . . . Art. 410

Access to Other Boxes (within ceilings)	410.118
As a Raceway	410.64

Ballast Type (Electric Discharge and LED)
   1000 Volts or Less . . . . Art. 410 Part XII
   Calculations, Inductive and LED Loads . . . . . 220.18(B)
   Conductors within 3" of a Ballast Must Be Rated Not Lower Than 90°C. 410.68
      Provision Table 310.4(A) 310.4
   Cord-Connected . . . . .410.62(B) & (C)
   For Signs and Outline Lighting . . . . Art. 600
   Luminaire Mounting. 410.136
   More than 1000 Volts . . . . Art. 410 Part XIII
   Thermal Protection. . . . . 410.130(E)
      High-Intensity Discharge Luminaires (Fixtures) 410.130(F)
   Voltage Limitations
      Branch Circuits 210.6
      Dwellings, Open-Circuit Voltage Exceeding 300 Volts. .410.135
      Lighting Equipment Installed Outdoors. 225.7
      Operating 1000 Volts or Less . . . . 410.130(A)
      Operating More than 1000 Volts . . . . . . 410.140(B)
   See Ballast Data Charts. . . . . . . *Ferm's Charts and Formulas*
Bathtub and Shower Areas . . . . 410.10(D)
Branch-Circuit Ratings to Luminaires . . . 210.19(A)(4) Ex. 1
   Maximum Load 220.18
   Overcurrent Protection. 210.20
   Permissible Loads. 210.23
   To Lampholders . . . . 210.21(A)
Breaker Rated "SWD" for Switching Fluorescent . . . 240.83(D)
Breaker Rated "HID" for Switching High Intensity Discharge . . . . 240.83(D)
   See (DIVQ) . . . . *UL Product iQ*
Calculations, General . . . . 220.14(D)
   Inductive and LED Loads, Maximum . . . . . . 220.18(B)
Canopies . . . . Art. 410 Part III
Clearance Required
   Bathtubs & Showers . . . . 410.10(D)
   Clothes Closet Light 410.16
   Hot Tubs and Spas, Indoors . . . . 680.43(B)
   Hot Tubs and Spas, Outdoors . . . . 680.22(B)
   Over Combustible Materials 410.12
   Recessed Luminaires, From Combustible Materials 410.116(A)(1) and (2)
   Recessed Luminaires, From Thermal Insulation . .410.116(B)
Clothes Closet (Restrictions) 410.16
Control Conductor Identification. 410.69
Cord-Connected . . . . 410.24(A)
   Adjustable Luminaires . . . . 410.62(B)

Fountain Luminaires. 680.51
Listed Electric-Discharge and LED Luminaires . . 410.62(C)
Storable Swimming Pool Luminaires. 680.33
Unit Equipment, Emergency Systems . . . . . . 700.12(F)
Unit Equipment, Legally Required Standby Systems 701.12(G)
See PENDANTS. . . . . *Ferm's Finder*
Covering of Combustible Material at Outlet Boxes 410.23
Crawl Space Outlets
   All Occupancies . . . . 210.70(C)
   Dwelling Units. . . . . 210.70(A)
   GFCI Protection . . . . 210.8(C)
Definition (Luminaire) . . . . Art. 100 Part I
Directly Controlled (Emergency Systems) 700.24
Disconnecting Means Required . . . . 410.130(G)
Electric-Discharge Lighting
   See (FIXTURES) LUMINAIRES, Ballast Type. . *Ferm's Finder*
Festoon Lighting
   See FESTOON LIGHTING. . . . . *Ferm's Finder*
(Fixture) Luminaire Wires
   See (FIXTURE) WIRE. . . . . *Ferm's Finder*
Taps
   See (FIXTURE) WHIPS. . . . . *Ferm's Finder*
Whips
   See (FIXTURE) WHIPS. . . . . *Ferm's Finder*
   See (ZIPR). . . . . *UL Product iQ*
Flush and Recessed Luminaires . . . . Art. 410 Part X
   Clearance Required. 410.116
   Supports to. 410.36
   Taps to Junction Box (Whip)
      See (FIXTURE) WHIPS . . . . *Ferm's Finder*
   Temperature Limits
      Combustible Materials. . . . . 410.115(A)
      Conductors in Outlet Boxes. 410.21
      Construction of Luminaires 410.119
      Near Combustible Materials. 410.11
      Where Recessed in Fire-Resistant Material . . .410.115(B)
Grounded (Neutral) Conductor Must Be Connected
   Polarization of Luminaires 410.50
   Screw Shell Lampholders 410.90
   Shell of Screw Shell Lampholders . . . . . 200.10(C) & (D)
Grounding . . . . Art. 250 Part VI
   Connected to an Equipment Grounding Conductor . 410.42
   Health Care, Patient Care Spaces. . . . . 517.13(B) Ex. 1 to 2
   Luminaires and Lighting Equipment. . . . . . Art. 410 Part V

Methods of Equipment Grounding	Art. 250 Part VII
Infrared Heating, Construction	422.48
Branch-Circuits, 40- and 50-Ampere Rating	210.23(C)
Branch-Circuit Requirements	424.3(A)
Overcurrent Protection	422.11(C)
Infrared Heating, Installation	425.14
Lamps Only (No Plug Fuses)	410.90
LED Luminaires, Closets	410.16
Grow Lights (plants)	Art. 410, Part XVI
Horticultural Lighting Equipment	Art. 410, Part XVI
Locations	Art. 410 Part II
Aircraft Hangars Class I Locations	513.4
Not Within Class I Locations	513.7
Commercial Garages Class I Locations	511.4
Over Class I Locations	511.7(B)
Corrosive	410.10(B)
Hazardous (Classified)	Art. 500
Anesthetizing Locations (Above)	517.61(B)
Anesthetizing Locations (Within)	517.61(A)
Class I, Division 1	501.130(A)
Class I, Division 2	501.130(B)
Class I, Zone 0, 1, and 2 Locations	Art. 505
Class II, Division 1	502.130(A)(1)
Class II, Division 2	502.130(B)
Class III, Division 1 & 2	503.130
Zone 20, 21, and 22 Locations	506.20
In Clothes Closets	410.16
In Ducts & Hoods, Commercial Cooking Hoods Only	410.10(C)
In Ducts for Dust, Loose Stock, or Vapor, Prohibited	300.22(A)
In Ducts or Plenums for Environmental Air	300.22(B)
In Show Windows	410.14
In Trees or Other Vegetation	410.36(G)
Marijuana and Agricultural Plants	Art. 410, Part XVI
Near Combustible Material	410.11
Other Space for Environmental Air	300.22(C)
Outdoors	225.7
Lampholders	410.96
Location Below Energized Conductors	225.25
Support by Trees	410.36(G)
Wet or Damp Locations	410.10(A)
Over 1000 Volts — Electric Discharge Type	Art. 410 Part XIII
Over 1000 Volts — Lockable Disconnecting Means	410.141(B)
Over Bathtubs	410.10(D)
Hydromassage Bathtubs	680.72
Mobile Homes	550.14(D)
Park Trailers	552.54(B)
Recreational Vehicles	551.53(B)
Over Combustible Material	410.11
Spas & Hot Tubs	
Indoors	680.43(B)
Outdoors	680.22(B)
Outdoors (General)	680.22(B)
Swimming Pools	680.22(B)
Underwater	680.23
Therapeutic Pools & Tubs	
General	680.60
Permanently Installed	680.61
Tubs (Hydrotherapeutic Tanks)	680.62(F)
Underwater	680.23
Wet & Damp	410.10(A)
See Hazardous Areas (Defined)	*Ferm's Charts and Formulas*
See HAZARDOUS (CLASSIFIED) LOCATIONS	Ferm's Finder
Low-Voltage at Swimming Pools	680.22(B)(6)
Low Voltage Lighting	Art. 411
Mechanical Strength of Luminaires and Parts	Art. 410 Part VII
Medium Base	
Branch Circuits	210.23(A)
Not Permitted in Clothes Closets	410.16(B)
Medium Base Lampholder	
(Not Over 120 Volts Between Conductors)	210.6(B)
Mogul Base Lampholder	
(Not Over 277 Volts to Ground)	210.6(C)
Outdoors	225.7
Multiwire Branch Circuits, Disconnects Required	410.130(G)(2)
Not Over 120 Volts (Dwellings)	210.6(A)
Not Over 600 Volts Between Conductors	210.6(D)
Outdoor Lighting	225.7
Outlet Box must be Accessible Under	
Surface-Mounted Electric-Discharge and LED Luminaire	410.24(B)
Suspended Ceiling, Securely Fastened	410.36(B)
Outline Lighting (and Electric Sign)	Art. 600
1000 Volts or Less	Art. 410 Part XII
More Than 1000 Volts	Art. 410 Part XIII
Overcurrent Protection	
Ampacity of Fixture Wires	Table 402.5
Branch-Circuit Ratings	210.19(A)(4)
For Conductors, Luminaire (Fixture) Wires	210.20(B)

For Luminaire (Fixture) Wires 240.5
For Lampholders. . . . . 210.21(A)
For Rated Ampacity . . . . Table 402.5
Outlet Device Ratings . . . . 210.21(A)
Outlet Devices . . . . 210.20(D)
Permissible Load 210.23
Protection of Conductors . . . . 240.5(B)(2)
Pendants
   Aircraft Hangars . . . . 513.7(B)
   Commercial Garages . . . . 511.7(A)(2)
   Conductors for Incandescent Lamps 410.54
   Hazardous (Classified) Locations
      Class I, Division 1 Locations . . . . 501.130(A)(3)
      Class I, Division 2 Locations . . . . 501.130(B)(3)
      Class II, Division 1 Locations . . . . 502.130(A)(3)
      Class II, Division 2 Locations . . . . 502.130(B)(4)
      Class III, Division 1 & 2 Locations . . . . . . .503.130(C)
   Hospitals (Hazardous Anesthetizing Locations) 517.61(B)(3) Ex. 2
   Not in Clothes Closets . . . . 410.16(B)
   Not in Theater Dressing Rooms 520.71
   Not Over Bathtubs. . . . . 410.10(D)
   Show Windows. 410.14
Polarization
   Identification of Terminals . . . . .200.10(C) & (D)
   Installation of Lampholders 410.90
   Of Luminaires (Fixtures) and Lampholders 410.50
Retrofit Kits
   Defined . . . . Art. 100
   Listing Requirements for Luminaires, Lampholders and Lamps 410.6
   Reconditioned Equipment 410.7
Roof Decking
   Installed Under. . . . . 410.10(F)
Signs & Outline Lighting . . . . 600.4(B)
Supports . . . . Art. 410 Part IV
   Boxes for . . . . 314.27(A)
   Ceiling-Suspended (Paddle) Fans, Including Lights 314.27(C)
   Class I, Division 1 Locations . . . . 501.130(A)(4)
   Class II, Division 1 Locations . . . . 502.130(A)(4)
   Swimming Pool Areas . . . . 680.22(B)
Taps
   Armored Cable (Type AC) . . . . 320.30(D)
   Conductor Sizes and Ampacity . . . . 210.19(A)(4)
   Flexible Metal Conduit. . . . . 348.30(A) Ex. 3
   Flexible Metallic Tubing . . . . 360.20(A) Ex. 2
   Length of. . . . . .410.117(C)
   Liquidtight Flexible Metal or # Conduit . . . 350.30(A) Ex. 3
   Liquidtight Flexible Nonmetallic Conduit . . . . . 356.30(2)
      Size . . . . .356.20(A)(2)
   Manufactured Wiring Systems . . . 604.100(A)(1) & (2) Ex. 1
   Metal-Clad Cable (Type MC) . . . . . 330.30(D)(2) and (3)
   Nonmetallic-Sheathed Cable . . . . .334.30(B)(2)
   Protection . . . . 240.5(B)(2)
Temperature Limits
   On Conductors in Outlet Boxes 410.21
   Special Provisions for Flush and Recessed 410.115
Track Lighting
   See LIGHTING TRACK . . . . Ferm's Finder
Used as a Raceway 410.64
Voltage Limitations
   Not Over 120 Volts Between Conductors (Dwellings)
   . . . . .210.6(A) & (B)
   Not Over 150 Volts Between Conductors
      Swimming Pool Underwater Lights . . . . . . 680.23(A)(4)
      Low Voltage Contact Limit, Definition 680.2
      Fountain Luminaires . . . . 680.51(B)
      GFCI Protection, Relamping . . . . .680.23(A)(3)
      General Compliance. . . . . .680.23(A)(8)
      Junction Box Location. . . . . 680.24(A)(2)
      Storable Pool Luminaires . . . . .680.33(A)&(B)
   Not Over 277 Volts to Ground. . . . . 210.6(C)
      Lighting Equipment Outdoors . . . . 225.7(C)
   Not Over 600 Volts Between Conductors . . . . . 210.6(D)
      Lighting Equipment Outdoors . . . . 225.7(D)
Wet & Damp Locations . . . . 410.10(A)
   To Be Weatherproof Type 410.96
Wire within 3" of a Ballast Must Be Rated Not Lower Than 90°C 410.64
Wired Luminaire Sections . . . . 410.137(C)
Wires, Luminaires (Fixture)
   See (FIXTURE) LUMINAIRE WIRES . . . . Ferm's Finder
Wiring of . . . . Art. 410 Part VI

LUMINAIRES FOR POOLS . . . . **Art. 680**
   See SWIMMING POOLS . . . . .Ferm's Finder
   See (WBDT) . . . . *UL Product iQ*

M

MACHINE ROOMS AND SPACES, ELEVATOR, ESCALATOR, ETC. . . . . Art. 620 Part VIII
Branch Circuit for Lighting and Receptacles 620.23
Flexible Cords Permitted . . . . . 620.21(A)(2)(d)(4) and (3)(e)
Guarding Equipment 620.71
Wiring in . . . . 620.21(A)(3)

MACHINE SCREWS (32 THREADS PER INCH)
Covers and Canopies 314.25
Receptacles 406.5
Switches. . . . . 404.10(B)

Note: The use of sheet metal screws or other screws not listed for devices, covers or canopies are not allowed.

MACHINE TOOL WIRE
Ampacity of . . . . 310.14(A)(1) Info. Note 2
See *Electrical Standard for Industrial Machinery, NFPA 79* NFPA 79
See (ZKHZ). . . . . UL Product iQ

MANDATORY RULES . . . . 90.5(A)

MAIN BONDING JUMPER 250.28
More Than One Service Enclosure . . . . 250.28(D)(2)
   Sizing. . . . . 250.28(D)
   When Used at Service Panels . . . . 250.24(B)
   Industrial Control Panels 409.108
   Motor Control Centers 430.95
   Switchboards and Panelboards . . . . 408.3(C)
   Table for Sizing . . . . Table 250.102(C)(1)

MANHOLES and OTHER ELECTRIC ENCLOSURES INTENDED FOR PERSONNEL ENTRY . . . .
Art. 110 Part V
Access to Vaults and Tunnels 110.76
Access to 110.75
Cabling Work Space 110.72
Conductor Bending Space 110.74
Conductors Racked, Subsurface Enclosures 110.74
   1000 Volts, Nominal, or Less . . . . 110.74(A)
   Over 1000 Volts, Nominal . . . . 110.74(B)
Equipment Work Space 110.73
General Requirements 110.70
   Using Appropriate Engineering Requirements and Practices . . 110.71 Info. Note

Note: *National Electrical Safety Code,* ANSI C-2 contains design and installation requirements applicable to this type of enclosure.

Over 1000 Volts, Conductors of Different Systems . . . 300.3(C)
Separation of Class 1 and Power Supply Circuits . . 725.48(B)(3)
Separation of Class 2 & 3 Circuits . . . . 725.136(A)&(F)

MANUFACTURED BUILDING . . . . Art. 545
Bonding and Grounding 545.11
Boxes 545.9
Component Interconnections 545.13
Definition of 545.2
Grounding Electrode Conductor Provisions for 545.12
Protection of Conductors and Equipment 545.8
Receptacle or Switch with Integral Enclosure 545.10
Relocatable Structures. . . . . Art. 545 Part II
Service Entrance Conductors
   As Service Conductors 545.5
      Installation 545.6
Service Equipment Location 545.7
Wiring Methods 545.4

MANUFACTURED HOMES . . . . Art. 550
Appliances 550.14
Arc-Fault Protections 550.25
Branch Circuits 550.12
Calculations. 550.18
Definition. 550.2
Disconnecting Means 550.11
Feeders 550.33
Ground-Fault Circuit Interrupters . . . . 550.13(B)
Grounding 550.16
Listed and Labeled . . . . 550.4(C)
Luminaires 550.14
Not Intended as a Dwelling Unit . . . . 550.4(A)
Power Supply 550.10
Receptacle Outlets 550.13
Service Equipment . . . . 550.32(A)
Testing 550.17
Wiring Methods 550.15

2020 Ferm's Fast Finder | 133

See MOBILE HOMES . . . . Ferm's Finder

MANUFACTURED WIRING SYSTEMS . . . . Art. 604
Construction, Cable or Conduit Types . . . . .604.100(A)
Definition of 604.2
Installation 604.7
Listing Requirements 604.6
Uses Not Permitted 604.12
Uses Permitted 604.10
   Of Flexible Cord As Part of Listed Assembly . 604.100(A)(3)
   See (QQVX) . . . . UL Product iQ

MANUFACTURER'S MARKING . . . . 110.21(A)

MARINAS & BOATYARDS . . . . Art. 555
Bonding Small Parts 555.13
Branch Circuits . . . . 555.34(B)(2)
Cable, Marina and Boatyard
   See (PDYQ) . . . . UL Product iQ
Disconnecting Means. 555.36
Dwellings (applicable) 555.1
Electrical Connections, Decks and Docks. 555.30
Equipment Grounding Conductor, Type of Conductor 555.37(B)
Feeder and Service Demand Factors. 555.6
Gasoline (Motor Fuel) Dispensing Stations
   Classification of Locations . . . . 514.3(C)
   Hazardous (Classified) Locations 555.11
Ground-Fault Circuit-Interrupter Protection . . . . 555.33(B)(1)
Ground-Fault Protection of Equipment- Boatyards 555.35
Ground-Fault Protection of Equipment- Floating Buildings 555.53
Ground-Fault Protection of Equipment- Marinas 555.35
Grounding
   Branch-Circuit Equipment Grounding Conductor 555.37(D)
   Equipment Grounding Conductor, Insulated Conductor . . .
   . . . . 555.37(B)
   Equipment Grounding Conductors, Size. . . . . . 555.37(C)
   Equipment to Be Grounded . . . . 555.37(A)
   Feeder Equipment Grounding Conductors . . . 555.37(E)
Leakage Current Measurement . . . . 555.35(B)
Motor Fuel Dispensing Stations
   Classification of Locations . . . . 514.3(C)
   Hazardous (Classified) Locations. 555.11
Occupancy Types .555.1
Receptacles, Requirements for 555.33
Service Equipment Location 555.4

Signage 555.10
Swimming (risk of) 555.10
Transformers 555.7
Wiring Methods 555.34
Wiring Over and Under Navigable Water. . . . . . 555.34(B)(1)
See FLOATING BUILDINGS . . . . Ferm's Finder
See Fire Protection Standard for Marinas and Boatyards, NFPA 303

MARKING AND LABELING REQUIREMENTS
Agricultural Buildings, Distribution Point . . 547.9(A)(10)&(D)
Arc-Flash Hazard Warning 110.16
Available Fault Current 110.24
Cable Trays, Over 600 Volts . . . . 392.18(H)
   Industrial Establishment . . . . 392.18(H) Ex.
Cords
   Optional . . . . 400.6(B)
   Standard . . . . 400.6(A)
Critical Operations Power Systems, Receptacle Identification . . . 708.10(A)(2)
Direct-Current Ground-Fault Detection 250.167
Directory, Interconnected Electric Power 705.10
Disconnecting Means, General . . . . 110.22(A)
Disconnecting Means, Engineered Series . . . . . . 110.22(B)
Disconnecting Means, Tested Series . . . . 110.22(C)
Disconnecting Means (Wind) . . . . 694.22(A)(4)
Electric-Discharge Lighting Systems of More Than 1000 Volts . .
. . . . 410.146
Electrolytic Cells, Isolating Transformer Receptacles . 668.21(C)
Field-Applied Hazard Marking . . . . 110.21(B)
Fire Alarm Systems, Branch Circuit Identification . 760.41(B) &
. . . . .760.121(B)
Fire Pump, Disconnecting Means. . . . . 695.4(3)(c)
Fuel Cells, Stand Alone 692.53
Fuel Cells, Stored Energy 692.56
Fuel Cells, Switch and Circuit Breakers 692.17
Generators
   Neutral Bonded to Frame, Manufacturer Marking . . . 445.11
   Neutral Bonded to Frame, Field Modified Marking . . 445.11
Health Care Facilities, Receptacle Identification . 517.31(E) and
. . . . 517.42(E)
High-Impedance Grounded Neutral . . . . 408.3(F)(3)
High Leg Marking 110.15
Identification, Grounded Conductor . . . . 210.5(A)
Identification of Power Sources (PV) 690.56
Identification of Power Sources (Wind) 694.54

Inverter Output Connection	705.12(B)(3)
Marinas and Boatyards, Other Than Shore Power Receptacles	555.33(B)
Marking, Boxes and Conduit Bodies	314.44
Marking, Circuit Breakers	240.83
Marking, Conductors	310.8
Marking, Field Applied Hazard Markings	110.21(B)
Marking, Fuel Cells	Art. 692 Part VI
Marking, Fuses	240.60(C)
Marking, Industrial Control Panels	409.110
Marking, Luminaries	410.74
Marking, Lamp Wattage for Luminaries	410.120
Marking, Manufacturer	110.21(A)
Marking, PV Systems	Art. 690 Part VI
Marking, Manhole Covers	110.75(E)
Marking, Panelboards	408.58
Marking, Product	110.21(A)
Marking, Switches	404.20
Marking, Transformer Disconnects	450.14
Marking, Wind Systems	Art. 694 Part VI
Mobile Home, Service Equipment, Receptacle	550.32(G)
Natural and Artificially Made Bodies of Water	682.14(A)
Outdoor Deicing and Snow Melting Equipment	426.13
Phase Converters, Conductors	455.6(B)
Recreation Vehicle, Supply Equipment	551.77(F)
Service Equipment, Disconnecting Means	230.70(B)
Service Equipment, Supply Side	230.82
Stand Alone System (PV)	710.10
Substations	490.48
Switches	404.20
Surface Metal Raceways	386.120
Surface Nonmetallic Raceways	388.120
Transformer	
General	450.11(A)
Source Marking (Reverse Feeding)	450.11(B)
Ungrounded Systems	250.21(C)
Wind Systems, Disconnecting Means	694.22(C)(2)

MAST-TYPE SUPPORTS

CATV Aerial Coaxial Cable Support	800.44(C)
Outside Feeders and Branch Circuits Support	225.17
Power Service-Drop Support	230.28
Radio and TV Equipment	810.12

MEANS OF EGRESS, ILLUMINATION

Hospitals	517.33(A)
Nursing Homes and Limited Care Facilities	517.43(A)
See Life Safety Code, NFPA 101	NFPA 101

MECHANICAL EXECUTION OF WORK

Cables and Conductors	110.12(C)
Circuits and Equipment Less Than 50 Volts	720.11
Class 1, 2 and 3 Remote Control, Signaling, Etc.	725.24
Communications Circuits	800.24
Community Antenna Television and Radio Distribution	800.24
Fire Alarm Systems	760.24
Optical Fiber Cables and Raceway	770.24
Workmanship	110.12

MEDIUM VOLTAGE CONDUCTORS AND CABLE   Art. 311

Ampacities	Art. 311 Part IV
Ampacity Tables	311.60
Conductors	311.12
Construction and Applications	311.10
Definitions	311.2
Direct-Burial Conductors	311.36
Installation	311.30
Listing	311.6
Marking	311.16
Shielding	311.44
Support	311.40
See (PITY)	*UL Product iQ*
Uses Permitted	311.32

MEETING ROOMS

Fixed Wall Receptacle Outlets	210.65(B)(1)
Floor Receptacle Outlets	210.65(B)(2)
Square Foot Requirements	210.65(A)

MESSENGER SUPPORTED WIRING . . . . Art. 396

Ampacity	310.14
Conductor Splices and Taps	396.56
Equipment Grounding Conductor	396.30(C)
Insulated Conductor, Definition	396.2
Messenger Grounding	396.60
Messenger Support	396.30
Neutral Conductor	396.30(B)
Uses Not Permitted	396.12
Uses Permitted	396.10
Cable Types	396.10(A)
Hazardous Locations	396.10(C)
In Industrial Establishments	396.10(B)

METAL BOXES
   See BOXES AND FITTINGS, Metallic Boxes . . . Ferm's Finder
   See (QCIT) . . . . UL Product iQ
   See (BGUZ). . . . . UL Product iQ

METAL DUSTS & POWDERS (Combustible)
   Class II Group Classification . . . . 500.6(B)(1)
   Class II Locations . . . . 500.5(C)
   Enclosures Specifically Approved for Locations . . . . 502.115(A)
   Requirements for Class II Locations . . . . Art. 502
   Transformers and Capacitors Prohibited . . . . . 502.100(A)(3)
   See HAZARDOUS (CLASSIFIED) LOCATIONS Ferm's Finder

METAL FRAME OF BUILDING
   Bonding . . . . 250.104(C)
   Equipment Grounding Conductor, Not Permitted as 250.136(A)
   Equipment Grounding Conductor, Restricted . . . . 250.121(B)
   Grounding Electrode . . . . 250.52(A)(2)
      For Separately Derived Systems . . . . 250.104(D)

METAL HALIDE LAMPS
   Containment requirements . . . . 410.130(F)(5)

METAL IN-GROUND SUPPORT STRUCTURE(S) 250.52(A)(2)

METAL PARTS IN VICINITY OF TRANSFORMERS 450.10(B)

METAL PIPING
   Bonding . . . . 250.30(A)(8)
   Gas Piping . . . . 250.104(B)
   Other Metal Piping . . . . 250.104(B)
   Water . . . . 250.104(A)

METAL RACEWAYS (MAINTAIN CONTINUITY)
      See CONDUIT, Continuity of Run . . . . Ferm's Finder
      See RACEWAYS . . . . Ferm's Finder

METAL SIDING, GROUNDING OF . . 250.116 (Info. Note)

METAL STRUCTURES, OVER 1000 VOLTS . . . . 250.194(B)

METAL SURFACE RACEWAY . . . . Art. 386
   See SURFACE METAL RACEWAY . . . . Ferm's Finder

METAL UNDERGROUND WATER PIPE
   As a Grounding Electrode . . . . 250.52(A)(1)

METAL WATER PIPE AND STRUCTURAL STEEL 250.68(C)

METAL WELL CASINGS
   Grounding Electrode . . . . 250.52(A)(8)
   Grounding Equipment . . . . 250.112(M)

METAL WIREWAYS, See WIREWAYS, METAL . . Ferm's Finder

METAL WORKING MACHINE TOOLS . . . . Art. 670
   Machine Tool Wire, Type MTW . . . . Table 310.4(A)
   See Electrical Standard for Industrial Machinery . . . . . NFPA 79
   See INDUSTRIAL MACHINERY . . . . Ferm's Finder
   See (ZKHZ) . . . . UL Product iQ

METAL-CLAD CABLE, TYPE MC . . . . Art. 330
   Ampacity 330.80
      Conductors Rated 0-2000 Volts 310.14
      Conductors Rated 2001 to 35,000 Volts 311.60
      Of Sizes 18 AWG and 16 AWG 330.80
      Where Installed in Cable Tray . . . . 330.80(A)
   Bending Radius 330.24
      Interlocked-Type Armor or Corrugated Sheath. . . 330.24(B)
      Shielded Conductors . . . . 330.24(C)
      Smooth Sheath . . . . 330.24(A)
   Exposed Work 330.15
   Grounding 330.108
   Hazardous Locations 330.130
   Installation . . . . 330 Part II
   Insulated Conductor Types 330.112
   Listing Requirements 330.6
   Support of 330.30
   Uses Not Permitted 330.12
   Uses Permitted 330.10
      Aircraft Hangars . . . . 513.7(A)
      Bulk Storage Plants. . . . . 515.7(A)
      Cable Trays, in 392.10
         Rated Over 1000 Volts . . . . 392.18(H)
      Class I, Division 1 Locations (Specific Restrictions) . . . . . .
      . . . . 501.10(A)(1)(c)
      Class I, Division 2 Locations . . . . 501.10(B)(1)(6)
      Class II, Division 1 Locations (Specific Restrictions) . . . . .
      . . . . 502.10(A)(1)(3)
      Class II, Division 2 Locations . . . . 502.10(B)(1)(3)
      Class III, Division 1 & 2 Locations . . . . . 503.10(A)(1)(4)
      Commercial Garages . . . . 511.7(A)(1)
      Cranes and Hoists 610.11
      Ducts or Plenums Used for Environmental Air. . . 300.22(B)
      Elevators, Dumbwaiters, Escalators, etc. 620.21

Equipment Grounding Conductor, As . . . . . . 250.118(10)

Health Care Facilities, Patient Care Spaces . . . . . 517.13(A)

    Outer Metal Armor to Be Identified As Ground Return Path . . . . 517.13(A)

    See Grounding Use (PJAZ) . . . . *UL Product iQ*

Information Technology Equipment, Raised Floors 645.5(E)(2)

Manufactured Wiring Systems . . . . .604.100(A)(2)(3)

Messenger Supported Wiring, As. . . . . . . Table 396.10(A)

Motion Picture and TV Studios 530.11

Motor Fuel Dispensing Facilities 514.4

Other Spaces Used for Environmental Air . . . .300.22(C)(1)

Outside Branch Circuits and Feeders. 225.10

Places of Assembly . . . . 518.4(A)

Safety-Control Equipment, Class 1, Wiring . . . . 725.31(B)

Service-Entrance Conductors. . . . . .230.43(13)

Spray Application, Dipping, and Coating Processes . 516.7(A)

Swimming Pool Motors . . . . .680.21(A)(1)

Theaters, Audience Areas of Motion Picture & TV . . 520.5(A)

See (PJAZ) . . . . *UL Product iQ*

METER FITTINGS

    See (PJVV) . . . . *UL Product iQ*

METER MOUNTING EQUIPMENT

    See (PJSR) . . . . *UL Product iQ*

METER SOCKETS . . . . **Art. 312**

    Conductors Entering Enclosures 312.5

    Damp, Wet, or Hazardous (Classified) Locations 312.2

    Deflection of Conductors 312.6

    Disconnects . . . . 230.82(3)

    Grounding 250.142

    Individual, Not Service Equipment 230.66

    Supply Side of Service Disconnect . . . . 230.82(2)

    Supply Side of Service Overcurrent Devices . . . . 230.94 Ex. 5

    See (PJYZ) . . . . *UL Product iQ*

METERING TRANSFORMER CABINETS . . . . . **Art. 312**

    See METER SOCKETS . . . . .Ferm's Finder

    See (PJXS) . . . . *UL Product iQ*

METERS

    Class I Locations 501.105

    Grounding 250.142

        Cases of, at 1000 Volts and Over 250.176

        Cases of, at 1000 Volts or less 250.174

Supply Side of Service Disconnect . . . . 230.82(2)

Supply Side of Service Overcurrent Devices. . . . . 230.94 Ex. 5

See (PJSR) . . . . *UL Product iQ*

Note: Type, location & mounting height should always be checked with the serving agency.

METRIC SYSTEM **90.9**

Metric Designators and Trade Sizes . . . . 300.1(C)

Numerical Designation for

    Electrical Metallic Tubing . . . . . . 358.20(B) (Info. Note)

    Flexible Metal Conduit. . . . . 348.20(B) (Info. Note)

    Flexible Metallic Tubing . . . . . . 360.20(B) (Info. Note)

    Intermediate Metal Conduit . . . . . 342.20(B) (Info. Note)

    Liquidtight Flexible Metal Conduit . 350.20(B) (Info. Note)

    Liquidtight Flexible Nonmetallic Conduit 356.20(B) (Info. Note)

    Nonmetallic Underground Conduit with Conductors 354.20(B) (Info. Note)

    Rigid Metal Conduit . . . . 344.20(B) (Info. Note)

    Rigid PVC Conduit . . . . 352.20(B) (Info. Note)

    RTRC Conduit, Fiberglass . . . . . . 355.20(B) (Info. Note)

Threaded Conduit, Hazardous (Classified) Locations 500.8(E)(2)

See METRIC SYSTEM, Measurements and Conversions. Ferm's Finder

MINERAL-INSULATED, METAL-SHEATHED CABLE Art. 332

Ampacity of 332.80

Ampacity of 310.14

Bends 332.24

Conductor Material. 332.104

Emergency Systems . . . . 700.10(D)(1)

Fire Pumps 695.6

Fittings . . . . 332.40(A)

Insulation Material 332.112

Listing Requirements 332.6

Outer Sheath 332.116

    For Equipment Grounding Purposes 332.108

Places of Assembly . . . . 518.4(A)

Single Conductors 332.31

Supports 332.30

Terminal Seals. . . . . 332.40(B)

Through Joists, Studs or Rafters . . . . 332.30(A)

Uses Not Permitted 332.12

Uses Permitted 332.10

Wet Locations . . . . 332.10(3)

See (PPKV) . . . . *UL Product iQ*

MOBILE HOMES . . . . Art. 550

Note: The term "mobile home" includes "manufactured homes" that are similar structure(s) designed to be used with or without a permanent foundation.

See Definitions of both terms in 550.2

Appliances
   Accessibility . . . . 550.14(B)
   Fastened During Transit . . . . 550.14(A)
   Grounding Frames of 550.16
Branch Circuits 550.12
Calculations
   Park Electrical Wiring System 550.30
     Demand Factors . . . . Table 550.31
   Supply-Cord and Distribution Panelboard Load . . 550.18(B)
Definitions 550.2
Disconnecting Means . . . . 550.11(A)
   Mounting Height, Outdoor Disconnecting Means 550.32(F)
Disconnecting Means and Branch-Circuit Protection 550.11(B)
Distribution System. 550.30
   Demand Factors 550.31
Feeder (Four Insulated Conductors) . . . . 550.10(I)(1)
   Capacity, Minimum Rating 100 Amperes . . . . . 550.33(B)
   Minimum Size Permitted 310.12
   Size. . . . . 215.2(A)(3)
   Type, Identification, Equipment Ground. . . . . . 550.33(A)
Ground-Fault Circuit-Interrupter Protection . . . . . 550.13(B)
   For Pipe Heating Cable Outlet, Where Installed . . 550.13(E)
   Receptacles At Service Equipment . . . . 550.32(E)
Grounding & Bonding 550.16
   Bonding of Non-Current-Carrying Metal Parts . . 550.16(C)
   Equipment Grounding Means . . . . 550.16(B)
   Insulated Grounded Conductor (Neutral) . . . . . 550.16(A)
Interconnection of Multiple-Section Mobile Home Units 550.19
Listing and Labeling . . . . 550.4(C)
Neutral Must Be Isolated from Enclosure in . . . . . 550.16(A)
Not Intended as a Dwelling Unit, Requirements for . . 550.4(A)
Outdoor Outlets, Luminaires (Fixtures), Air-Cooling Equipment, etc. 550.20
Power Supply . . . . 550.10(B)
   Equipment Grounding. 550.16
   Feeders. 550.33
Receptacle Outlets, Where Required or Permitted . . . 550.13
   At Service Equipment . . . . 550.32(D)
   Pipe Heating Cable Outlet . . . . 550.13(E)
Relocatable Structures. . . . . .see Art. 545, Part II
Services and Feeders. . . . . Art. 550 Part III
Service Equipment, Manufactured Home. . . . . . . 550.32(B)
Service Equipment . . . . 550.32(A)
   Ground-Fault Circuit-Interrupter Protection. . . . 550.32(E)
   Marking of 125/250-Volt Receptacle Used in . . . 550.32(G)
   Minimum Size Permitted . . .310.12, 230.79(C), 550.32(C)
   Rating of. . . . . 550.32(C)
   Size. . . . . 215.2(A)(3)
Testing
   Continuity, Operational, and Polarity Checks . . . 550.17(B)
   Dielectric Strength . . . . 550.17(A)
Wiring Methods & Materials 550.15
   Boxes, Fittings, and Cabinets . . . . 550.15(I)
   Component Interconnections . . . . 550.15(K)
   For Ranges, Clothes Dryers, or Similar Appliance . 550.15(E)
   Metal and Nonmetallic Cable Protection through Studs .550.15(C)
   Nonmetallic Cable Protection Where Exposed . . 550.15(B)
   Ratings of Switches. . . . . 550.15(G)
   Under Chassis Wiring . . . . 550.15(H)

MOBILE X-RAY EQUIPMENT . . . . Art. 660
Connection to Supply Circuit. . . . . 660.4(B)
Definition. 660.2
Grounding 660.48
Manual Control Device. 660.21

MODULAR DATA CENTERS
Cable Routing Assemblies. . . . . 646.3(F)
Definitions 646.2
Emergency Lighting Circuits 646.17
Equipment 646.10
Field-Wiring Compartments 646.8
Flexible Power Cords and Cables 646.9
Installation and Use. 646.14
Lighting
   Emergency 646.16
   General Illumination. 646.15
Listing 646.4
Nameplate Data. 646.5
Other Articles. 646.3
Overcurrent Protection 646.6
Plenums. . . . . 646.3(B)
Receptacles 646.12
Scope. 646.1
Short-Circuit Current Rating 646.7

Supply Conductors .646.6

Transformers 646.11

Working Space for ITE 646.20

Workspace

    Battery Installations 646.21

    Entrance to and Egress 646.19

    General. 646.18

    Service and Maintenance 646.22

MODULE, SOLAR PHOTOVOLTAIC (Definition of) . . 690.2

Module use in Solar Photovoltaic Devices . . . . Art. 690

MOGUL BASE LAMPHOLDERS

Branch-Circuit Voltage Limitations . . . . 210.6(C)(3)

Ratings 410.103

MOTELS AND HOTELS

AFCI Requirements. . . . . 210.12(C)

General Lighting Load . . . . Table 220.12

Ground-Fault Circuit-Interrupter Requirements . . . . 210.8(A)

    For Other Than Dwelling Units . . . . 210.8(B)

Lighting Load Demand Factors . . . . Table 220.42

Receptacle Load Demand Factors, Other Than Dwelling Unit . . . . Table 220.44

Receptacle Outlets, General . . . . 210.60(A)

Receptacle Outlets, Required 210.52

Receptacle Outlets, Placement of . . . . 210.60(B)

Tamper-Resistant Receptacles . . . . 406.12(2)

See Life Safety Code, NFPA 101 . . . . NFPA 101

MOTION PICTURE PROJECTORS . . . . Art. 540

Audio Signal Equipment, etc. 540.50

Definitions 540.2

    Nonprofessional Projector 540.2

    Professional-Type Projector. 540.2

Nonprofessional Projectors . . . . Art. 540 Part III

    Listing of Equipment. 540.32

    Projection Room Not Required If Safety Film Used . . 540.31

Professional Type Equipment . . . . .Art. 540 Part II

Emergency System Control . . . . 540.11(C)

Listing of Equipment 540.20

Location of Associated Electrical Equipment . . . . . 540.11(A)

Projection Room Required 540.10

Work Space 540.12

MOTION PICTURE & TELEVISION STUDIOS . . Art. 530

Cellulose Nitrate Film Storage Vaults . . . . . . .Art. 530 Part V

Definitions 530.2

Dressing Rooms. . . . . Art. 530 Part III

Feeder Conductor Sizing, General . . . . 530.19(A)

    Demand Factors Permitted . . . . Table 530.19(A)

    Demand Factors, Portable Feeders . . . . 530.19(B)

Grounding 530.20

Lamps, Portable 530.16

Lamps, Portable Arc Lamp 530.17

Live Parts, Enclosing and Guarding 530.15

Overcurrent Protection 530.18

Stage or Set . . . . .Art. 530 Part II

    Plugs and Receptacles 530.21

    Single-Pole Separable Connectors 530.22

    Stage Lighting and Effects Control 530.13

Wiring

    Permanent 530.11

    Portable 530.12

Substations . . . . Art. 530 Part VI

See Also: THEATERS, AUDIENCE AREAS OF MOTION . . PICTURE AND TELEVISION STUDIOS AND

SIMILAR LOCATIONS . . . . Ferm's Finder

MOTOR CONTROL CENTERS . . . . . . Art. 430 Part VIII

Available Fault Current 430.99

Busbars and Conductors 430.97

Dedicated Equipment Space . . . . 110.26(E)

Definition of . . . . Art. 100 Part I

Grounding 430.96

Overcurrent Protection 430.94

Service-Entrance Equipment, Use As 430.95

Spaces About 110.26

Spacings. . . . . Table 430.97(D)

Wire Bending Space, Minimum . . . . 430.97(C)

See (NJAV) . . . . UL Product iQ

MOTOR CONTROLLERS . . . . Art. 430 Part VII

Auxiliary Devices

    See (NKCR) . . . . UL Product iQ

Combination Fuseholder and Switch As 430.90

Combination Motor Controllers

    See (NKJH) . . . . UL Product iQ

Copper Conductors. 430.9(B)

Definition, of (Controller) 430.2

Design 430.82

Disconnect Means Permitted in Same Enclosure As. . . 430.103
Enclosure Types (Environmental Conditions). .110.28 and Table
Float & Pressure-operated
   See (NKPZ). . . . . *UL Product iQ*
Location of Disconnecting Means . . . . 430.102(A)
Magnetic
   See (NLDX) . . . . *UL Product iQ*
   See (NLRV) . . . . *UL Product iQ*
Motor Not in Sight of Controller 430.102
Need Not Open All Conductors Unless Also Disconnect 430.84
Number of Motors Served by 430.87
Number of Overload Units 430.37
Over 1500 Volts
   See (NJHU) . . . . *UL Product iQ*
Permitted to Serve As Overload Protection 430.39
Portable Motor 1/3 Horsepower or Less . . . . . . 430.81(B)
Ratings 430.83
   Stationary Motors, 2 Horsepower or Less . . . . 430.83(C)
   Torque Motors . . . . 430.83(D)
   Voltage Rating . . . . 430.83(E)
Stationary Motor, 1/8 Horsepower or Less . . . . . 430.81(B)
Torque Requirements . . . . 430.9(C)
Wire Bending Space 430.10
See (NJOT) . . . . *UL Product iQ*

MOTOR FUEL DISPENSING
Circuit Disconnects
   Attended Self-Service Motor Fuel . . . . 514.11(B)
   Emergency Electrical Disconnect. . . . . 514.11(A)
   Unattended Self-Service Motor Fuel . . . . 514.11(C)
Classification of Locations 514.3
   Boatyards and Marinas. . . . . 514.3(C)
   Classified Areas. . . . . 514.3(B)
   Unclassified Areas . . . . 514.3(A)
Definitions . . . . Art. 100, Part III
Figures for Area Classification
   Adjacent to Dispensers. . . . . Figure 514.3
   Adjacent to Storage Tanks . . . . Figure 514.3(B)
Grounding and Bonding 514.16
Maintenance 514.13
Sealing
   At Boundary . . . . 514.9(B)
   At Dispenser . . . . 514.9(A)
Tables

   Class I Locations . . . . Table 514.3(B)(1)
   Electrical Equipment. . . . . Table 514.3(B)(2)
Underground Wiring. .514.8
Wiring and Equipment
   Above Class I Locations 514.7
   In Class I Location .514.4
See GASOLINE (MOTOR FUEL) DISPENSING Ferm's Finder

MOTORS . . . . Art. 430
Adjustable Speed Drive Systems . . . . Art. 100 Part I
Air-Conditioning Units. . . . . 430.1 Info. Note 1
Appliances, Motor-Driven 422.3
   Ampacity and Motor Rating Determination 430.6
Automatic Restarting 430.43
Branch-Circuit Overload Protection . . . . . . Art. 430 Part III
   Continuous-Duty Motors 430.32
   Intermittent-Duty Motors 430.33
   Number of Overload Units Required . . . . . . Table 430.37
Branch-Circuit Short-Circuit & Ground-Fault Protection Art. 430 Part IV
   Individual Motor. 430.52
      Maximum Rating or Setting of . . . . Table 430.52
   Several Motors or Loads on One Branch-Circuit . . . 430.53
      Not Over 1 Hp Each . . . . 430.53(A)
      Other Group Installations . . . . 430.53(C)
      Single Motor Taps . . . . 430.53(D)
      Smallest Motor Protected . . . . 430.53(B)
Use of Tables 430.148 to 150 to Determine Value 430.6(A)(1)
Calculations – See Example D8 . . . . . . Informative Annex D
Capacitors 430.27
   Conductor Ampacity. . . . . 460.8(A)
   Disconnecting Means . . . . 460.8(C)
   Overcurrent Protection. . . . . 460.8(B)
   Rating or Setting of Motor Overload Device. 460.9
   See CAPACITORS. . . . . Ferm's Finder
Code Letters (Locked Rotor Indicating) . . . . . Table 430.7(B)
Compressor Motors. . . . . Art. 430
   Hermetic Refrigerant Motor-Compressor . . . . . . Art. 440
   See AIR-CONDITIONING & REFRIGERATING EQUIPMENT. . . . . Ferm's Finder
Conductors for Motors . . . . Art. 430 Part II
   Conductors for General Wiring 310.1
   Constant Voltage Direct-Current – Power Resistors . . 430.29
   Controllers and Control Circuit Devices . . . . 430.9(B)
   Determining Ampacity of . . . . 430.6(A)(1)

Feeder Demand Factor Permitted 430.26
Feeder Tap Conductors. 430.28
Integral Parts of Equipment . . . . 300.1(B)
Multimotor and Combination-load Equipment 430.25
Several Motors or a Motor(s) & Other Load(s). . . . 430.24
Single Motor 430.22
Conductors for Small Motors
   18 AWG Copper . . . . 430.22(G)(1)
   16 AWG Copper . . . . 430.22(G)(2)
   Note: Always use tables to size conductors and use nameplate to size overload protection.
Control Centers . . . . Art. 430 Part VIII
Control Circuits . . . . Art. 430 Part VI
Class 1, 2, and 3 Circuits . . . . 725.3(F)
   Copper Conductors At Terminals . . . . 430.9(B)
   Disconnection 430.75
   Electrical Arrangement of 430.74
   For Fire Pumps 695.14
   Grounding
      Arrange So Accidental Ground Won't Start 430.73
      Circuits Permitted but Not Required to Be Grounded. . . . 250.21(A)(3)
      Ungrounded Circuits Permitted 685.14
   Overcurrent Protection. 430.72
      For Specific Conductor Applications . . . . . . 240.4(G)
      Maximum Rating of. . . . . Table 430.72(B)
      Orderly Shutdown. 430.44
      System Coordination 240.12
   Transformer . . . . 430.75(B)
      In Controller Enclosure 430.74
      Requirements . . . . Art. 450
      Separately Derived Systems 250.30
Controllers . . . . Art. 430 Part VII
   See MOTOR CONTROLLERS . . . . Ferm's Finder
   See NEMA Sizes . . . . Ferm's Charts and Calculations
   Cord- and Plug-Connected . . . . 430.42(C)
   Disconnecting Means for. . . . . 430.109(F)
   Controller for 1/3 Horsepower or Less . . . . . . 430.81(B)
Cranes & Hoists . . . . Art. 610
DC Motors
   Full-Load Current . . . . Table 430.247
   Power Resistors. 430.29
      Conductor Rating Factors . . . . Table 430.29
Design B, C, and D

Conversion Table for Selecting Disconnecting Means . . Table 430.251(B)
Disconnecting Means . . . . 430.109(A)(1)
Instantaneous Circuit Breaker Trip Setting 430.52(C)(3) Ex. 1
   Note: Also applies to Design B motors
Listed Self-Protected Combination Controller . . 430.52(C)(6)
Note: Also applies to Design B motors
Marking on Motors . . . . 430.7(A)(9)
Maximum Rating or Setting . . . . Table 430.52
Motor Controllers . . . . 430.83(A)(1)
Motor Short-Circuit Protector . . . . 430.52(C)(7)
Disconnecting Means . . . . Art. 430 Part IX
   A Group of Motors Served by a Single Disconnect . . . 430.112 Ex.
      For Combination Loads . . . . 430.110(C)
   Ampere Rating of 430.110
   Cord- and Plug-Connected. . . . . 430.109(F)
      Rating of . . . . 430.42(C)
   Damp or Wet Locations 404.4
   Disconnect Both Motor & Controller 430.101
   Disconnect in Sight from Motor and Driven Machinery . . . . 430.102(B)
   Disconnect Must Be in Sight from Controller . . 430.102(A)
   Grounded Conductor 430.105
   Motor Not in Sight from Controller 430.102(B) Ex. to (1) and (2)
      In Sight from, Definition of . . . . Art. 100 Part I
   Power from Two Sources 430.113
   Readily Accessible 430.107
   Single Disconnect for Each Motor 430.112
   Switch or Circuit Breaker As Both Controller & Disconnect. . 430.111
   To Be Indicating 430.104
   Type, Every Switch in Circuit to Comply 430.108
      Ampere Rating and Interrupting Capacity. 430.110
Electronically Protected, Defined 430.2
Feeder Taps. 430.28
   Overcurrent Protection. . . . . 240.21(F)
   Single Motor Tap . . . . 430.53(D)
Feeders . . . . 430 Part II
   Feeder Demand Factor 430.26
   Feeder Taps. 430.28
   Multimotor and Combination-Load Equipment 430.25
   Several Motors or a Motor(s) and Other Load(s) 430.24
   Short-Circuit and Ground-Fault Protection . . Art. 430 Part V
      Rating or Setting – Motor Load 430.62
      Rating or Setting – Power and Light Loads 430.63

Full-Load Running Current 430.6
    Direct-Current Motors. . . . . Table 430.247
    Single-Phase Alternating Current. . . . . . . Table 430.248
    Three-Phase Alternating Current . . . . Table 430.250
    Two-Phase Alternating Current . . . . Table 430.249
Ground-Fault Protection, Branch-Circuit . . . Art. 430 Part IV
Ground-Fault Protection, Feeder . . . . .Art. 430 Part V
Grounding . . . . Art. 430 Part XIII
    Enclosures for Controllers 430.244
    Equipment Grounding and Conductors . . . Art. 250 Part VI
    Means for Equipment Grounding Connection . . 430.12(E)
    Method of Grounding 430.245
    Methods of Equipment Grounding . . . . .Art. 250 Part VII
    Stationary Motors 430.242
Group Installations
    Ampacity of Conductors 430.24
    Number of Controllers. 430.87
    Overload Protection 430.42
    Short-Circuit and Ground-Fault Protection 430.53
    Single Disconnect Permitted for Group . . 430.102(A) Ex. 2
Hazardous (Classified) Locations
    Class I Locations 501.125
    Class II Locations 502.125
    Class III Locations 503.125
High Voltage (Over 1000 Volts, Nominal) . . . Art. 430 Part XI
    See OVER 1000 VOLTS, NOMINAL . . . . . Ferm's Finder
Highest Rated or Smallest Rated 430.17
In Sight from (Definition of) . . . . Art. 100
Industrial Machinery . . . . Art. 670
    See INDUSTRIAL MACHINERY . . . . Ferm's Finder
Location of Motors 430.14
    Protected Against Dust 430.16
    Protected Against Liquids 430.11
    See HAZARDOUS (CLASSIFIED) LOCATIONSFerm's Finder
Locked-Rotor Currents . . . . Table 430.251(A) and 430.251(B)
    Code Letters (Locked Rotor Indicating) . . . Table 430.7(B)
Motor Control Centers . . . . Art. 430 Part VIII
    See MOTOR CONTROL CENTERS . . . . . Ferm's Finder
Orderly Shut Down of 430.44
    Generator Considered Vital to Operation . . . 445.12(E) Ex.
    Integrated Electrical Systems . . . . Art. 685
    Power Loss Hazard . . . . 240.4(A)
    System Coordination 240.12
Over 1000 Volts, Nominal . . . . Art. 430 Part XI

Engineering Supervision, Protective Device Settings430.225(B)(1)
    See OVER 1000 VOLTS, NOMINAL . . . . . . Ferm's Finder
Overcurrent Protection
    Branch Circuits . . . . Art. 430 Part IV
        Individual Motor 430.52
        Maximum Rating of . . . . Table 430.52
        Several Motors on One Branch Circuit 430.53
    Feeders . . . . .Art. 430 Part V
        Rating or Setting – Motor Load 430.62
        Rating or Setting – Power and Light Loads 430.63
    Use Tables 430.247 through 250 for Determining Value of . .
    . . . . 430.6(A)(1)
Overload Protection . . . . Art. 430 Part III
    Adjustment for Capacitors 460.9
    Continuous-Duty Motor. 430.32
    Intermittent-Duty Motor 430.33
    Number of Overload Units
        Fuses. 430.36
        Other Than Fuses 430.37
        Number and Location . . . . Table 430.37
    Number of Conductors Opened, by Other Than Fuses
    or Thermal Protectors 430.38
    Use of Nameplate Values for Separate Overload . 430.6(A)(2)
Overtemperature Protection 430.126
Phase Converters . . . . Art. 455
Protection of Live Parts . . . . .Art. 430 Part XII
    Guards for Attendants 430.233
    Required If 50 Volts or More. 430.232
Restarting, Automatic. 430.43
Setting of Branch-Circuit Protective Device 430.6
    Maximum Rating or Setting . . . . Table 430.52
    Short-Circuit and Ground-Fault . . . . Art. 430 Part IV
Short-Circuit & Ground-Fault Protection
    Branch Circuit . . . . Art. 430 Part IV
        Maximum Rating or Setting of . . . . Table 430.52
    Feeder . . . . Art. 430 Part V
    Use of Tables 430.247 through 250 to Determine 430.6(A)(1)
Small Motor Conductors
    18 AWG Copper . . . . .430.22(G)(1)
    16 AWG Copper . . . . .430.22(G)(2)
Starters . . . . .Art. 430 Part VII
Tables — Motor Information
    Duty Cycle Service . . . . Table 430.22(E)
    Feeders 430.28

Full Load Current
    Three-Phase AC Motors. . . . . Table 430.250
    Two-Phase AC Motors. . . . . Table 430.249
    Single-Phase AC Motors. . . . . Table 430.248
Locked-Rotor Current, Conversion Single-Phase . Table 430.251(A)
Locked-Rotor Current, Conversion Polyphase . . Table 430.251(B)
Locked-Rotor Indicating Letter . . . . Table 430.7(B)
Maximum Rating or Setting of Protective Devices Table 430.52
Number of Overload Units Required. . . . . . Table 430.37
Other Articles . . . . Table 430.5
Secondary Conductor . . . . Table 430.23(C)
Single Motor Taps . . . . 430.53(D)
Terminal Housings 430.12
Three Overload Devices, Three-Phase . . . . . . Table 430.37
Valve Actuator Motor (VAM) Assemblies, Definition of . .430.2
Ventilation . . . . 430.14(A)
Water Pumps, Motor Operated . . . . .250.112(L)
Welders, Arc, Motor-Generator . . . . .Art. 630 Part II
Wire-Bending Space . . . . 430.10(B)

MOTORS & GENERATORS FOR USE IN CLASS I, GROUPS C & D; CLASS II, GROUPS E, F & G
    See (PSPT) . . . . *UL Product iQ*
    See (PTDR). . . . . *UL Product iQ*
    See(PTKQ) . . . . *UL Product iQ*
    See (PUCJ) . . . . *UL Product iQ*
    See (PTHE) . . . . *UL Product iQ*

MOUNTING OF EQUIPMENT (General) . . . . . 110.13(A)
Mounting of Snap Switches 404.10
Outlet, Device, Pull, and Junction Boxes 314.23
Switches and Circuit Breakers (readily accessible) 404.8

MOUNTING POSTS AND PEDESTALS FOR DISTRIBUTION EQUIPMENT
    See (PUPR) . . . . *UL Product iQ*

MOVING WALKS . . . . Art. 620
    See ELEVATORS . . . . Ferm's Finder

MULTI-FAMILY DWELLINGS
Definition. . . . . Art. 100
Disconnecting Means, in Separate Enclosures . . . 230.40 Ex. 2
    Access to Occupants . . . . 230.72(C)
    Appliances . . . . Art. 422 Part III
        Cord- and Plug-Connected 422.33

    With Unit Switches . . . . 422.34(A)
    Electric Heat, with Unit Switches . . . . .424.19(C)(1)
    Location 230.70
    Maximum Number for Sets of Service-Entrance Conductors .
    . . . . 230.71(A)
    Service Overcurrent Devices, Access to . . . . . 240.24(B)
Overcurrent Protection
    Access to Devices Protecting Conductors Supplying . . 240.24(B)
    Access to Occupants, Service . . . . 230.72(C)
    Location in Circuit 240.21
    Locked Service Overcurrent Devices 230.92
    Service Equipment, Rating . . . . 230.90(A)
Type NM Cable (Romex) and NMC Not Permitted . . . 334.12
Type NM Cable (Romex) and NMC Permitted 334.10
Services
    Access to Devices Protecting Conductors Supplying . . 240.24(B)
    Access to Disconnecting Means . . . . 230.72(C)
    Additional Permitted . . . . 230.2(B)(1)
    Location of Disconnecting Means 230.70
    Locked Service Overcurrent Devices 230.92
    Maximum Number . . . . 230.71(A)
    Number of Service-Entrance Conductor Sets . . 230.40 Ex. 4
    See CALCULATIONS . . . . Ferm's Finder
    See SERVICES . . . . Ferm's Finder

MULTIGROUNDED NEUTRAL SYSTEM
Solidly Grounded . . . . 250.184(C), Ex.

MULTIMODE INVERTER
Microgrid Systems . . . . 705.70(3) Info Note
Photovoltaic Systems (defined) . . . . .Art. 100, Part I

MULTIOUTLET ASSEMBLY . . . . Art. 380
Calculations . . . . 220.14(H)
    See CALCULATIONS . . . . Ferm's Finder
Cord and plug connected . . . . 380.12(7)
Definition. . . . . Art. 100 Part I
Insulated Conductors . . . . 380.23(A)
Pull Boxes. . . . . 380.23(B)
Through Partitions 380.76
Uses Not Permitted 380.12
Uses Permitted 380.10
    See (PVGT) . . . . *UL Product iQ*

MULTIOUTLET BRANCH-CIRCUIT
Ampacity . . . . .210.19(A)(2)

Branch-Circuit Requirements – Summary 210.24
Greater Than 50 Amperes, Industrial Premises . . . . 210.3 Ex.
Next Higher Standard Overcurrent Device Not Permitted . . . .
. . . . 240.4(B)(1)
Outlet Devices 210.21
Permissible Loads 210.23

MULTIPLE CIRCUITS

Equipment Bonding Jumper on Load Side of Service 250.102(D)
Multiple Branch Circuits, Simultaneously Disconnect . . .210.7
Multiple Circuit Connections, Grounding Means . . . 250.144
Multiwire Branch Circuit Considered Multiple Circuits 210.4(A)
Rating. 210.3
Size of Equipment Grounding Conductors . . . . . 250.122(C)

MULTIPLE-OCCUPANCY BUILDING

Disconnecting Means, in Separate Enclosures . . . 230.40 Ex. 2
    Access to Occupants . . . . 230.72(C)
    Appliances . . . . Art. 422 Part III
        Cord- and Plug-Connected 422.33
        With Unit Switches . . . . 422.34(D)
    Electric Heat, with Unit Switches . . . . .424.19(C)(4)
    Location 230.70
    Maximum Number for Sets of Service-Entrance
        Conductors . . . . 230.71(A)
    Service Overcurrent Devices, Access to . . . . . . 240.24(B)
Overcurrent Protection
    Access to Devices Protecting Conductors Supplying 240.24(B)
    Access to Occupants, Service . . . . 230.72(C)
    Location in Circuit 240.21
    Locked Service Overcurrent Devices 230.92
    Service Equipment, Rating. . . . . 230.90(A)
Type NM Cable (Romex) Not Permitted 334.12
Type NM Cable (Romex) Permitted 334.10
Services
    Access to Devices Protecting Conductors Supplying . . 240.24(B)
    Access to Disconnecting Means . . . . 230.72(C)
    Additional Permitted . . . . 230.2(B)(1)
    Location of Disconnecting Means . . . . 230.70(A)
    Locked Service Overcurrent Devices 230.92
    Maximum Number . . . . 230.71(A)
    Number of Service-Entrance Conductor Sets . . 230.40 Ex. 4
    See CALCULATIONS . . . . Ferm's Finder
    See SERVICES . . . . Ferm's Finder

MULTIPLE RACEWAYS

Installation — Equipment Bonding Jumper . . 250.102(C)(2), 250.122(F)
Minimum Size of Bonding Jumper, Load Side . . . 250.122(D)
Minimum Size of Bonding Jumper, Supply Side . . 250.102(C)
Size Equipment Bonding Jumper, Supply Side . . . 250.102(C)
Size of Equipment Bonding Jumper, Load Side . . 250.102(D)

MULTIPLE SEPARATELY DERIVED SYSTEMS 250.30(A)(6)

MULTIWIRE BRANCH CIRCUITS

Branch Circuits Permitted to Be Multiwire . . . . 210.4(A)
Definition of . . . . Art. 100
Emergency Systems, Branch Circuits 700.19
Grouping of Conductors in Panelboard. . . . . 210.4(D)
    Disconnecting Means . . . . 210.4(B)
    Disconnecting Means, Luminaires . . . . 410.130(G)
    Disconnecting Means, Location of Luminaires 410.130(G)(3)
Neutral Continuity Must Not Be Dependent on Device
    Connections . . . . 300.13(B)
Not Permitted, Freestanding-Type Partitions,
    Cord- and Plug-Connected . . . . 605.9(D)
Requirements for in Class I, Division 1 Locations . . . 210.4(B)
Requirements for in Class II, Division 1 Locations . . . 210.4(B)
Tie Bars Required (Line-to-Line Loads)
    Devices on Same Yoke 210.7
    Disconnecting Means Temporary Circuits . . . . . . 590.4(E)
    Dwelling Units, Simultaneous Disconnection . . . . 210.4(B)
    Permission to Supply Line-to-Line Loads. . . . 210.4(C) Ex. 2
Ungrounded Conductors Tapped from Grounded 210.10

N

NATURAL AND ARTIFICIALLY MADE BODIES OF WATER Art. 682

Bonding and Grounding . . . . Art. 682 Part III
Definitions of 682.2
Electrical Connections 682.12
Electrical Equipment and Transformers 682.10
Equipotential Plane 682.33
    Areas Not Requiring . . . . 682.33(B)
    Areas Requiring . . . . 682.33(A)
    Bonding of . . . . 682.33(C)

Ground-Fault Circuit Interrupter Protection 682.15

Submersible or Floating Equipment. 682.14

NEAT AND WORKMANLIKE

Audio Signal Processing, Amplification, and Reproduction . 640.6(A)

CATV and Radio Distribution 800.24

Class 1, 2, and 3 Circuits 725.24

Communication Circuits 800.24

Fire Alarm Systems 760.24

General 110.12

Less than 50 Volts. 720.11

Network-Powered Broadband Communications Systems . 800.24

Optical Fiber Cables and Raceways 770.24

Premises-Powered Broadband Communications Systems 840.24

NEON

Neon Secondary-Circuit Wiring, 1000 Volts or Less, Nominal 600.31

Neon Secondary-Circuit Wiring, Over 1000 Volts, Nominal 600.32

Neon Tubing 600.41

Definition of 600.2

NETWORK POWERED BROADBAND COMMUNICATIONS SYSTEMS . . . . Art. 830

Abandoned Cables 800.25

Access to Electrical Equipment Behind Panels 800.21

Burial Depth . . . . Table 830.47(C)

Cable Routing Assemblies . . . . 800.110(C)

Cable Ties and Accessories, Nonmetallic 800.24

Cables Outside and Entering Buildings. . . . . . Art. 800 Part II

Definitions 830.2

Ducts, Wiring Within . . . . 800.3(B)

Grounding Devices, Required to be Listed 830.180

Grounding Methods . . . . Art. 830 Part IV

Grounding and Bonding at Mobile Homes 830.106

Installation Methods Within Buildings . . . . . Art. 800 Part IV

Listing Requirements . . . . 800.113(A)

Mechanical Execution of Work 830.24

Minimum Cover Requirements. . . . . Table 830.47(C)

Output Circuits. 830.3(B)

Overhead (Aerial) Cables 830.44

Power Limitations. 830.15

Protection (Electrical) . . . . Art. 830 Part III

Raceway Types . . . . 830.110(A)

Underground Cables Entering Buildings 830.47

NEUTRAL OR GROUNDED CONDUCTOR

Bare (Permitted)

Ampacity of Bare Conductors . . . . 310.15(D)

Table Value . . . . Table 310.21

Outside Branch Circuits and Feeders. . . . . .225.4 Ex.

Overhead Service Conductors . . . . 230.22 Ex.

Service Entrance . . . . 230.41 Ex.

Solidly Grounded Systems 1kV and Over . . . . 250.184(A)(1) Ex. 1 & 2 & 3

Underground Service Lateral . . . . 230.30 Ex.

Bonded to Service Equipment . . . . 250.24(C)

Separately Derived Systems. . . . . 250.30(A)(1)

Two or More Buildings Supplied by Feeder or Branch Circuit . . . . 250.32(B)

Within Service Disconnect Enclosure 250.28

Brought to Service Equipment . . . . 250.24(C)

Calculation of (Feeder or Service Load) 220.61

Considered Current-Carrying . . . . 310.15(E)

Examples. . . . . Informative Annex D

Change in Size 220.61

Common Neutral Not Permitted (Generally)

Class I, Division 1 Locations . . . . 210.4(C)

Class I, Zone 1 Locations . . . . 210.4(C)

Class II, Division 1 Locations . . . . 210.4(C)

Freestanding-Type Partitions, Cord- and Plug-Connected . . . . 605.9(D)

Zone 20 and 21 Locations . . . . 210.4(C)

Common Neutral Permitted

Feeders. 215.4

Lighting Equipment Installed Outdoors . . . . 225.7(B)

Multiwire Branch Circuits 210.4

Definition of. . . . . Art. 100 Part I

Considered Current-Carrying. . . . . 310.15(E)

Examples. . . . . Informative Annex D

Contained within the Same Raceway, Cable Tray,

Trench, Cable or Cord . . . . 300.3(B)

Exceptions

Auxiliary Gutters . . . . 300.3(B)(4)

Paralleled Installations . . . . 300.3(B)(1) Ex.

Switch Loops . . . . 404.2(A) Ex.

Underground Installations. . . . . 300.5(I) Ex. 1 & 2

Metal (Ferrous) Raceways . . . . 300.20(A)

Over 1000 Volts 300.35

Underground Feeder and Branch-Circuit Cable . . 340.10(2)

Underground Installations . . . . 300.5(I)

Definition of Neutral Conductor . . . . Art. 100 Part I

Definition of Neutral Point . . . . Art. 100 Part I

Disconnection of, at Main Service 230.75

For More than one Branch Circuit, Neutral Conductor. . .200.4

Grouping Multiwire Circuits, Ungrounded Conductors 210.4(D)

Grouping with Same Circuit . . . . 200.4(B)

Harmonics (Nonlinear Load)

    Definition of . . . . Art. 100 Part I

    Feeder or Service Neutral . . . . 220.61(C) IN 2

    Flexible Cords and Cables . . . . 400.5(A)

    Multiwire Branch Circuits . . . . 210.4(A) IN

    Neutral Considered Current-Carrying . . . . 310.15(E)

High Impedance Grounding 250.36

Identification & Marking of . . . . Art. 200

    Conductors for General Wiring . . . . 310.6(A)

    For Branch Circuits . . . . 210.5(A)

    Means of Identifying Grounded Conductors 200.6

Impedance Grounded Neutral Systems 250.187

Installation . . . . 200.4(A)

Insulated, Required

    1000 Volt Minimum Solidly Grounded Neutral Systems

        1 kV and Over. . . . . 250.184(A)

    High Impedance Grounded System . . . . 250.36(B)

Messenger Supported Wiring . . . . 396.30(B)

Multiple Circuits . . . . 200.4(B)

Neutral Conductors (more than one branch circuit) 200.4

Neutral (Current-Carrying Conductor) . . . . 310.15(E)

    Flexible Cords and Cables (Ultimate Insulation Temp) . . 400.5(B)

Neutral (Feeder Load) 220.61

    Neutral (to Every Service) Must Not Be Smaller Than the

        Required Grounding Electrode Conductor . .250.24(C)(1)

Not to Be Dependent on Device Connections for

    Continuity on Multiwire Branch Circuits . . . . . 300.13(B)

Overcurrent Device Not Permitted in the Grounded Conductor . . . . 230.90(B)

    Exceptions

        Fuse in Grounded Conductor 3-Wire, 3-Phase AC . . 430.36

        Other Than Fuses . . . . Table 430.37

        Simultaneously Opened . . . . 230.90(B)

        Unless Conditions Met. . . . . 240.22(1) & (2)

    Reduction in Size 220.61

    Re-Identifying at Switches . . . . 200.7(C)(1)

Size of Conductor

    Minimum 250.66

    Not Smaller Than Grounding Electrode Conductor . . . . . . 250.24(C)(1)

    Overhead Service Conductors . . . . 230.23(C)

    Parallel Conductors . . . . .250.24(C)(2)

        Existing Installations, Under Engineering Supervision. . . . . .310.10(G)(1) Ex. 2

    Phase Conductors over 1100 kcmil Copper . . .250.24(C)(1)

    Service Entrance . . . . 230.42(C)

    Table for Sizing. . . . . Table 250.102(C)(1)

    Underground Service . . . . 230.31(C)

Switch in Grounded Conductor Not Permitted. . . . . 404.2(B)

    Exceptions

        Circuit Breaker If Simultaneously Opens All

            Conductors . . . . 230.90(B)

        For Circuits through or to Motor Fuel Dispensers. . 514.11(A)

        In Service Disconnecting Means. 230.75

        Motor Controller 430.105

        Switches or Circuit Breakers (General). . . . . 404.2(B) Ex.

    Service Overcurrent Device . . . . 230.90(B)

    Wind Electric Systems 694.20

Use of for Grounding Equipment

    Load Side .250.142(B)

        Agricultural Buildings . . . . 547.9(B)(3)

        Frames of Ranges, Dryers, Etc., Existing Installations . .250.140

        In Separate Buildings . . . . 250.32(B)

        Mobile Homes 550.16

        Recreational Vehicle Site Supply, Not Permitted 551.76(C) & (D)

        Recreational Vehicles . . . . 551.54(C)

        Separately Derived Systems 250.30

    Supply Side. . . . . 250.142(A)(3)

Wind Electric Systems 694.20

NEUTRAL CONDUCTOR

Definition of . . . . Art. 100

NEUTRAL POINT

Definition of . . . . Art. 100

NIGHTCLUBS . . . . Art. 518

NIPPLES & NIPPLE FILL (CONDUIT OR TUBING)

60 Percent Fill Allowed, Not Exceeding 600 mm (24 in.) Chapter 9 Table 1, Note 4

Approved for the Condition . . . . 300.6(C)

Class I Division 2- Seals. . . . . 501.15(B)(1)

Class I Locations- Seals . . . . 501.15(A)(3)

Corrosion Protection . . . . 300.6(A)
 Supplementary . . . . 300.6(B)
Derating Factors Not Applicable, Raceways Not Exceeding 600 mm (24 in.) . . . . 310.15(C)(1)(b)
Pole Luminaire . . . . 410.30(B)(2)
Zone 1 . . . . 505.16(B)(2)(2)(c)

NOMINAL VOLTAGE
See VOLTAGE . . . . Ferm's Finder

NONAUTOMATIC, DEFINED . . . . Art. 100

NONCOINCIDENT LOADS 220.60
Room Air Conditioners and Other Equipment on Same Circuit . . . . 440.62(C)

NONELECTRICAL EQUIPMENT, GROUNDING OF 250.116

NONFERROUS METALS
Aluminum Rigid Conduit 344.100
Electrical Metallic Tubing . . . . 358.10(B)
Faceplates, Receptacles . . . . 406.6(A)
Faceplates, Switches . . . . 404.9(C)
Fixed Outdoor Electric Deicing and Snow Melting Equipment 426.26
Grounding Electrodes . . . . 250.52(A)(7)
Grounding Electrode Conductor Enclosures . . . . . 250.64(E)
Markings, Corrosion Resistant 344.120
Metal-Clad Cable 330.31
Red Brass Conduit 344.100
Rigid Metal Conduit (Aluminum) 344.100
Stainless Steel Conduit 344.100
Underfloor Raceways 390.12
Wiring Methods . . . . 300.3(B)(3)

NONGROUNDING-TYPE RECEPTACLES, REPLACEMENT OF . . . . 406.4(D)

NONINCENDIVE CIRCUIT
Class I, Division 2 Locations
  Meters, Instruments, and Relays . . . . 501.105(B)(1), Ex.(3)
  Signaling, Alarm, Communications, Remote Control 501.150(B)(1) Ex.(3)
  Wiring Methods . . . . 501.10(B)(3)
Class II, Division 2 Locations
  Signaling, Alarm, Communications, Remote Control, Meters, etc. . . . . 502.150(B)(1) Ex.
  Wiring Methods . . . . 502.10(B)(3)

Class III, Division 1 Locations
  Wiring Methods . . . . 503.10(A)(4)
Definitions of Nonincendive Words and Terms,
  Specific to Article 506, Zone 20, 21, and 22 Locations. . . . . Art. 100, Part III
Protection Techniques. . . . . 500.7(F)(G)(H)
Zone 20, 21, and 22 Locations. . . . . 506.8(F)

NONINTERCHANGEABLE
Cartridge Fuseholders . . . . 240.60(B)
Type S Fuses . . . . 240.53(B)

NONINSTANTANEOUS TRIP (Arc Energy Reduction) 240.87

NONLINEAR LOADS
Definition of . . . . Art. 100
Feeder or Service Neutral . . . . 220.61(C) IN
Flexible Cords and Cables. . . . . 400.5(A)
Multiwire Branch Circuits . . . . 210.4(A) IN 1
Neutral Considered Current-Carrying . . . . . . 310.15(B)(E)

NONMETALLIC AUXILIARY GUTTERS . . . . . . Art. 366
Ampacity of Conductors . . . . 366.23(B)
Definition of 366.2
Listing Requirements 366.6
Marking 366.120
Number of Conductors . . . . 366.22(B)
Uses Permitted . . . . 366.10(B)
  See AUXILIARY GUTTERS . . . . Ferm's Finder

NONMETALLIC BOXES
See BOXES & FITTINGS, Nonmetallic . . . . . . Ferm's Finder

NONMETALLIC CONDUIT
See CONDUIT . . . . Ferm's Finder
See High Density Polyethylene Conduit – Type HDPE . Art. 353
  See CONDUIT, High Density Polyethylene Conduit . Ferm's Finder
See Liquidtight Flexible Nonmetallic Conduit . . . . . Art. 356
  See LIQUIDTIGHT FLEXIBLE NONMETALLIC CONDUIT . . . . Ferm's Finder
See Nonmetallic Underground Conduit with Conductors. . . Art. 354
  See NONMETALLIC UNDERGROUND CONDUIT WITH CONDUCTORS . . . . Ferm's Finder
See Reinforced Thermosetting Resin Conduit . . . . . Art. 355
  See REINFORCED THERMOSETTING RESIN CONDUIT Ferm's Finder

See Rigid PVC Conduit . . . . Art. 352
    See RIGID PVC CONDUIT . . . . Ferm's Finder

NONMETALLIC EXTENSIONS . . . . Art. 382
Bends 382.26
Boxes and Fittings 382.40
Concealable Exposed Runs . . . . 382.15(B)
Construction Specifications (Concealed Type Only) . . Art. 382 Part III
Definition of 382.2
Devices, Receptacles and Housings 382.42
Exposed Runs. 382.15
Listing Requirements 382.6
Securing and Supporting 382.30
Splices and Taps. 382.56
Uses Not Permitted 382.12
Uses Permitted 382.10
See (PXXT) . . . . UL Product iQ

NONMETALLIC-SHEATHED CABLE: TYPE NM and NMC
. . . . Art. 334
Accessible Attics 334.23
Ampacity Shall Be That of 60°C Conductors 334.80
    Derating Permitted from 90°C Ampacity. 334.80
    In Contact With Draft or Firestopping or Thermal Insulation .
. . . . 334.80
    Sealed Within Thermal Insulation, Caulk, or Sealing Foam 334.80
Bending Radius 334.24
Conductors, Sizes and Type 334.104
Conductors Shall Be Rated 90°C 334.112
Construction Types, Sheath 334.116
    Type NM. . . . . 334.116(A)
    Type NMC. 334.116(B)
Definition. 334.2
Devices of Insulating Materials . . . . 334.40(B)
Devices with Integral Enclosures . . . . 334.40(C)
Devices without a Separate Outlet Box . . . . 334.30(C)
Entering Boxes 314.17
    Cabinets, Cutout Boxes, and Meter Socket Enclosures . 312.5
Exposed Work 334.15
Installation of . . . . Art. 334 Part II
Interconnectors . . . . 334.40(B)
Listing Required 334.6
Marking of . . . . 334.112 IN
Nonmetallic Outlet Boxes . . . . 334.40(A)
Protection of

At Crawl Hole (Attics & Roof Spaces) 334.23
Cables and Raceways Installed In or Under Roof Decking . . . 300.4(E)
Closely Follow Surface . . . . 334.15(A)
Exposed Work 334.15
Metal Cabinets, Cutout Boxes, Meters 312.5
Parallel to Framing Members and Furring Strips . . . 300.4(D)
Shallow Grooves, in . . . . 300.4(F)
Through Bored Holes and Notches in Wood Framing
    Members . . . . 300.4(A)(1) & (2)
Through Floors. 334.15(B)
Through Metal Framing Members . . . . . 300.4(B)(1) & (2)
Through or Parallel to Wood or Metal Framing Members . 334.17
Unfinished Basements and Crawl Spaces . . . . . 334.15(C)
SE Cable, Interior Installations Comply with Article 334 . . . . .
. . . . 338.10(B)(4)
Securing and Supporting 334.30
    Every 1.4 m (4 1/2 ft) 334.30
    In Unfinished Basements and Crawl Spaces . . . . 334.15(C)
    Within 300 mm (12 in.) of Metal Box 334.30
    Within 200 mm (8 in.) of Nonmetallic Box . . 314.17(C) Ex.
    Exceptions
        Concealed Work in Finished Buildings . . . 334.30(B)(1)
        Wiring Devices without Separate Box . . . . . . 334.30(C)
        Within Accessible Ceilings (Whips) . . . . . . 334.30(B)(2)
UF Cable Used for Interior Wiring Shall Comply with the Requirement of Article 334 . . . . 340.10(4)
Unfinished Basements and Crawl Spaces . . . . 334.15(C)
Uses Permitted (General, Types NM, NMC, and NMS) 334.10
    Electric Discharge and LED Luminaires,
        Connection of . . . . 410.24(A)
    In Cable Trays, Identified for . . . . 334.10(4)
    In Nursing Homes and Limited Care Facilities . . 517.10(B)(2)
    In Business Offices, Corridors, Waiting Rooms in Clinics,
        Medical and Dental Offices & Outpatient Facilities
. . . . 517.10(B)(1)
    In Places of Assembly, Not Fire-rated Construction . 518.4(B)
    In Theaters, Motion Picture & Television Studios, etc.
        Not Fire-Rated Construction . . . . 520.5(C)
        (Note: Refer to 334.10(3) for building construction types permitted.)
    Multifamily Dwellings of Types III, IV, and V Construction . .
. . . . 334.10(2)
    One-and Two-Family Dwellings . . . . 334.10(1)
    Other Structures of Types III, IV, and V Construction 334.10(3)

See Informative Annex E (for determination of building types)

See Standard on Types of Building Construction). . NFPA 220

See . . . . also applicable building code)

Spas, Hot Tubs and Permanently Installed Immersion Pools . . 680.42(C)

Swimming Pool Motors, Interior One-Family Dwellings and Another Building or Structure Associated with . . 680.21(A)(1)

Temporary Installations (No Height Limitation) . . 590.4(B)

Uses Not Permitted . . . . 334.12(A)(1)-(10)

Uses Not Permitted, Types NM . . . . 334.12(B)

Uses Permitted (Specific)

    Type NM. . . . . 334.10(A)

    Type NMC. . . . . 334.10(B)

White Conductor 200.7

    See MULTIFAMILY DWELLINGS . . . . Ferm's Finder

    See MULTIPLE-OCCUPANCY BUILDING . . Ferm's Finder

    See (PWVX) . . . . UL Product iQ

NONMETALLIC-SHEATHED CABLE INTERCONNECTORS 334.40(B)

NONMETALLIC UNDERGROUND CONDUIT WITH CONDUCTORS . . . . Art. 354

Aboveground Use, Encased in Concrete . . . . . . 354.10(5)

Bends

    How Made 354.24

    Number of 354.26

Bushings 354.46

Conductor Terminations 354.50

Construction of 354.100

Definition of 354.2

Installation of . . . . .Art. 354 Part II

Insulation Temperature Limitations . . . . .355.10(I)

Listing Requirements 354.6

Marking 354.120

Maximum and Minimum Size . . . . 354.20(A) & (B)

Number of Conductors 354.22

Trimming 354.28

Uses Not Permitted 354.12

Uses Permitted 354.10

See (QQRK) . . . . UL Product iQ

NONMETALLIC WIREWAYS

Dead Ends 378.58

Definitions 378.2

Expansion Fittings 378.44

Extensions 378.70

Grounding 378.60

Insulated Conductors 378.23

Listing 378.6

Marking 378.120

Number of Conductors 378.22

Parallel Conductors 378.20

Securing and Supporting 378.30

Size of Conductors 378.21

Splices and Taps 378.56

Uses Not Permitted 378.12

Uses Permitted 378.10

NON-POWER-LIMITED FIRE ALARM CIRCUIT (NPLFA) . . . . Art. 760 Part II

Definition of NPLFA .760.2

See FIRE ALARM SYSTEMS . . . . Ferm's Finder

NONREMOVABLE

Type S Fuse Adapters . . . . 240.54(C)

NONTAMPERABLE

Circuit Breakers 240.82

Type S Fuses . . . . 240.54(D)

NUMBER OF

Bends

    EMT, Electrical Metallic Tubing 358.26

    ENT, Electrical Nonmetallic Tubing 362.26

    FMC, Flexible Metal Conduit 348.26

    HDPE, High Density Polyethylene Conduit 353.26

    IMC, Intermediate Metal Conduit 342.26

    LFMC, Liquidtight Flexible Metal Conduit 350.26

    LFNC, Liquidtight Flexible Nonmetallic Conduit . . . 356.26

    NUCC, Nonmetallic Underground Conduit with Conductors 354.26

    PVC, Rigid Polyvinyl Chloride Conduit 352.26

    RTRC, Reinforced Thermosetting Resin Conduit . . . 355.26

    RMC, Rigid Metal Conduit 344.26

    See BENDS, Number of Bends Permitted in . . . Ferm's Finder

Circuits Required . . . . 210.11(A)

    Central Heating Equipment (Other Than Fixed Electric) 422.12

    Computation of Loads to Determine 220.10

    Dwelling Units . . . . 210.11(C)(1)-(3)

    Elevator Hoistway Pit Lighting and Receptacles . . 620.24(A)

    Elevator Machine Room/Machinery Space . . . . . 620.23(A)

    Manufactured and Mobile Homes . . . . 550.12(A)-(E)
    Marinas and Boatyards . . . . 555.33(A)(3)
    Recreational Vehicles . . . . 551.42(A)-(D)
    Signs . . . . 600.5(A)
Conductors in
    See CONDUCTORS, Number of, in . . . . . . Ferm's Finder
Lighting Outlets Required
    About Electric Equipment . . . . 110.26(D)
        Over 1000 Volts . . . . 110.34(D)
    All Occupancies . . . . 210.70(C)
    Dwellings . . . . 210.70(A)
    Guest Rooms and Guest Suites . . . . 210.70(B)
Overcurrent Devices (Panelboards) 408.36
    Maximum Number in 408.54
Receptacle Outlets Required
    All Occupancies, Wherever Flexible Cords with
        Attachment Plugs Are Used . . . . 210.50(B)
    Dwellings 210.52
    Electrical Service Areas 210.64
    Meeting Rooms 210.65
Service Disconnecting Means 230.71
    Grouping of 230.72
Service-Entrance Conductor Sets 230.40
Services to a Building 230.2
Supplies to More Than One Building or Structure 225.30
    Disconnecting Means for Each Supply 225.33
    Grouping of Disconnects 225.34

NURSING HOMES AND LIMITED CARE FACILITIES
Definition, Limited Care Facility 517.2
Definition, Nursing Home 517.2
Essential Electrical System 517.40
    Capacity of System . . . . 517.42(C)
    Equipment Branch 517.44
    Life Safety Branch 517.43
    Power Sources . . . . 517.41(A)
    Receptacle Identification . . . . 517.42(E)
    Separation from Other Circuits . . . . 517.42(D)
    Transfer Switches . . . . 517.42(B)
Wiring and Protection . . . . Art. 517 Part II
    Spaces Used Exclusively for Patient Sleeping Rooms 517.10(B)(2)
See HEALTH CARE FACILITIES . . . . Ferm's Finder

O

OCCUPANCIES
Assembly . . . . Art. 518

OFFICE FURNISHINGS . . . . Art. 605
Application
    Covered . . . . 605.1(A)
    Not Covered . . . . 605.1(B)
Cords Permitted
    Freestanding-Type, Cord- and Plug-Connected . . . 605.9(A)
    Lighting Accessories . . . . 605.6(B)
Covered by Article . . . . 605.1(A)
Definition 605.2
Fixed-Type Office Furnishings 605.7
Freestanding Type Office Furnishings
    Cord- and Plug-Connected 605.9
    Not Fixed 605.8
General Information 605.3
Hazardous (Classified) Locations . . . . 605.3(B)
Interconnections 605.5
Lighting Accessories 605.6
    Connection . . . . 605.6(B)
    Listed 605.6
    Receptacle Outlet in, Not Permitted . . . . 605.6(C)
    Support . . . . 605.6(A)
Multiwire Circuits Not Permitted in . . . . 605.9(D)
Not Covered by Article . . . . 605.1(B)
Office Furnishing Interconnections 605.5
Power Supply
    Fixed-Type Office Furnishings 605.7
    Freestanding-Type Office Furnishing, Cord- and
    Plug-Connected 605.9
    Freestanding-Type Office Furnishing, Not Fixed 605.8
Receptacles
    Located Not More Than 300 mm (12 in.) from Partition   605.9(B)
    Maximum Number in Office Furnishings or Groups   605.9(C)
    Supplying Power to on Separate Circuit . . . . 605.9(B)
Uses Permitted . . . . 605.3(A)
Wireways 605.4
See (QAWZ) . . . . UL Product iQ
See (QAXB) . . . . UL Product iQ

OFFICE TRAILERS 545.1
Definition of Relocatable Structures- Info Note 545.2

See MOBILE HOMES . . . . Ferm's Finder

OPEN BOTTOM EQUIPMENT
Mechanical Continuity of Raceways and Cables 300.12

OPEN KNOCKOUT (Wiring through) . . . . 300.4(G)
Boxes 314.17

OPEN WIRING . . . . Art. 398
Clearance from Piping, Exposed Conductors, etc. 398.19
Conductor Supports 398.30
Conductor Supports, Mounting . . . . 398.30(D)
Conductors, Type. 398.104
Definition. 398.2
Devices 398.42
Dry Locations. . . . . 398.15(A)
Entering Spaces Subject to Wetness, Dampness or
   Corrosive Vapors . . . . 398.15(B)
Exposed to Physical Damage . . . . 398.15(C)
For Service Entrance . . . . 230.43(1)
   Entering Buildings or Other Structures. 230.52
   Mounting Supports . . . . 230.51(C)
   Protection, Aboveground . . . . 230.50(B)
For Temporary Wiring, Prohibited (General) . . . 590.4(B)&(C)
   Exceptions
      Branch Circuits, Emergencies and Tests and
      90-day Maximum for Holiday Lighting, etc. 590.4(C) Ex.
      Feeders for Emergencies and Tests Only . . . . 590.4(B) Ex.
In Accessible Attics 398.23
Through Walls, Floors, Wood Cross-Members, etc. . . . 398.17
Tie Wires . . . . 398.30(E)
Uses Not Permitted 398.12
Uses Permitted 398.10
See OUTSIDE BRANCH CIRCUITS & FEEDERS . Ferm's Finder

OPENINGS ADEQUATELY CLOSED
Conductors Entering Boxes, etc. . . . . 314.17(A)
Spread of Fire or Products of Combustion 300.21
Unused Openings, Boxes and Conduit Bodies . . . . 110.12(A)
   Cabinet Cutout Boxes and Meter Socket Enclosures . 312.5(A)

OPENINGS, APPROVED FOR DRAINAGE (BOXES, ETC.) . 314.15

OPERATING ROOM RECEPTACLES . . . . 517.19(C)
OPTICAL FIBER CABLES AND RACEWAYS . . . . Art. 770
Applications of Listed Optical Fiber Cables 770.154
Cable Marking . . . . Table 770.179
Cable Routing Assemblies . . . . 770.110(C)
Cable Substitutions . . . . Table 770.154(b)
Cable Trays . . . . 770.110(D)
Circuit Integrity Cable . . . . 770.179(E)(1)
Class 1, Division 1 . . . . 501.10(A)(1)(3)
Class 1, Division 2 . . . . 501.10(B)(1)(7)
Class 2, Division 1 . . . . 502.10(A)(1)(4)
Class 2, Division 2 . . . . 502.10(B)(1)(8)
Definitions . . . . Art 100
Ducts for Dust, Loose Stock, or Vapor Removal . . . . 770.3(B)
Electrical Circuit Protective System, Defined 770.2
Field Assembled. . . . . 770.179(F)
Fire Resistance, Listing for . . . . 770.179(A)
Fire-Resistive Cables . . . . 770.179(E)(2)
Fire Spread 770.26
Grounding 770.93, 770.100
   Intersystem Bonding Termination . . . . 770.100(B)(2)
   Length of Conductor. . . . . 770.100(A)(4)
   Listed or Part of Listed Equipment. 770.180
   Metallic Entrance Conduit 770.49
   Non-Current-Carrying Conductive Members 770.106
Hazardous Locations . . . . 770.3(A)
Innerduct, Defined 770.2
Installation of Optical Fibers and Electrical Conductors 770.133
Listing, Marking, and Installation. 770.113
Listing Requirements 770.179
Mechanical Execution of Work 770.24
Overhead (Aerial) 770.44
Substitutions 770.154
Types 770.2
Underground Cables Entering Building 770.47
Unlisted Cables Entering Buildings 770.48
See (QAYK) . . . . UL Product iQ
See (QAZM) . . . . UL Product iQ

OPTIONAL STANDBY SYSTEMS . . . . Art. 702
Capacity and Rating of 702.4
Circuit Wiring 702.10
Definition. 702.2
Emergency Disconnect . . . . 702.7(A)
Equipment Load Selection 702.4
Generator Sets

Outdoor, Permanently Installed . . . . 702.12(A)
Outdoor, Portable . . . . 702.12(B)
Grounding, Portable Generators 702.11
    Nonseparately Derived System . . . . 702.11(B)
    Separately Derived System . . . . 702.11(A)
Outdoor Generator Sets. 702.12
    Permanently Installed . . . . 702.12(A)
    Portable- 15kW or Less . . . . 702.12(B)
    Portable- Greater the 15 kW . . . . 702.12(A)
    Portable- Power Inlets, 100 Amperes or Greater . . 702.12(C)
Power Inlet . . . . 702.7(C)
Signals 702.6
Signs . . . . 702.7(A)
    At Disconnect . . . . 702.7(A)
    At Grounding Location . . . . 702.7(B)
    At Power Inlet . . . . 702.7(C)
Transfer Equipment. 702.5
    Automatic . . . . 702.4(B)(2)
    Manual. . . . . 702.4(B)(1)
    Short-Circuit Current Rating 702.5

ORDERLY SHUTDOWN
Coordination (Selective) . . . . Art. 100 Part I
Electrical System Coordination 240.12
Ground-Fault Protection of Equipment (Branch Circuit)
. . . . 210.13 Ex. 1
Ground-Fault Protection of Equipment (Feeder) . . 215.10 Ex. 1
Ground-Fault Protection of Equipment (General) 240.13
Ground-Fault Protection of Equipment, (Not Applicable) . . . .
. . . . 230.95 Ex. 1
Ground-Fault Protection of Equipment (Service) . . 230.95 Ex.
Integrated Electrical Systems . . . . 685 Part II
Insulation Level- 173 Percent . . . . .311.10(C)(3)
Motor Overload Protection 430.44
Motor Overtemperature Protection . . . . .430.126(C)
Overcurrent Protection for Generators Deemed Essential 445.12 Ex.
Overcurrent Protection, Cranes and Hoists . . . . . . 610.53(B)
Overload Protection, Elevators, etc. . . . .620.61(B) Info. Note
Permanent Amusement Attractions 522.25
Power Loss Hazard . . . . 240.4(A)

ORGANS, PIPE . . . . **Art. 650**
Conductors. 650.6
Definitions. 650.2
Overcurrent Protection. 650.8

Protection from Accidental Contact 650.9
Sources of Energy 650.4

OUTDOOR
Lighting
    Equipment Installed Outdoors 225.7
    Lampholders 410.96
    Location of Outdoor Lamps 225.25
    Outdoor Lampholders 225.24
    Supported by Trees. . . . . 410.36(G)
    Wet and Damp Locations . . . . 410.10(A)
    See FESTOON LIGHTING . . . . .Ferm's Finder
Receptacles
    Damp and Wet Locations . . . . 406.9(A)&(B)
    GFCI Requirements . . . . 210.8(A)(3)
    Required, One- and Two-Family Dwellings . . . . 210.52(E)

OUTDOOR, OVERHEAD CONDUCTORS, OVER 1000 VOLTS
. . . . .Art. 399
Defined 399.2

OUTLET
Definition of . . . . Art. 100 Part I
    Lighting Outlet, Definition of . . . . Art. 100 Part I
        See LIGHTING, Lighting Outlets. . . . . . .Ferm's Finder
    Receptacle Outlet, Defined. . . . . Art. 100
        See RECEPTACLE . . . . .Ferm's Finder

OUTLET BOX **314.27**
Ceiling . . . . .314.27(A)(2)
Ceiling Fan . . . . 314.27(C)
Extra Duty . . . . 406.9(B)
Floor . . . . 314.27(B)
Separable Attachment Fittings . . . . 314.27(E)
Utilization Equipment . . . . 314.27(D)
Vertical Surface . . . . .314.27(A)(1)

OUTLINE LIGHTING . . . . **Art. 600**
Definition. . . . . Art. 100 Part I
Electric-Discharge Systems 1000 Volts or Less . . . . Art. 410 Part XII
Electric-Discharge Systems More than 1000 Volts . . Art. 410 Part XII
See SIGNS, ELECTRIC AND OUTLINE LIGHTING . . . .
. . . . .Ferm's Finder

OUTPUT CIRCUITS
Amplifiers . . . . .640.9(C)
Characteristics, Interconnected Power Production Source. . .705.14

Fuel Cell Systems . . . . Art. 692
Induction and Dielectric Heating Equipment 665.5
Solar Photovoltaic Systems . . . . Art. 690

OUTSIDE
Fire Pump Supply Conductors . . . . 695.6(A)
Service Conductors Considered 230.6
See OUTDOOR . . . . Ferm's Finder

OUTSIDE BRANCH CIRCUITS & FEEDERS . . . Art. 225
Attached to Buildings or Structures. 225.11
Clearance, Not over 1000 Volts
    Final Spans. . . . . 225.19(D)
    From Buildings. 225.19
    From Finish Grade, Sidewalks, or Platforms 225.18
    From Nonbuilding or Nonbridge Structures . . . . 225.19(B)
    Horizontal Clearances . . . . 225.19(C)
    Over Roofs . . . . 225.19(A)
    Over Swimming Pools 680.9
    Zone for Fire Ladders . . . . 225.19(E)
Clearance, Over 1000 Volts
    Over Buildings and Other Structures. 225.61
    Over Roadways, Walkways, Rail, Water, and Open Land. . 225.60
    See Life Safety Code, NFPA 101. . . . . NFPA 101
Common Neutral, for Lighting . . . . 225.7(B)
Communications Circuits . . . . 805 Part I
Conductor Covering 225.4
Conductor Size and Support 225.6
    Overhead Spans . . . . 225.6(A)
        Vegetation Such As Trees Not for Support 225.26
    Festoon Lighting . . . . 225.6(B)
Disconnecting Means
    Access 225.35
    Construction 225.38
    General 225.34
    Identification 225.37
    Over 1000 Volts, Location . . . . 225.52(A)
    Over 1000 Volts, Not Readily Accessible . . . . . . 225.52(A)
    Over 1000 Volts, Pre-Energization and Operating Tests . . . . .
    . . . . .225.56, 110.41
    Overcurrent Protective Devices, Access 225.40
    Rating 225.39
Type 225.36
Entering a Building or Structure 225.11
Exiting a Building or Structure 225.11

Fire Alarm Circuits 760.32
Lampholders 225.24
More Than One Building or Other Structure . . .Art. 225 Part II
    Access to Overcurrent Protective Devices. 225.40
    Disconnecting Means 225.31
        Access to Occupants 225.35
        Construction 225.38
        Grouping 225.34
        Identification 225.37
        Location 225.32
        Number 225.33
        Rating 225.39
        Suitable for Use as Service Equipment 225.36
    Grounding 250.32
    Number of Supplies 225.30
    Over 1000 Volts . . . . Art. 225 Part III
Open-Conductor Spacings 225.14
Size of Conductor .225.5
Wiring on Outside of Buildings or Other Structures . . . 225.10
    Raceways on Exterior of Buildings or Other Structures 225.22
    Using Service Entrance Cable 225.21
        Exterior Installations. . . . . 338.10(B)

OVENS & RANGES
Commercial
    Branch-Circuit Ratings (General) . . . . 210.19(A)(4)
    Calculated Load 220.56
    Disconnecting Means (General) 422.30
        Cord- and Plug-Connected . . . . 422.33(A)
        Permanently Connected . . . . 422.31(B)
        Unit Switch(es) As Disconnecting Means . . . . 422.34(D)
    Individual Branch Circuit Ratings . . . . 422.10(A)
    Overcurrent Protection. . . . . .422.11(A)(D)(F)
Flexible Cord .422.16(B)(3)
Grounding of Frames 250.140
    Use of Grounded Conductor . . . . 250.142(B) Ex.1
    In Mobile Homes 550.16
    In Park Trailers . . . . 552.55(C)
    In Recreational Vehicles . . . . 551.54(C)
Household Appliances
    Branch-Circuit Ratings (General) . . . . 210.19(A)(3)
    Calculation of Load 220.55
    Branch-Circuit Computations . . . . 220.14(B)
    Demand Factors . . . . Table 220.55

Optional Demand Factors, Mobile Homes . 550.18(B)(4)(5)
Optional Demand Factors, Park Trailers . . . 552.47(B)(4)(5)
Disconnecting Means (General) 422.30
    Cord- and Plug-Connected . . . . 422.33(A)&(B)
    Permanently Connected . . . . 422.31(B)
    Unit Switch(es) as Disconnecting Means . 422.34(A)(B)(C)
Individual Branch-Circuit Ratings . . . . 422.10(A)
Minimum Ampacity and Size . . . . 210.19(A)(3)
    Feeder or Service Neutral Load 220.61
    Neutral Conductor . . . . 210.19(A)(3), Ex. 2
    Tap Conductors . . . . 210.19(A)(3) Ex. 1
Overcurrent Protection . . . . 422.11(A)&(B)
Receptacle Ratings . . . . 210.21(B)(4)
See CALCULATIONS . . . . Ferm's Finder

OVER 1000 VOLTS, NOMINAL (High Voltage)

Aboveground Wiring Methods 300.37
Ampacity
    Conductors to 2000 Volts 310.14
        Allowable Ampacities . . . . . . Table 310.16 through 21
    Conductors 2001 to 35,000 Volts 311.60
        By Tables . . . . Table 311.60(C)67 through 86
        Definitions . . . . Art. 100, Part II
        Modifications to Table Ambients and Burial Depths. . . . .
        . . . . .311.60(D)(2)
        Under Engineering Supervision . . . . 311.60(B)
    See Informative Annex B for examples of formula applications
    See . . . . IEEE *Standard Power Cable Ampacity Tables*. . . . .
Bends (Conductor Radius) 300.34
Boxes, Junction and Pull . . . . Art. 314 Part IV
    General. 314.70
    Size of
        Angle or U Pulls. . . . . 314.71(B)
        Removable Sides. . . . . 314.71(C)
        Straight Pulls . . . . 314.71(A)
        Suitable Covers. . . . . 314.72(E)
Branch Circuits . . . . 210.19(B)
Busways . . . . Art. 368 Part II
Cable
    Medium Voltage Type MV . . . . Art. 311
    Portable . . . . Art. 400 Part III
        Splices 400.36
        Radius of Bends 300.34
Cable Tray . . . . Art. 392

Capacitors . . . . Art. 460 Part II
Clearance of Live Bare Parts 490.24
Clearances of Outside Branch Circuits and Feeders . Art. 225 Part III
    See OUTSIDE BRANCH CIRCUITS AND FEEDERS . . .
    . . . . Ferm's Finder
Clearances of Service Conductors
    See Life Safety Code, NFPA 101. . . . . NFPA 101
Conductors (General). 310.1
    Braid-Covered Insulated Conductors – Open Installation 300.39
    Direct Burial Conductors . . . . 310.10(E)
        Above 2000 Volts 311.36
    Insulation and Jacket Thickness Nonshielded Solid Types RHH and RHW
        Dielectric Insulated Conductors Rated 2400 Volts   311.36 Ex.
    Insulation Shielding 300.40
    Insulation Thickness, Nonshielded Dielectric Insulated Conductors
        Rated 2400 Volt . . . . Table 311.10(B)
    Insulation Thickness Shielded Solid Dielectric
        Insulated Conductors 2001 to 35,000 Volts Table 311.10(C)
        Radius of Bends 300.34
        Shielding, Above 2000 Volts. 311.44
    Type MV, Conductor Application and Insulation . . . . . .
    . . . . Table 311.60(C)(67) thru (86)
Conductors of Different Systems 300.32
Conductor Installations. . . . . 110.74(B)
Definitions, General . . . . Art. 100 Part II
    High Voltage .490.2
Disconnecting Means
    Equipment . . . . Art. 490
    Mobile and Portable Equipment . . . . 490.51(D)
    Outside Branch Circuits and Feeders . . . . Art. 225 Part III
    Service 230.205
        Isolating Switches 230.204
        Location . . . . 230.205(A)
        Overcurrent Devices As 230.206
        Permitted to be Located Not Readily Accessible . . 230.205
        Remote Control . . . . 230.205(C)
        Type . . . . 230.205(B)
        Doors . . . . .110.31(A)(3)
Electrode Type Boilers . . . . Art. 490 Part V
    Branch-Circuit Requirements 490.72
    Grounded Neutral Conductor . . . . 490.72(E)
    Bonding 490.74
    Supply System 490.71

Enclosure for Electrical Installations 110.31
   Conductor Installations 110.74
   Enclosed Equipment Accessible to Unqualified Persons 110.31(D)
   Indoor Installations
      Accessible to Qualified Persons Only . . . . . 110.31(B)(2)
      Accessible to Unqualified Persons . . . . 110.31(B)(1)
   Locked Rooms or Enclosures . . . . 110.34(C)
   Outdoor Installations
      Accessible to Qualified Persons Only . . . . . 110.31(C)(2)
      Accessible to Unqualified Persons . . . . 110.31(C)(1)
   See Definition of Qualified Person . . . . Art. 100
Entrance and Access to Work Space. 110.33
Equipment Over 1000 Volts
   Backfeed 490.25
   Circuit Breakers 490.45
   Circuit-Interrupting Devices 490.21
      Circuit Breakers . . . . 490.21(A)
      Distribution Cutouts and Fuse Links, Expulsion Type . . .
      . . . . 490.21(C)
      Load Interrupters . . . . 490.21(E)
      Oil-Filled Cutouts . . . . 490.21(D)
      Power Fuses and Fuseholders . . . . 490.21(B)
   Danger Signage
      Accessibility to Energized Parts 490.35
      Backfeed Installations 490.25
      Cable Connections. Portable 490.55
      Enclosures, Portable 490.53
      Fuseholders . . . . 490.21(B)(6)
   Definition Applying to Article 490.2
   Door Stops and Cover Plates 490.38
   Electrode-Type Boilers . . . . .Art. 490 Part V
   Enclosures in Wet or Damp Locations . . . . 490.3(B)
   Fused Interrupter Switches 490.44
   Gas Discharge from Interrupting Devices 490.39
   Grounding 490.36
   Isolating Means 490.22
   Metal-Enclosed Power Switchgear and Industrial Control Assemblies . . . . Art. 490 Part III
   Minimum Space Separation 490.24
   Mobile and Portable . . . . Art. 490 Part IV
   Oil-Filled Equipment . . . . 490.3(A)
   Specific Provisions . . . . Art. 490 Part II
   Substations 490.48
   Switchgear Used as Service Equipment 490.47
   Visual Inspection Windows 490.40
   Voltage Regulators 490.23
   Warning Signage
      Isolating Switches 490.22
      Substations . . . . 490.48(B)
Fences (Metal), Grounding and Bonding . . . . . . 250.194(A)
General Installations . . . . Art. 110 Part III
Grounding . . . . Art. 250 Part X
   Cable Shields. 490.47
   Connected to an Equipment Grounding Conductor . . 490.36
   Derived Neutral Systems 250.182
   Equipment 250.190
   Grounding Service-Supplied Alternating-Current Systems . . .
   . . . . 250.186
   Impedance Grounded Neutral Systems 250.187
   Of Equipment, Fences and Enclosures, etc. 250.190
   Of Systems Supplying Portable or Mobile Equipment 250.188
   Solidly Grounded Neutral Systems 250.184
Illumination about Electrical Equipment . . . . 110.34(D)
Impedance Grounded Neutral Systems 250.187
Locking, Circuit Breaker Capability 490.46
Locks . . . . 110.31(A)(4)
Manholes and Other Enclosures Intended for Personnel Entry . .
. . . . .Art. 110 Part V
Mobile & Portable Equipment . . . . Art. 490 Part IV
Moisture or Mechanical Protection for Metal-Sheathed Cables . .
. . . . 300.42
Motors . . . . Art. 430 Part XI
Motor Disconnecting Means 430.227
Motor Grounding, Connected to EGC 430.245
Outside Branch Circuits and Feeders . . . . . . Art. 225 Part III
   Isolating Switches 225.51
Overcurrent Protection . . . . Art. 240 Part IX
   Feeders and Branch Circuits 240.100
      Additional Requirements for Feeders 240.101
   Services. . . . . Art. 230 Part VIII
      Enclosed Devices . . . . 230.208(B)
      Overcurrent Device as Disconnecting Means . . . 230.206
      Protection Requirements 230.208
   Transformers . . . . 450.3(A)
Protection against Induction Heating 300.35
Resistors & Reactors . . . . Art. 470 Part II
Services . . . . Art. 230 Part VIII
Signs, Warning
   Distribution Cutouts and Fuse Links, Expulsion Type . . . .

  . . . . .490.21(C)(2)
 Fused Interrupter Switches . . . . 490.44(B)
 High-Voltage Fuses. . . . . 490.21(B)(7) Ex.
 Load Interrupters. 490.21(E)
 Mobile and Portable Equipment Enclosures 490.53
 Power Cable Connections to Mobile Machines. 490.55
 Pull and Junction Boxes . . . . 314.72(E)
 Rooms and Enclosures . . . . 110.34(C)
 Substations. . . . . 490.48(B)
 Vaults and Equipment Rooms . . . . 490.48(B)(1)
Structures (Metal), Grounding and Bonding . . . . .250.194(B)
Substations 490.48
Surge Arresters (see Overvoltage Protection) . . . . . . Art. 242
  Services (Lightning Arresters) 230.209
Switching Mechanism, Lock Remains in Place . . . . 490.44(C)
Temporary Wiring, Guarding 590.7
Transformers . . . . Art. 450
 Overcurrent Protection of . . . . 450.3(A)
 Specific Provisions Applicable to Different Types Art. 450 Part II
 Vaults . . . . 110.31(A)(5)
Tunnel Installations. . . . . Art. 110 Part IV
Underground Installations 300.50
 Backfill. . . . . 300.50(E)
 Industrial Establishments. . . . . 300.50(A)(2)
 Minimum Cover Requirements . . . . Table 300.50
 Other Nonshielded Cables . . . . 300.50(A)(3)
 Protection from Damage . . . . 300.50(C)
 Raceway Seal . . . . 300.50(F)
 Shielded Cables and Nonshielded Cables in Metal-Sheathed Cable
  Assemblies . . . . 300.50(A)(1)
 Splices . . . . 300.50(D)
Vaults
 Conductors Considered Outside Building . . . . . . 230.6(3)
 Doors . . . . .110.31(A)(3)
 Floors . . . . 110.31(A)(2)
 General Requirements for . . . . Art. 450 Part III
 Installations in 110.31
 Locks . . . . .110.31(A)(4)
 Separation from Low-Voltage Equipment . . . 110.34(B) Ex.
 Services Over 35,000 Volts 230.212
 Specific Provisions Applicable to Different Types of
  Transformers . . . . .Art. 450 Part II
 Transformers . . . . 110.31(A)(5)
 Walls and Roof . . . . 110.31(A)(1)

Warning Signs, Conductor Access in Conduit and Cable
 Systems 300.45
 Wet Locations . . . .
  Above Grade 300.38
  Underground . . . . 300.50(B)
Wiring Methods . . . . Art. 300 Part II
Working Space
 About Equipment 110.32
 Clear Space . . . . 110.34(A)
 Entrance and Access to 110.33

OVERCURRENT DEVICES
Not In Bathrooms . . . . 240.24(E)
Not In Clothes Closets . . . . 240.24(D)
Not Over Steps . . . . 240.24(F)

OVERCURRENT PROTECTION . . . . **Art. 240**
Access to (Location of) . . . . .Art. 240 Part II
 Circuit Breakers Used as Switches . . . . 404.8(A)
 In Circuit 240.21
 In or on Premises. 240.24
 More Than One Building or Structure 225.40
 Multiple-Occupancy Building . . . . 230.72(C)
  By Each Occupant . . . . 240.24(B)
  Locked Service Overcurrent Devices. 230.92
Air-Conditioning & Refrigeration Equipment . . . . .240.4(G)
 Branch-Circuit Short-Circuit & Ground-Fault . . Art. 440 Part III
 Location of 440.14
 Motor-Compressor & Branch-Circuit Overload Art. 440 Part VI
Ampere Rating 240.6
 Adjustable-Trip Circuit Breakers . . . . 240.6(B)
 Fuses and Fixed-Trip Circuit Breakers . . . . 240.6(A)
 Restricted Access Adjustable-Trip Circuit Breakers . . 240.6(C)
 Table-Standard Ampere Rating . . . . Table 240.6(A)
Appliances 422.11
Branch-Circuit Conductors and Equipment 210.20
 Overcurrent Protective Device, Branch-Circuit
 Definition of . . . . Art. 100
 Overcurrent Protective Device, Supplementary
 Definition of . . . . Art. 100
Branch-Circuit Taps . . . . 210.19(A)(4) Ex. 1
 General Requirements . . . . 240.21(A)
 Protection of Conductors 240.4
Breakers (Up Position Must Be On Position) 240.81
 Used As Switches 404.7

See BREAKERS & FUSES . . . . Ferm's Finder
Busways 368.17
   Busway Taps . . . . 240.21(E)
   Feeders or Branch Circuits 368.17
   Feeders . . . . 368.17(A)
   Rating for Branch Circuit . . . . 368.17(D)
   Reduction in Size . . . . 368.17(B)
Capacitors . . . . 240.4(G)
   Conductors . . . . 460.8(B)
   For Units 460.25
   See Power Factor Correction Capacitors*Ferm's Charts and Formulas*
Cartridge Fuses . . . . Art. 240 Part VI
   Classification of 240.61
   See FUSES . . . . Ferm's Finder
Class 1 Remote Control and Signaling Circuits 725.41
Class 2 and 3 Remote Control and Signaling Circuits . 725.121
   Conductors 240.4
   Limitations Chapter 9 . . . . Tables 11(A)&(B)
   Taps 240.21
Conductors from Generator Terminals . . . . 240.21(G)
Control Circuits
   Cranes 610.53
   Elevators . . . . 620.61(A)
   Motors 430.72
   Others (General) 240.4
      Specific Applications . . . . 240.4(G)
Cords (Flexible) 240.5
   Ampacity to Be Used for Determining . . . .
   . . . . Table 400.5(A)(1) & Table 400.5(A)(2)
Cranes & Hoists . . . . Art. 610 Part VI
   Control Circuits, Creating Hazard . . . . 610.53(B)
Definition of
   Overcurrent Protective Device, Branch-Circuit . . . . Art. 100
   Current-Limiting Overcurrent Protective Device 240.2
   Overcurrent . . . . Art. 100 Part I
Deicing & Snow-Melting Equipment 426.4
   Nonheating Leads . . . . 210.19(A)(4) Ex.1(e)
Devices (Fuses or Circuit Breakers) in Parallel 240.8
Devices Rated 800 Amperes or Less . . . . 240.4(B)
Devices Rated Over 800 Amperes . . . . 240.4(C)
Direct Current Systems (General) 240.1
Disconnecting Means Ahead of Fuses 240.40
Electrical System Coordination 240.12
   Ground-Fault Protection, Health Care Facilities . . . .

. . . . 517.17(B) and (C)
Electric Vehicle Charging Systems 625.41
Electric Welders . . . . 240.4(G)
   Arc Welders . . . . Art. 630 Part II
      For Conductors . . . . 630.12(B)
      For Welder . . . . 630.12(A)
   Resistance Welders . . . . Art. 630 Part III
      For Conductors . . . . 630.32(B)
      For Welders . . . . 630.32(A)
Electroplating 669.9
Elevators, Dumbwaiters, Escalators, Moving Walks,
   Wheelchair & Stairway Chair Lifts . . . . . Art. 620 Part VII
Emergency Systems . . . . Art. 700 Part VI
Feeder Taps Not Over 3 m (10 ft) . . . . 240.21(B)(1)
Feeder Taps Not Over 7.5 m (25 ft) . . . . 240.21(B)(2)
   Transformer (Primary Plus Secondary) . . . . . 240.21(B)(3)
Feeder Taps Over 7.5 m (25 ft) (High Bay Manufacturing) . . .
. . . . 240.21(B)(4)
Feeder Taps Outside, Unlimited Length . . . . . . 240.21(B)(5)
Feeders 240.4
   General 215.3
   Taps . . . . 240.21(B)
Fire Protective Signaling Systems . . . . 240.4(G)
   NPLFA Circuit Conductors 760.43
      Limitations . . . . Chapter 9 Tables 12(A) and 12(B)
      Location of 760.45
Fire Pumps . . . . 695.4(B)(1)(a)(1)
   Feeder Sources . . . . 695.5(C)(2)
   Overload Protection . . . . 695.6(C)
   Transformers Supplying . . . . 695.5(B)
(Fixture) Luminaire Wire 240.5
   Ampacity To Be Used for Determining . . . . . . Table 402.5
   General 402.14
   Supplementary Protection 240.10
   Overcurrent Protective Device, Supplementary, Definition of .
. . . . Art. 100 Part I
Fuel Cell Systems 692.9
   Conductor Ampacity, Relative to . . . . 692.8(B)
Fuses, *See* FUSES . . . . Ferm's Finder
Generators 445.12
Heating Equipment (Space) 424.22
Heating Equipment for Pipelines & Vessels 427.4
Heating Equipment, Induction & Dielectric 665.11
Industrial Machinery . . . . 670.4(B) & (C)

Irrigation Machines (Electrically Driven) . . . . Art. 675
Lampholders . . . . 210.21(A)
 Not for Plug Fuses 410.90
 Summary Requirements 210.24
Location of, in Circuit 240.21
Location of, in or on Premises
 Circuit Breakers Used As Switches . . . . 404.8(A)
 For Service Equipment . . . . Art. 230 Part VII
 General 240.24
 Location Related to Service Disconnecting Means . . . 230.91
 Not Exposed to Physical Damage . . . . 240.24(C)
 Not Permitted in Clothes Closets or in Bathrooms 240.24(D)&(E)
 Protection of Specific Circuits 230.93
Locked, Sealed or Not Readily Accessible
 More than One Building or Structure 225.40
 Service Overcurrent Device 230.92
 Specific Circuits 230.93
Metal Working Machine Tools, etc. . . . . 670.4(C)
Motion Picture Studios 530.18
 Feeders . . . . 530.18(B)
 Lighting, Other Than Stage Set. . . . . 530.18(G)
 Location Boards . . . . 530.18(D)
 Plugging Boxes . . . . 530.18(E)
 Stage or Set Cables . . . . 530.18(A)
 Substations, DC Generators 530.63
Motor Circuits & Feeders
 Branch-Circuit Short-Circuit & Ground-Fault Art. 430 Part IV
 Feeder Short-Circuit & Ground-Fault . . . . . Art. 430 Part V
 Motor and Branch-Circuit Overload Protection Art. 430 Part III
 Over 1000 Volts 430.225
Motor & Motor Control . . . . 240.4(G)
 Control Circuits 430.72
Motor-Operated Appliances (General) . . . . 240.4(G)
 Specific . . . . 422.11(B) through (G)
Multiple-Occupancy Buildings
 See MULTIPLE-OCCUPANCY BUILDINGS . Ferm's Finder
 See OVERCURRENT PROTECTION, Access to (Location of)
 . . . . Ferm's Finder
Next Higher Size Permitted . . . . 240.4(B)
 Individual Motor Circuit . . . . . . 430.52(C)(1) Ex. 1 & 2
No Overcurrent Device in the Grounded Conductor
 Fuses Used as Overload Protection 430.36
 General Restriction 240.22
 Service Equipment . . . . 230.90(B)

Not Located Over Steps of a Stairway . . . . 240.24(F)
Not Required Fire Pump Power Wiring, Short Circuit Only . . .
. . . . 695.6(C)
 Fire Pump Transformer Secondary . . . . 695.5(B)
 Generators . . . . 445.12 Ex.
 Power Loss Hazard . . . . 240.4(A)
 Selection . . . . 695.4(B)(2)
Orderly Shutdown
 Ground-Fault Protection of Equipment . . . . . . 240.13(1)
 Integrated Electrical Systems 685.10
 Motors 430.44
 Power Loss Hazard . . . . 240.4(A)
 Selective Coordination (Emergency Systems) 700.32
 Selective Coordination (Legally Required Standby Systems) . .
 . . . . 701.32
 Services, GFPE . . . . 230.95 Ex.
 System Coordination 240.12
Note: If immediate automatic shutdown of equipment would increase personnel hazard and equipment damage, it is permissible to connect the overload protective device to an alarm system instead of causing immediate interruption of the circuit, so corrective action or an orderly shutdown can be initiated.
Organs (Pipe) 650.8
Outlet Devices 210.21
 Rated Less Than 800 Amperes . . . . 240.4(B)(1)
Outside Branch Circuits and Feeders 225.3
Over 1000 Volts, Nominal . . . . Art. 240 Part IX
 Enclosed Devices . . . . 230.208(B)
 Equipment Requirements . . . . Art. 490
 Services, Overcurrent Device As Disconnecting Means 230.206
 Transformers . . . . 450.3(A)
Panelboards 408.36
 Maximum Number in Panelboards 408.54
Phase Converters (General) . . . . 240.4(G)
 Supply Conductors and Converter 455.7
Pipe Organs 650.8
Places of Assembly, Power Outlets 518.5
Plug Fuses . . . . Art. 240 Part V
 Not in Lampholders 410.90
Power Loss Hazard . . . . 240.4(A)
Protection of Conductors 240.4
Readily Accessible Place
 Accessibility . . . . 240.24(A)
 Services 230.92
 Switches and Circuit Breakers . . . . 404.8(A)

Receptacles 210.21
    Conductors Feeding, Under 800 Amperes . . . . . . 240.4(B)
    Summary Requirements 210.24
Remote Control, Signaling, Power Limited . . . . 240.4(G)
Resistance-Type Boilers 424.72
Selective Coordination
    Critical Operations Power Systems (COPS) 708.54
    Elevators, Etc. 620.62
    Emergency Systems 700.32
    Legally Required Standby Systems 701.32
Selectivity, GFPE, Health Care Facilities . . . . 517.17(C)
Sensitive Electronic Equipment . . . . 647.4(A)
Separate Buildings or Structures 225.3
Separately Derived Systems . . . . 240.4(F)
    Conductor Protection . . . . 240.21(C)
    Panelboard Supplied through a Transformer . . . . 408.36(B)
    Sensitive Electronic Equipment, 60 Volts to Ground 647.4(A)
    Supervised Industrial Installations, Feeder Taps. . . 240.92(B)
    See SEPARATELY DERIVED SYSTEMS . . . . Ferm's Finder
Service Conductors
    Location of. 230.91
    Over 1000 Volts As Service Disconnect 230.206
    Relative Location of Device 230.94
    Service Equipment 230.90
Services . . . . Art. 230 Part VII
Signs & Outline Lighting, Maximum Rating for . . . . 600.5(B)
Small Conductors, 16 and 18 AWG . . . . 240.4(D)
Solar Photovoltaic Systems 690.9
Standard Ampere Ratings 240.6
    Adjustable-Trip Circuit Breakers . . . . 240.6(B)
    Fuses and Fixed-Trip Circuit Breakers . . . . 240.6(A)
    Restricted Access Adjustable-Trip Circuit Breakers . . 240.6(C)
Supervised Industrial Installations . . . . . . Art. 240 Part VIII
    Definition 240.2
    Location in Circuit
        Feeder and Branch-Circuit Conductors . . . . . 240.92(A)
        Outside Feeder Taps. . . . . 240.92(D)
        Protection by Primary Overcurrent Device . . . 240.92(E)
        Transformer Secondary Conductors, Separately
        Derived Systems . . . . 240.92(C)
Supplementary Overcurrent Protection
    Appliances with Resistance-Type Elements Over 48 Amperes .
    . . . . 422.11(F)
    Fixed Electric Space-Heating 424.22

Not As Substitute for Branch-Circuit Devices 240.10
    Tapped Motor Control Circuits . . . . 430.72(A)
Switchboards & Panelboards 408.36
    Protection of Instrument Circuits 408.52
Tap Conductors . . . . 240.4(E)
    Battery Conductor Taps . . . . 240.21(H)
    Busway Taps . . . . 240.21(E)
    Definition of 240.2
    From Generator Terminals . . . . 240.21(G)
    Motor Circuit Taps. . . . . 240.21(F)
    Service Conductor Taps . . . . 240.21(D)
    Transformer Secondary taps . . . . 240.21(C)
Theaters
    Dimmers 520.25
    Portable Equipment Other Than Switchboards . . . 520.62(B)
    Portable Switchboards on Stage Overcurrent Protection 520.53
    Receptacle Circuits . . . . 520.62(B)
    Road Show Connection Panel . . . . 520.50(C)
Thermal Devices Not Designed for Short-circuit Use 240.9
Transformers 450.3
    1000 Volts or Less . . . . 450.3(B)
    Autotransformers, 1000 Volts or Less . . . . 450.4(A)
    Ground Reference for Fault Protection Devices . . 450.5(B)(2)
    Grounding Autotransformers Three-Phase 4-Wire 450.5(A)(2)
    Over 1000 Volts . . . . 450.3(A)
    Panelboard Supplied from . . . . 408.3(C)
    Secondary Conductors (General) . . . . 240.4(F)
    Secondary Conductors (Specific) . . . . 240.21(C)
    Secondary Ties . . . . 450.6(B)
    Supplying Transformer, Primary Plus Secondary Not Over
    7.5 m (25 ft) . . . . 240.21(B)(3)
    Voltage (Potential) Transformers . . . . 450.3(C)
Ungrounded Conductors 240.15
Vertical Position, Enclosures 240.33
Wet Locations 240.32
    Cabinets and Cutout Boxes 312.2
    Used As Switches 404.4
X-Ray Equipment
    Health Care Facilities 517.73
    Industrial, Nonmedical and Nondental Use 660.6

OVERLOAD
Definition . . . . Art. 100
Example- How Calculated . Informative Annex D, Example D8

Protection

    Appliances- Motor Operated . . . . 422.11(G)

    Cranes and Hoists 610.43

    Devices (other than fuses) . . . . . 430.37 and Table 430.37

    Elevators, Etc. . . . . 620.21(B)

    Fire Pumps- not have. . . . . 695.6(C)

    Motor . . . . Art. 430 Part III

    Power Loss Hazard (created) . . . . 240.4(A)

    Service Conductors. 230.90

Thermal Devices (not for protection of conductors). 240.9

P

P CABLE (TYPE P CABLE) (ARMORED AND UNARMORED)
. . . . .Art. 337

Ampacity of. 337.80

Armor. 337.116

Bending Radius 337.24

Conductors 337.104

Definition. 337.2

Equipment Grounding Conductor 337.108

Installation . . . . Art. 337 Part II

Insulation 337.112

Jacket 337.115

Listing Requirements 337.6

Marketing. 337.120

Scope 337.1

Securing and Supporting 337.30

Shield 337.114

Single Conductors 337.31

Uses Not Permitted 337.12

Uses Permitted 337.10

PADDLE (CEILING-SUSPENDED) FANS

Bathtubs and Shower Areas above. . . . . 410.10(D)

Outlet Box, Support of . . . . 314.27(C)

Separable Attachment Fitting . . . . 314.27(E)

Spare Separately Switched Ungrounded Conductors . 314.27(C)

Spas and Hot Tubs, Indoor . . . . 680.43(B)

Support of 422.18

Swimming Pools . . . . 680.22(B)

See also CEILING-SUSPENDED (PADDLE) FANS . Ferm's Finder

PAINT SPRAYING . . . . Art. 516
See SPRAY APPLICATION, DIPPING, AND COATING
. . . . Ferm's Finder

See (QEFY) . . . . .UL Product iQk

PANELBOARDS, SWITCHGEAR & SWITCHBOARDS Art. 408

Accessibility 240.24

    Containing Switches or Circuit Breakers 404.8

    Of Energized Parts over 1000 Volts. 490.35

    To Occupants . . . . 230.72(C)

Bare Metal Parts, Minimum Spacing 408.56

Barriers . . . . 312.11(D)

    Conductors in Same Vertical Section. . . . . . . 408.3(A)(2)

    In Service Switchboards and Switchgear . . . . . 408.3(B)

Bending Space for Wire Within Panelboard Enclosure

    Back Wire-Bending Space . . . . 408.55(C)

    Side Wire-Bending Space. . . . . 408.55(B)

    Top and Bottom Wire-Bending Space . . . . . . 408.55(A)

Bonding, Health Care 517.14

Breakers (Up Position On) 240.81

In Enclosures 404.7

Cable Entering . . . . 312.5(C)

    Deflection of Conductors 312.6

Clearances

    Around Switchboards and Switchgear . . . . . . 408.18(B)

    Clear Spaces . . . . 110.26(B)

    From Ceilings, Switchboards and Switchgear. . . 408.18(A)

Clearance of Bare Live Parts

    Auxiliary Gutters, In . . . . .366.100(E)

    Bus Enclosures 408.5

    Cabinets and Cutout Boxes, in . . . . 312.11(A)(3)

    Minimum Spacings Bare Metal Parts 408.56

    Over 1000 Volts 110.34

        Minimum Space Separation 490.24

    Under 1000 Volts 110.26

Conductor Bending Space Within Panelboard Enclosure

    Back Wire-Bending Space . . . . 408.55(C)

    Side Wire-Bending Space. . . . . 408.55(B)

    Top and Bottom Wire-Bending Space . . . . . . 408.55(A)

Conductors Feeding through (General) . . . . 312.8(A)

    Isolated Equipment Grounding Conductor . . . 250.146(D)

    Not Required to Terminate in . . . . 408.40 Ex.

Conduit Risers Entering Bottom Shall Not Exceed 75 mm (3 in.)
. . . . .408.5

Damp or Wet Locations

Cabinets and Cutout Boxes 312.2
Containing Overcurrent Devices. 240.32
Panelboards 408.37
Switchboards and Switchgear. 408.16
Dead Front (Required) 408.38
Dedicated Equipment Space . . . . 110.26(E)
Over 1000 Volts . . . . 110.34(F)
Definition of (Panelboard, Switchboard, Switchgear)Art. 100 Part I
Enclosure 408.38
Field Identification Required 408.4
Face-Up orientation (Prohibited) 408.43
Grounding . . . . 408.3(C)
Of Panelboards 408.40
Supplying Swimming Pool Equipment 680.25
Grounding Terminal Bar (Required) 408.40
Health Care Facilities 517.14
Critical Care (Category 1) Spaces . . . . 517.19(D)
High-Leg Identification. . . . . 408.3(F)
High-Leg Marking 110.15
High-Leg Phase Arrangement . . . . 408.3(E)
Identification of Disconnecting Means 110.22
Circuits and Modifications. 408.4
Illumination . . . . 110.26(D)
Over 1000 Volts . . . . 110.34(D)
Marking of Panelboards. 408.58
See Marking Guide for Panelboards . . . . . . *UL Product iQ*
Maximum Number of Overcurrent Devices 408.54
Minimum Spacing, Bare Minimum Parts. 408.56
Mounting of (General Equipment) 110.13
Damp and Wet Locations 312.2
Position in Wall 312.3
No Delta Breakers . . . . 408.36(C)
Open Bottom Equipment, Mechanical Continuity of Raceways and Cables. 300.12
Orientation (face-up prohibited) 408.43
Overcurrent Protection 408.36
Power Monitoring Equipment . . . . 312.8(B)
Readily Accessible
Air-Conditioning or Refrigeration Equipment 440.14
Location in or on Premises (General). 240.24
Motor Disconnecting Means 430.107
Service Equipment . . . . 230.70(A)(1)
Switches and Circuit Breakers . . . . 404.8(A)
Repairing Noncombustible Surfaces 312.4

Signs
Caution, Meet These Requirements . . . . 110.21(B)
Danger, Meet These Requirements . . . . 110.21(B)
Field-Applied. . . . . 110.21(A)
Warning, Meet These Requirements . . . . 110.21(B)
Source of Power, Marking. . . . . 408.4(B)
Spare Circuits Identified as Spares. 408.4
Splices in . . . . 312.8(A)
Split-Bus Panels (As Service Equipment) . . . . . 408.36 Ex. 1
Support of (General Equipment) . . . . 110.13(A)
Damp, Wet, or Hazardous Locations. 312.2
Position in Wall 312.3
Taps to Conductors . . . . 312.8(A)
Terminals . . . . .408.3(D)
Wet or Damp Locations
Cabinets and Cutout Boxes 312.2
Containing Overcurrent Devices 240.32
Panelboards 408.16
Switchboards. 408.16
Switchgear 408.16
Wire Bending Space
At Terminals . . . . 312.6(B)
For Conductors Entering Bus Enclosures 408.5
In Panelboards 408.55
Back Wire-Bending Space . . . . 408.55(C)
Side Wire-Bending Space . . . . 408.55(B)
Top and Bottom Wire-Bending Space . . . . . 408.55(A)
Working Clearance and Space . . . . 110.26(A)
Access and Entrance to. . . . . 110.26(C)
Existing Dwelling Units . . . . 110.26(A)(3) Ex. 1
Face-Up Position (prohibited) 408.43
Headroom . . . . 110.26(A)(3)
Over 1000 Volts 110.32
Entrance and Access to 110.33
Clear Space. 110.34
See (QEUY) . . . . *UL Product iQ*
See Switchboards (WEIR). . . . . *UL Product iQ*

PANEL, (SOLAR PHOTOVOLTAIC SYSTEMS)
Definition of 690.2

PANIC HARDWARE, LISTED
Over 1000 Volts . . . . .110.33(A)(3)
Storage Batteries 480.9
Under 1000 Volts . . . . .110.26(C)(3)

PARALLEL
Alternate Power Sources, Additional Service Permitted 230.2(A)(5)
    Interconnected Electric Power Production Sources . . Art. 705
Breakers & Fuses Not Permitted in 240.8
    Fused Switches 404.27
Cables and Raceways to Framing Members . . . . . . 300.4(D)
    Type NM Cable 334.17
Conductors 1/0 AWG and Larger . . . . 310.10(G)
    Ampacity Adjustment (Derating) . . . . 310.10(G)4
    Cable Tray Installations . . . . 392.20(C)
    Conductors of Same Circuit . . . . 300.3(B)(1)
    Conductors of Same Circuit Underground . . 300.5(I) Ex. 1-2
    Equipment Bonding Conductors . . . . 310.10(G)(5)
    Equipment Grounding Conductors . . . . .250.122(F)
    Exceptions to Minimum Size Rule . . . .310.10(G)(1) Ex. 1-2
    General Requirements for Paralleled Conductors . . 310.10(G)
    Run in Separate Raceways or Cables (Paralleled) .310.10(G)(3)
Elevator Traveling Cables . . . . 620.12(A)(1)
Equipment Bonding Jumpers . . . . 250.102(D)
Equipment Grounding Conductors . . . . .250.122(F)
Frequencies 360 Hz and Higher . . . . . . 310.10(G)(1) Ex. 1
Grounded Neutral Conductors, Existing Installations
    Under Engineering Supervision . . . . . 310.10(G)(1) Ex. 2
Raceways
    Conductor and Installation Characteristics . . 310.10(G)(2)
    Conductors in Same . . . . 300.3(B)(1)
    Conductors in Same, Underground . . . . . 300.5(I) Ex. 1-2
    Same Physical Characteristics . . . . 310.10(G)(2)
Service-Entrance Conductors . . . . 250.24(C)(2)
Stage Switchboard Feeders . . . . 520.27(A)
Supply-Side Bonding Jumper . . . . 250.102(C)(2)
Transformers
    Parallel Operation Permitted 450.7
    Secondary Ties 450.6

PARK TRAILERS . . . . Art. 552
Attachment Plug . . . . 552.44(C)
Bonding 552.57
Branch Circuits
    Determined- Number 552.46
    Protection 552.42
Calculations 552.47
Combination Electrical Systems 552.20
Conductors and Boxes 552.49
Cord 552.44
    Attachment Plugs . . . . 552.44(C)
    Cord Length . . . . 552.44(B)
    Labeling . . . . 552.44(D)
Definition of 552.2
Distribution Panelboard 552.45
    Grounded Conductor Insulated from Enclosure . . 552.45(A)
Grounding 552.55
Interior Equipment Grounding 552.56
Labels 552.5
Low-Voltage Systems 552.10
Luminaires (Lighting Fixtures) 552.54
Nominal 120- or 120/240-Volt Systems 552.40
Outdoor Outlets, Luminaires (Fixtures), Equipment . . . 552.59
Power Supply 552.43
Receptacle Configuration . . . . Figure 552.44(C)(1)
Receptacle Outlets Not Permitted . . . . 552.41(F)
Receptacle Outlets Required 552.41
    Ground-Fault Circuit-Interrupter Protection . . . 552.41(C)
    Pipe Heating Cable Outlet . . . . 552.41(D)
    Outdoor Receptacle Outlets . . . . 552.41(E)
Switches 552.52
Tests Required (Factory) 552.60
Wiring Methods 552.48

PART-WINDING MOTORS 430.4
Code Letter Markings . . . . 430.7(B)(5)
Single Motor Conductors . . . . 430.22(D)

PATH, GROUNDING
Effective Ground Fault Current Path, Defined . . . . . Art. 100
    For Grounded Systems . . . . 250.4(A)(5)
    For Ungrounded Systems . . . . 250.4(B)(4)
To Grounding Electrode At Service . . . . 250.24(D)

PATIENT BED LOCATION
Critical Care (Category 1) Space 517.19
Definition of 517.2
General Care (Category 2) Space 517.18
Multiwire Branch Circuit Prohibited 517.18

PATIENT CARE SPACES
Definitions within Patient Care Space
    Basic Care (Category 3) Space 517.2
    Critical Care (Category 1) Space 517.2
    General Care (Category 2) Space 517.2

Support (Category 4) Space 517.2
Critical Care (Category 1) Space 517.19
General Care (Category 2) Space 517.18
Wet Procedure Locations 517.20

PATIENT CARE VICINITY (Definition of) **517.2**
Grounding and Bonding (Optional) . . . . 517.19(D)

PEDIATRIC LOCATIONS
Listed Tamper Resistant Receptacles or Covers Required 517.18(C)

PENDANTS
See FIXTURES (LUMINAIRES) LIGHTING, Pendants   Ferm's Finder

PENINSULA COUNTERTOP
Peninsula Countertop Locations, Receptacle Outlets . 210.52(C)

PERFORMANCE TESTING
Ground Fault Protection of Equipment. . . . . 230.95(C)
SCADA Systems . . . . Informative Annex G

PERMANENT AMUSEMENT ATTRACTIONS
Conductors
    Ampacity. 522.22
    Overcurrent Protection. 522.23
    Size. 522.21
    Type 522.20
Control Circuits . . . . .Art. 522 Part II
Control Systems . . . . Art. 522
Definitions 522.2
Ungrounded Control Circuits 522.25
Wet Locations. 522.28
Wiring Methods, Control Circuits . . . . . . Art. 522 Part III
See   Article 525 for Carnivals, Circuses, Fairs, and Similar Events. . . . . .Ferm's Finder

PERMANENTLY INSTALLED GENERATORS
Grounding 250.35

PERMANENT WARNING SIGNS (REQUIRED)
See WARNING SIGNS. . . . . .Ferm's Finder

PERMISSIBLE LOADS
Individual Branch Circuits 210.22
Multiple-Outlet Branch Circuits 210.23

PERSON, QUALIFIED (Definition of) . . . . **Art. 100**

See QUALIFIED PERSON, WORK BY . . . . . . Ferm's Finder

PHASE CONVERTERS . . . . **Art. 455**
Capacitors 455.23
Conductors
    Ampacity. . . . . 455.6(A)
    Manufactured Phase Marking . . . . 455.6(B)
Definitions 455.2
Disconnecting Means 455.8
Equipment Grounding Connection 455.5
Marking (Nameplate Information) 455.4
Overcurrent Protection 455.7
Rotary-Phase Converter, Definition of 455.2
Single-Phase Loads Not Connected to Manufactured Phase 455.9
Specific Provisions Applicable for Different Types of . Art. 455 Part II
    Disconnecting Means, Static-Phase Converter 455.20
    Power Interruption, Rotary-Phase Converter. 455.22
    Start-Up, Rotary-Phase Converter 455.21
Static-Phase Converter, Definition of 455.2
Terminal Housings 455.10

PHOTOVOLTAIC SYSTEMS (SOLAR) . . . . **Art. 690**
Arc-Fault Circuit Protection (DC) 690.11
Auxiliary Grounding Electrode . . . . 690.47(B)
Calculations for Maximum Current of Circuit . . . . . 690.8(A)
Charge Control 690.72
Circuit Requirements . . . . .Art. 690 Part II
Circuit Sizing 690.8
Connection to Other Sources 690.59
Connectors 690.33
Correction Factors . . . . Table 690.31(A)(a)
Defined . . . . Article 100
Definitions Applicable to System 690.2
Disconnecting Means . . . . Art. 690 Part III
    Disconnect PV Equipment. 690.15
    Location . . . . 690.13(A)
    Marking . . . . 690.13(B)
    Maximum Number of . . . . 690.13(C)
    Not in Grounded Conductor 690.15
    Rapid Shutdown 690.12
    Ratings . . . . 690.13(D)
    Type of Disconnect. . . . . 690.13(E)
Disconnect Type . . . . 690.15(D)
Energy Storage Systems. 690.71
Equipment- Not Located in Bathrooms. . . . . 690.4(E)

Functional Grounded PV System 690.2
General Requirements 690.4
Ground-Fault Protection . . . . .690.41(B)
Grounding . . . . .Art. 690 Part V
Grounding Electrode System 690.47
Identification of Power Systems. 690.56
Interrupting Rating . . . . .690.15(B)
Large-Scale Photovoltaic (PV) Electric Power Production Facility
. . . . Art. 691
   Applicable PV Systems (no less than 5000 kW) 691.1
   Arc-Fault Mitigation 691.10
   Conformance of Construction to Engineered Design . .691.7
   Definitions. 691.2
   Direct Current Operating Voltage 691.8
   Disconnection of Photovoltaic Equipment 691.9
   Engineered Design .691.6
   Equipment Approval 691.5
   Fence Grounding. .691.11
   Special Requirements. 691.4
Listed Equipment .690.4(B)
Locations Not Permitted . . . . .690.4(E)
Marking. .Art. 690 Part VI
Modules- Alternating-Current (ac) 690.6
Multiple Inverters .690.4(D)
Overcurrent Protection 690.9
Qualified Personnel . . . . .690.4(C)
Rapid Shutdown 690.12
   Labels . . . . .690.56(C)
Serving a Building with an Electrical Supply System . .690.4(A)
Stand-Alone Systems 690.10
Storage Batteries . . . . Art. 690 Part VIII
Temperature Correction Factors . . . . Table 690.31(A)(a)
Wiring Methods and Materials . . . . Art. 690 Part IV
See SOLAR PHOTOVOLTAIC SYSTEMS . . . . Ferm's Finder
See (QHWJ) . . . . UL Product iQ
See (QIGU) . . . . UL Product iQ

PHYSICAL PROTECTION
Agricultural Buildings. . . . . .547.5(E)
Bushing. . . . . .300.5(H)
Conductors, Raceways, and Cables 300.4
Cables, Raceways and Boxes Installed Under Roof Decking . . 300.4(E)
Electrical Metallic Tubing (EMT) 358.2
Fixed Electric Heating Equipment for Pipelines and Vessels . . 427.11

Furring Strips . . . . .300.4(D)
Information Technology Equipment . . . . .645.5(D)
Insulated Fittings . . . . .300.4(G)
Intermediate Metal Conduit (IMC). 342.2
Notches in Wood . . . . 300.4(A)(2)
Parallel to Framing Members . . . . .300.4(D)
Park Trailers- Low-Voltage Wiring . . . . .552.10(C)(1)
Raceway (Minor Damage) . . . . 300.4 Info Note
Rigid Metal Conduit (RMC) 344.2
Safety-Control Equipment . . . . .725.31(B)
Shallow Grooves . . . . .300.4(F)
Transformer Secondary Conductors. . . . . .240.92(C)(3)

PIN AND SLEEVE TYPE PLUGS, RECEPTACLES AND CABLE CONNECTORS
See (QLGD) . . . . UL Product iQ

PIPE ELECTRODES . . . . .250.52(A)(5)
Connection of Grounding Electrode Conductor to . . 250.66(A)
Installation of . . . . 250.53(G)
Physical Protection 250.10
Resistance of . . . . 250.53(A)(2) Ex.
See GROUNDING ELECTRODES . . . . Ferm's Finder

PIPE ORGANS . . . . Art. 650
Abandoned Cable. .650.7
Conductors 650.6
   Installation 650.7
Definitions 650.2
Electronic Organ Equipment . . . . 650.3(A)
Electronic Organs and Other Musical Instruments 650.1
Grounding 650.5
Optical Fiber Cables . . . . 650.3(B)
Overcurrent Protection 650.8
Protection from Accidental Contact. 650.9
Sources of Power 650.4

PIPELINES AND VESSELS, HEATING . . . . Art. 427
Continuous Load 427.4
Control and Protections. . . . . .Art. 427 Part VII
Definitions 427.2
Disconnecting Means 427.55
Identification 427.13
Induction Heating . . . . .Art. 427 Part V
Impedance Heating . . . . Art. 427 Part IV
Overcurrent Protection 427.57

Protected from Physical Damage 427.11

Resistance Heating Elements . . . . Art. 427 Part III

Skin-Effect Heating . . . . Art. 427 Part VI

Thermal Protection 427.12

See FIXED ELECTRIC HEATING EQUIPMENT FOR PIPE-LINES AND VESSELS . . . . Ferm's Finder

PIPING SYSTEMS, BONDING **250.104**

PLACES OF ASSEMBLY . . . . **Art. 518**

Examples of . . . . 518.2(A)

General Classifications 518.2

    Multiple Occupancies . . . . 518.2(B)

    Theatrical Areas . . . . 518.2(C)

Portable Switchboards and Power Outlets 518.5

Temporary Wiring, Exhibition Halls . . . . 518.3(B)

Wiring Methods (General) . . . . 518.4(A)

    For Use in Nonrated Construction . . . . 518.4(B)

    For Use in Spaces with Finish Rating . . . . 518.4(C)

PLANTS

Bulk Storage . . . . Art. 515

Cleaning and Dyeing . . . . 500.5(B)(1), Info. Note 1(6)

    Class I Location . . . . Art. 501

Clothing Manufacturing . . . . . . 500.5(D)(1), Info. Note 1

    Class III Location . . . . Art. 503

Woodworking . . . . 500.5(D)(1), Info. Note 1

    Class III Location . . . . Art. 503

PLATE GROUNDING ELECTRODES . . . . . . **250.52(A)(7)**

Connection of Grounding Electrode Conductor to . 250.66(A)

Installation of 250.53

PLAQUES *See* **LABELS** . . . . Ferm's Finder

PLENUM

Cable Ties and Cable Accessories (Nonmetallic) . .300.22(C)(1)

Communication Circuits within Plenum . . . . . . 805.170(C)

Communication Circuit Support 800.24

Definition of . . . . Art. 100 Part I

Low Smoke and Heat Release Properties . . . . 300.22(C)(1) IN

Wiring in 300.22

    Cable Tray Systems . . . . .300.22(C)(2)

    Community Antenna Television and Radio Distribution Systems 800.24

    Network-Powered Broadband Communications Systems . 800.24

    Equipment . . . . .300.22(C)(3)

Information Technology Equipment . . . . 300.22(D)

Wiring Methods . . . . .300.22(C)(1)

See DUCTS & HOODS . . . . .Ferm's Finder

POLARITY

Adapters .406.10(B)(3)

Cord- and Plug-Connected Appliances 422.40

Flat Conductor Cable: Type FCC . . . . 324.40(B)

Identification of Terminals 200.10

Lampholders, Screw-Shell Type 410.90

Of Connection 200.11

Of Luminaires (Fixtures) 410.50

Portable Luminaires . . . . 410.82(A)

PORTABLE LUMINAIRES & LIGHTING EQUIPMENT (FORMERLY HAND LAMPS) . . . . **410.82, 511.4(B)(2)**

PORTABLE POWER CABLES

Flexible Cords and Cables . . . . Art. 400

See (QPMU) . . . . UL Product iQ

POTENTIAL TRANSFORMERS . . . . **450.3(C)**

POTTING COMPOUNDS FOR USE AT POOLS

Flush Deck Boxes . . . . 680.24(A)(2)(c)(1)

Swimming Pool, Wet-Niche Luminaires (Fixtures) 680.23(B)(2)(b)

See (WCRY) . . . . UL Product iQ

POWER & CONTROL TRAY CABLE (TYPE TC) . . **Art. 336**

Ampacity 336.80

Bends 336.24

Between Cable Tray and Equipment . . . . 336.10(7)

Construction . . . . Art. 336 Part III

Definition of 336.2

Dwelling Units- Use . . . . 336.10(9)

Installation .Art. 336 Part II

Listing Requirements 336.6

Jacket, Flame Retardant 336.116

Marking 336.120

Uses Not Permitted 336.12

Uses Permitted 336.10

See (QPOR) . . . . UL Product iQ

POWER CONVERSION EQUIPMENT

Branch-Circuit Short-Circuit and Ground-Fault Protections

    Several Motors or Loads 430.131

    Single Motor Circuits 430.130

POWER DISTRIBUTION BLOCKS . . . . **314.28(E)**
  Conductors Not to Obstruct . . . . 376.56(B)(5)
  In Metal Raceways . . . . 376.56(B)
  Installation . . . . 376.56(B)(1)
  Live Parts . . . . 376.56(B)(4)
  Size of Enclosure . . . . 376.56(B)(2)
  Splices and Taps. . . . . 376.56(A)
  Wire Bending Space . . . . 376.56(B)(3)

POWER ELECTRONIC DEVICES . . . . **430.52(C)(5)**

POWER FACTOR
  Use of in Examples . . . . .Informative Annex D

POWER INLET
  Optional Standby Systems 702.12
  Signs Required . . . . 702.7(C)

POWER-LIMITED CIRCUIT CABLE
  See (QPTZ) . . . . *UL Product iQ*

POWER-LIMITED FIRE ALARM (PLFA) CIRCUITS
  Fire Alarm Systems . . . . Art. 760 Part III
  See FIRE ALARM SYSTEMS. . . . . .Ferm's Finder

POWER-LIMITED TRAY CABLE (TYPE PLTC) . . .**725.135**
  Cable Trays . . . . 725.135(H)
  Class I, Division 2 Locations . . . . 501.10(B)(1)(3)
  Class I, Zone 2 Locations . . . . 505.15(C)(1)(4)
  Class II, Division 2 Locations . . . . 502.10(B)(1)(4)
  Cross-Connect Arrays. . . . . 725.135(I)
  Definition. 725.2
  Fabricated Ducts Used for Environmental Air . . . .725.135(B)
  Industrial Establishments . . . . 725.135(J)
  Listing and Marking of . . . . .725.135(A)
  Multifamily Dwellings . . . . .725.135(L)
  One- and Two-Family Dwellings . . . . 725.135(M)
  Other Building Locations . . . . 725.135(K)
  Plenums . . . . 725.135(C)
  Risers
    Fireproof Shafts . . . . 725.135(F)
    Metal Raceways . . . . 725.135(E)
    One-and Two-Family Dwellings . . . . 725.135(G)
    Vertical Runs . . . . 725.135(D)

POWER MONITORING EQUIPMENT . . . . **312.8(B)**

POWER OUTLETS & FITTINGS
  Assembly Occupancies 518.5
  Definition of . . . . Art. 100 Part I
  Electric Vehicle Supply Equipment 625.2
  Electrified Truck Parking Spaces (permitted disconnect) . . .626.31(B)
  Enclosure type number (Marked) 110.28
  Marina and Boatyard Power Outlets 555.2
    Disconnecting Means (permitted) . . . . 555.17(B)
    Manual Operation 555.11
    Shore Power Receptacles . . . . 555.33(A)
  Mobile Homes- Service Equipment. . . . . 550.32(C)
  Recreational Vehicle Site Supply Equipment 551.2
    Grounding Electrode (not required) . . . . 551.76(A)
  Temporary Wiring . . . . 590.4(C) and (E)
  Theaters, Motion Pictures, TV Studios 520.51
  See Informative Annex A- Product Standard Name
  See (QPYV). . . . . *UL Product iQ*

PREASSEMBLED CABLE IN NONMETALLIC CONDUIT .
. . . .**Art. 354**
  See NONMETALLIC UNDERGROUND CONDUIT WITH
  CONDUCTORS. . . . . .Ferm's Finder
  See (QQRK) . . . . *UL Product iQ*

PREMISES-POWERED BROADBAND COMMUNICATION SYSTEMS . . . . **Art. 840**
  Abandoned Cable. 840.25
  Definitions 840.2
  Ducts, Wiring Within . . . . 800.3(B)
  Grounding
    Grounding Devices, Required to be Listed 800.180
    Grounding, Metallic Entrance Conduits 800.49
    Mobile Homes 800.106
    Primary Protector and Cable Bonding and Grounding 800.100
    Not Leaving the Building 840.101
  Installation Requirements . . . . .800.3(D)
  Installation Methods Within Buildings . . . . . .Art. 840 Part V
  Listing Requirements 840.170
    Accessory Equipment. . . . . 840.170(H)
    Communication Equipment . . . . .840.170(C)
    Network Terminal . . . . .840.170(A)
  Leaving a Building . . . . 840.94 and 840.102
  Listing 840.160
  Mechanical Execution of Work 840.24
  Optical Fiber Cables
    Overhead. 770.44

Underground	840.47
Other Articles	840.3
Overhead (Aerial) Coaxial Cable)	800.44
Overhead (Aerial) Communication Wires & Cables	800.44
Premise Power over Communication Cables	840.160
Powering Circuits	840.160
Power Source (Limitations)	840.170(G)
Raceways and Cable Routing Assemblies	800.110
Underground Wires and Cables	840.47
Unlisted Wires and Cables	840.48

PREMISES WIRING (SYSTEM)

Branch Circuits (more than one nominal voltage)	210.5(C)(1)
Connection to Grounded System	200.3
Critical Operations Power Systems (COPS)	708.1
Definition of	Art. 100 Part I
Electric Vehicle Supply Equipment (to vehicle)	625.2
Feeders (more than one nominal voltage)	215.12(C)(1)
Floating Building (has a system)	555.50
Fuel Cell System (feeder to the)	692.8(B)
Grounded Conductors	200.1
Microgrid System	705.2
Optional Standby Systems	702.1
Photovoltaic System (disconnect from)	690.13
Pool Lifts (connected to)	680.82
Service Conductors (simultaneously disconnect)	230.74
Service Point	Art 100 Part 1
Site Isolation Device- Agricultural	547.9(A)(3)
Stand-Alone Systems	710.15
Surge Arrestors- Over 1000 Volts	242 Part III
Surge-Protective Devices- 1000 Volts and Less	242 Part II
System Grounding Connections	
Alternating-Current System	250.24(A)
Ungrounded System	250.24(E)

PRESSURE CONNECTORS

Agricultural Building Concrete Embedded Elements	547.10(B)
Disconnection of Grounded Conductor	230.75
Listed	
Connecting to Grounding Electrodes	250.70
Grounding Conductors and Equipment	250.8
Separately Installed, Temperature Rating	110.14(C)(2)
Service Conductors to Terminals	230.81
Swimming Pool Equipotential Bonding Grid	680.26
Terminals, Electrical Connections	110.14(A)

See (ZMVV)	UL Product iQ

PRINTING PROCESSES . . . . **Article 516**

PRODUCT CERTIFICATION OF EQUIPMENT **110.3**

PRODUCT SAFETY STANDARDS . . . **Informative Annex A**

PROTECTION

Combustible Material, Appliances	422.17
Corrosion	
Boxes, Metal (Over 1000 Volts)	314.72(A)
Cable Trays	392.10(C)
Conductors	310.10
Electrical Metallic Tubing	358.10(B)
Enamel Coated	300.6(A)(1)
Ferrous Metal Equipment	300.6(A)
Protection (suitable)	300.6
General Equipment	300.6
Intermediate Metal Conduit	342.10(B)
Supports, Hardware, Etc., Wet Locations	342.10(D)
Metal-Clad Cable: Type MC	330.12
Mineral-Insulated, Metal-Sheathed Cable: Type MI	332.10(9) and (10)
Nonmetallic-Sheathed Cable: Type NMC	334.10(B)(1)
Construction, Type NMC	334.116(B)
Nonmetallic Wireways	378.10(2)
Rigid Metal Conduit	344.10(B)
Nonferrous Corrosion-Resistant, Marking	344.120
Protected by Enamel Only	344.10(A)(4)
Supports, Hardware, Etc., Wet Locations	344.10(D)
Rigid PVC Conduit	352.10(B)
Strut-Type Channel Raceway	384.100(B)
Underfloor Raceways	390.12
Underground Feeder and Branch-Circuit: Type UF	340.10(3)
Hazardous (Classified) Locations, Techniques	500.7
Class I, Zone 0, 1, and 2 Locations	505.8
Liquids, Motors	430.11
Live Parts	110.27
Generators	445.14
Transformers	450.8(C)
Motor Overload	Art. 430 Part III
Overcurrent, *See* OVERCURRENT PROTECTION	*Ferm's Finder*
Physical Damage, from	
Armored Cable: Type AC	320.15
Busways	368.10(C)(2) and 368.12(A)

Cable Trays 392.12

Conductors 300.4

    Over 1000 Volts . . . . 300.50(C)

Electrical Equipment . . . . 110.27(B)

Electrical Metallic Tubing . . . . 358.12(1)

Flat Cable Assemblies: Type FC . . . . 322.10(3)

Flexible Metal Conduit . . . . 348.12(7)

Flexible Metallic Tubing . . . . 360.12(5)

Grounding Conductor

    CATV Systems . . . . 800.100(A)(6)

    Communication Circuits . . . . 800.100(A)(6)

    Network-Powered Broadband Communications Systems . . . . 800.100(A)(6)

    Radio and Television Equipment . . . . 810.21(D)

Grounding Electrode Conductor . . . . 250.64(B)

    Burial Depth . . . . 250.64(B)(4)

Ground Clamps and Fittings 250.10

Instrumentation Tray Cable: Type ITC . . . . 727.4(5)

Lamps, Electric Discharge Lighting More Than 1000 Volts 410.145

Lighting Track . . . . 410.151(C)(1)

Liquidtight Flexible Metal Conduit 350.12

Liquidtight Flexible Nonmetallic Conduit . . . . 356.12(1)

Luminaire (Fixture) Wiring 410.48

Metal-Clad Cable: Type MC . . . . 330.12(1)

Metal-Sheathed Cables over 1000 Volts 300.42

Metal Wireways . . . . 376.12(1)

Mineral-Insulated Metal-Sheathed Cable: Type MI 332.12(1)

    Underground Runs . . . . 332.10(10)

Multioutlet Assembly . . . . 380.12(2)

Network-Powered Broadband Communications Systems

    Attached to Buildings . . . . 800.3(F)

    Underground Circuits Entering Buildings . . . . . 830.47(C)

    Wiring within Buildings . . . . 830.47(C)

Nonmetallic-Sheathed Cable . . . . 334.15(B)

Nonmetallic Wireways . . . . 378.12(1)

Non-Power-Limited Fire Alarm Cables (NPLFA) . 760.53(A)(1)

Open Conductors and Cables (Services) 230.50

Open Wiring on Insulators . . . . 398.15(C)

Overcurrent Devices . . . . 240.24(C)

Power and Control Tray Cable: Type TC . . . . . 336.12(1)

Power-Limited Fire Alarm Cables (PLFA) . . . 760.130(B)(1)

Raceways .300.5(D)

    Over 1000 Volts . . . . 300.50(C)

Rigid PVC Conduit . . . . 352.10(F)

RV Park, Underground Branch Circuits and Feeders 551.80(B)

Safety Control Equipment Circuits . . . . 725.31(B)

Space-Heating Systems . . . . 424.12(A)

Surface Metal Raceways . . . . 386.12(1)

Surface Nonmetallic Raceways . . . . 388.12(2)

Techniques- Hazardous (Classified) Locations 500.7

Transformers . . . . 450.8(A)

Underground Branch-Circuit & Feeder Cable: Type UF . . . . 340.12(10)

Underground Installations

    Conductors, Cables, and Raceways Over 1000 Volts . . 300.50(C)

    Conductors, Cables, and Raceways . . . . 300.5(D)

    Protection from Ground Movement . . . . 300.5(J)

    Service-Entrance Conductors 230.50

PROTECTOR, COMMUNICATION SYSTEMS

Antenna 810.6

Application . . . . 805.90(A)

Bonding 805.93

Critical Operations Power Systems (COPS) . . . . . 708.14(4)

Grounding 805.93

Hazardous (Classified) Locations (not located in) . . . 805.90(C)

Installation 805.50

Listed 805.170

Location . . . . 805.90(B)

Mobile Homes, At 800.106

Modular Data Centers . . . . 646.3(C)

Network-Powered Broadband Systems 830.90

Primary Protector Requirements – Listed 805.170

Primary Protector 805.90

Secondary Protector Requirements . . . . 805.90(D)

Short as Practicable- Length . . . . 805.93(A)

Unlisted Cables 805.48

See ANSI/UL 497B. Protectors for Data Communications

PUBLIC ADDRESS SYSTEMS . . . . Art. 640

Class 1, 2, & 3 Remote Control, Signaling Systems . . Art. 725

Communication Circuits . . . . Art. 805

PULL BOXES & JUNCTION BOXES

See BOXLESS DEVICES . . . . Ferm's Finder

See JUNCTION AND PULL BOXES . . . . Ferm's Finder

PULLOUT SWITCHES DETACHABLE TYPES

See (WGEU) . . . . *UL Product iQ*

PUMPS
Water, Motor-Operated. . . . . 250.112(L)

PUMP HOUSES
See GROUNDING, Fixed Equipment . . . . Ferm's Finder
See MOTORS . . . . Ferm's Finder
See Submersible Pump Cable (YDUX) . . . . . . UL Product iQ
See UNDERGROUND WIRING . . . . Ferm's Finder

PV SYSTEMS . . . . Art. 690
See PHOTOVOLTAIC SYSTEMS (SOLAR) . . . Ferm's Finder
See Large-Scale Photovoltaic (PV) Electric Power Production Facility
. . . . Ferm's Finder

PVC (ELECTRICAL NONMETALLIC TUBING) . . Art. 362
See ELECTRICAL NONMETALLIC TUBING . . Ferm's Finder

PVC (RIGID POLYVINYL CHLORIDE CONDUIT) Art. 352
See RIGID PVC CONDUIT. . . . . Ferm's Finder
See (DZYR) . . . . UL Product iQ

Q

QUALIFIED PERSON, Definition of . . . . . . Art. 100 Part I

QUALIFIED PERSON, WORK MUST BE PERFORMED BY OR ACCESSIBLE ONLY TO
Aboveground Wiring Methods, Over 1000 Volts 300.37
AC Systems 50 to 1000 Volts Not Required to Be Grounded . . . 250.21(A)(3)b
Amusement Attractions- Control Systems 522.7
Arc-Flash Hazard Warning 110.16
Available Fault Current, Marking Requirements . 110.24(B) Ex.
Branch Circuits from Autotransformers. . . . . . . 210.9 Ex. 2
Branch Circuits Over 600 Volts. . . . . 210.19(B)(2)
Branches from Busways by Cord and Cable Assemblies . . . . .
. . . . 368.56(B)(2) Ex.
Cable Trays
    Airfield Lighting . . . . 392.10(E)
    Grounding and Bonding- Metal Tray. . . . . 392.60(A)
    Support of Raceways and Cables . . . . 392.18(G)
    Temporary Wiring . . . . 518.3(B) Ex.
    Wiring Methods in Industrial Establishments . . . 392.10(B)
    Warning Notices . . . . 392.18(H) Ex.
Capacitors, Accidental Contact . . . . 460.3(B)

Carnivals, Circuses and Fairs (Guarding) . . . . 525.10(A)
Cartridge Fuses 240.40
Circuit Breakers Indoors, over 1000 Volts. . . . 490.21(A)(1)(a)
Class I, Zone 0, 1, and 2 Locations . . . . 505.7(A)
    Wiring Methods, Zones 1 and 2 . . . . . . 505.15(1)(b) & (c)
    Flexible Cords, Zones 1 and 2 . . . . 505.17
Communication System Test Equipment . . . . 805.18 Ex.
Cords and Cables (Splices)- over 600 Volts 400.36
Cranes and Hoists (Disconnect) . . . . 610.31 Ex.
Critical Operations Data Systems. . . . . 645.10(B)(2)
Critical Operations Power Systems (COPS). . . . . . . 708.5(B)
Depth of Working Space, 600 Volts, Nominal or Less . . . . . .
. . . . 110.26(A)(1)(c)
Direct-Current Switchboards, Not Required Deadfront 530.64(A)
Disconnecting Means, Air-Conditioning Equipment 440.14 Ex. 1
Enclosures, Rooms, or Areas, Over 1000 Volts 110.31
Energy Storage Systems
    Batteries 706.3
    Charge Control 706.3
    Fuses 706.3
Equipment Over 1000 Volts
    Circuit Breakers . . . . 490.21(A)(1)
    Enclosures 490.53
    Guarding. 490.32
Feeder Taps- over 25 feet long. . . . . 240.21(B)(4)(1)
Feeders from Autotransformers . . . . 215.11 Ex. 2
Feeders over 600 Volts . . . . 215.2(B)(3)
Fixed Electrostatic Equipment, Restricted Access 516.10(A)(8)(3)
Flexible Cords and Cables- Industrial 400.17
Fuel Cell Systems . . . . 692.4(C)
Fuseholders, Over 1000 Volts. . . . . 490.21(B)(6) Ex.
Grounded Conductor of Multiconductor Cables . 200.6(E) Ex. 1
Ground-Fault Protection for Receptacles 210.8(B)(4) Ex. 2 to (4)
Ground-Fault Protection of Equipment. . . . . 230.95(C)
Guarding of High-Voltage Parts within a Compartment . 490.32
Guarding of Live Parts . . . . 110.27(A)
High-Impedance Grounded Neutral Systems . . . . . 250.36(1)
Identification of Equipment Grounding Conductors 250.119(B)
Indoor Installations, Over 1000 Volts. . . . . 110.31(B)(2)
Installation of Optical Fibers and Electrical Conductors
    Composite Optical Fiber Cables and Over 1000 Volts . 770.133(B)
    Nonconductive Optical Fiber Cables and Over 1000 Volts. . .
    770.133(B)
Instrument Transformer Cases 250.172

Instrumentation Tray Cable: Type ITC, Uses Permitted. . . 727.4
Integrated Electrical Systems . . . . 685.1(2)
Interconnected Electric Power Production Sources 705.8
Interior Metal Water Pipe As Grounding Electrode. . .250.68(C)(1) Ex.
Isolating Switches, Services Over 1000 Volts . . . . 230.204(C)
Large-Scale Photovoltaic Electric Power Production Facility . . .
   Disconnect for Isolation 691.4
Location of Switchboards 408.20
Locked Electrical Equipment Room or Enclosures
   1000 Volts or Less . . . . 110.26(F)
   Over 1000 Volts . . . . 110.34(C)
Low-Voltage Suspended Ceiling Power Distribution Systems
. . . . 393.14 Info. Note
Manholes & Enclosures Intended for Personnel Entry 110.70 Ex.
Means of Identification of Terminals . . . . 200.9 Ex.
Medium Voltage Cable: Type MV, Installation of 311.30
Messenger-Supported Wiring, Industrial Establishments 396.10(B)
Metallic Cable Trays as Equipment Grounding Conductor . . . .
. . . . 392.60(A)
Motor Controllers, Speed Limitation . . . . 430.89 Ex. (*2*)
Multioutlet Branch Circuits Greater Than 50 Amperes 210.3 Ex.
Neon Tubing . . . . 600.41(D)
Non-Power-Limited Fire Alarm (NPLFA) Circuits
   Branch Circuit . . . . 760.41(B)
Oil-Insulated Transformers Installed Indoors . . . . 450.26 Ex. 5
Open Wiring Supports, Industrial Establishments . . 398.30(C)
Outdoor Installations, Over 1000 Volts. . . . . . .110.31(C)(2)
Panelboard Enclosures Other Than Deadfront . . . . 408.38 Ex.
Portable Stage and Studio Equipment, Temporary Use . . .530.6
Portable Switchboards on Stage 520.53
Power Cable Connections to Mobile Machines Over 1000 Volts .
. . . . 490.55
Power and Control Tray Cable: Type TC, Uses Permitted 336.10(7)
Power-Limited Fire Alarm (PLFA) Circuits
 Branch Circuits . . . . .760.121(B)
Power-Limited Tray Cable (PLTC) . . . . 725.135(J)
Receptacles for 60/120-Volt Technical Power . . . . 647.7(A) Ex.
Reconditioned Equipment- Industrial . . . . .110.21(A)(2) Ex.
Restricted Access Adjustable-Trip Circuit Breakers . 240.6(C)(3)
Selective Coordination
   Critical Operations Power Systems (COPS) 708.54
   Emergency Systems 700.32
   Interconnected Electric Power Production Systems. . . .705.8
   Legally Required Standby Systems 701.30
Sensitive Electronic Equipment, General . . . . 647.3(2)

Services Exceeding 1000 Volts
   Isolating Switches 230.204
Solar Photovoltaic Systems . . . . .690.4(C)
Splices (Cords)- over 600 Volts 400.36
Substations 490.48
   Definition. . . . . Art. 100
Supervised Industrial Installations . . . . . . Art. 240, Part VIII
   Definition . . . . 240.2(1)
Switchboards and Panelboards- live parts 408.20
Temporary Wiring for Receptacles, Industrial. . . . 590.6(A) Ex.
Temporary Wiring Over 600 Volts 590.7
Temporary Wiring, Use of Single Conductors for Feeders. . . . .
. . . . 590.4(B) Ex.
Terminations (Cords) - over 600 Volts 400.36
Theaters, Motion Pictures and TV . . . . 520.54(K)
Time Switches, Flashers, and Similar Devices . . . . 404.5 Ex.
Transformer Overcurrent Protection Over 1000 Volts . . . . . .
. . . . .Table 450.3(A), Note 3
Transformer Secondary Conductors
   Industrial Installations- Taps not over 25 feet. . .240.21(C)(3)
   Taps over 25 feet . . . . .240.21(C)(4)
Transformer Vault Door Locks . . . . 450.43(C)
Transformers, Accessibility 450.13
Transformers, Exposed Energized Parts . . . . .450.8(C)
Underground Installations
   Industrial Establishments. . . . . Table 300.50 Note 3
   Over 1000 Volts 300.50
Wind Electric Systems 694.7
   Fuses 694.26
Working Clearances, Elevator Equipment 620.5
X-Ray Equipment (Disconnect). . . . . .660.5 Ex.
Zone 20, 21, and 22 Locations Wiring Methods 506.15

R

RACEWAYS ABOVEGRADE, WET LOCATIONS
  1000 Volts or Less 300.9
  Over 1000 Volts 300.38
Arrange to Drain
   On Exterior Surfaces of Buildings or Other Structures . 225.22
   Service Suitable for Use in Wet Locations 230.53
Arrange So Water Will Not Enter . . . . 230.54(G)
Bonding

Class I Locations . . . . 501.30(A)
Class II Locations . . . . 502.30(A)
Class III Locations . . . . 503.30(A)
Service and Other Raceways . . . . 250 Part V

Busways . . . . Art. 368
See BUSWAYS . . . . Ferm's Finder

Cellular Concrete Floor . . . . Art. 372
See CELLULAR CONCRETE FLOOR RACEWAYS . Ferm's Finder

Cellular Metal Floor . . . . Art. 374
See CELLULAR METAL FLOOR RACEWAYS Ferm's Finder

Class I Locations . . . . 501.30(A)
Class II Locations . . . . 502.30(A)
Class III Locations . . . . 503.30(A)

Complete Runs . . . . 300.18(A)

Continuity of Run
   Bonding in Hazardous Locations 250.100
   Bonding Loosely Jointed Metal Raceways 250.98
   Bonding of Other Enclosures . . . . 250.96(A)
   Bonding of Service Raceways . . . . 250.92(B)
   Bonding Over 250 Volts 250.97
   Electrical Continuity Metal Raceways and Enclosures . 300.10
   Electrical Metallic Tubing 358.42
   Intermediate Metal Conduit 342.42
   Isolated Grounding Circuits . . . . 250.96(B)
   Mechanical Continuity 300.12
   Method of Bonding at Service . . . . 250.92(B)
   Rigid Metal Conduit 344.42
   Threaded Conduit Hazardous Locations . . . . . . 500.8(E)

Definition of Bonding . . . . Art. 100

Different Systems in Same Enclosure . . . . 300.3(C)
   1000 Volts, Nominal or Less . . . . 300.3(C)(1)
   Cable Trays
      Cables Over 1000 Volts . . . . 392.20(B)
      Multiconductor Cables 1000 Volts or Less . . . 392.20(A)
   CATV and Radio Distribution Systems . . . . Art. 820
   Class 1 Remote Control, Signaling Circuits 725.48
   Class 2 & 3 Remote Control, Signaling Circuits . . . 725.133
   Communications Circuits . . . . Art. 805
   Elevators, Dumbwaiters, etc. . . . . Art. 620
   Grounded Conductors . . . . 200.6(D)
   Intrinsically Safe Systems . . . . Art. 504
   Network-Powered Broadband Communications Syst. . . Art. 830
   Nonelectrical Systems Prohibited 300.8
   Nonpower-Limited Fire Alarm (NPLFA) Circuits   760.48(A)
   Optical Fiber Cables 770.133
   Optional Standby Wiring 702.10
   Over 1000 Volts, Nominal . . . . 300.3(C)(2)
      Requirements for 300.32
   Power-Limited Fire Alarm (PLFA) Circuits 760.136
   Prohibition of with Service Conductors 230.7
   Seal
      Outside Branch Circuits and Feeders 225.27
   Solar Photovoltaic Systems . . . . 690.31(B)
   Surface Metallic Raceways 386.70
   Surface Nonmetallic Raceways 388.70
   Wiring Legally Required Standby Wiring 701.10

Electrical Metallic Tubing . . . . Art. 358
See ELECTRICAL METALLIC TUBING . . . Ferm's Finder

Electrical Nonmetallic Tubing . . . . Art. 362
See ELECTRICAL NONMETALLIC TUBING Ferm's Finder

Emergency Circuits (Independent) . . . . 700.10(B)

Expansion Joints
   Bonding Loosely Jointed Metal Raceways 250.98
   Coefficient of Expansion, Steel Conduit   300.7(B), Info. Note
   Earth Movement . . . . 300.5(J) IN
   In Rigid PVC Conduit 352.44
   Nonmetallic Auxiliary Gutters 366.44
   Provided Where Necessary . . . . 300.7(B)

Exposed to Different Temperatures 300.7

Field-Cut Threads . . . . 300.6(A) Info. Notes

Flexible Metal Conduit . . . . Art. 348
See FLEXIBLE METAL CONDUIT . . . . . . Ferm's Finder

Flexible Metallic Tubing . . . . Art. 360
See FLEXIBLE METALLIC TUBING . . . . . Ferm's Finder

Grounding and Bonding of
   Class I Locations . . . . 501.30(A)
   Class II Locations . . . . 502.30(A)
   Class III Locations . . . . 503.30(A)
   Equipment Grounding . . . . Art. 250 Part VI
   Methods of . . . . Art. 250 Part VII
   Service, Enclosures and Other Raceways . . Art. 250 Part IV

High Density Polyethylene Conduit Type (HDPE) . . Art. 353

Indoor, Wet Locations . . . . 300.6(D)

Induced Currents in
   Enclosures for Grounding Electrode Conductors . . 250.64(E)
   Group Conductors . . . . 300.20(A)
   Protection against Induction Heating 300.35

2020 Ferm's Fast Finder | 171

Inserting Conductors in
- Complete Runs. . . . . 300.18(A)
- Conductors of the Same Circuit . . . . 300.3(B)
- Equipment Grounding Conductor in Same Raceway 250.134(B)
- Number and Size. 300.17
- Stranded Conductors. . . . . 310.3(C)

Installation of
- Complete Runs. . . . . 300.18(A)
- Welding . . . . 300.18(B)

Installed in Shallow Grooves . . . . 300.4(F)
Insulating Fittings (Bushings) . . . . 300.4(G)
- In Lieu of Box or Termination Fitting . . . . . 300.16(B)

Intermediate Metal Conduit . . . . Art. 342
- *See* INTERMEDIATE METAL CONDUIT . . Ferm's Finder

Liquidtight Flexible Metal Conduit . . . . Art. 350
- *See* LIQUIDTIGHT FLEXIBLE METAL CONDUIT . . . . Ferm's Finder

Liquidtight Flexible Nonmetallic Conduit . . . . . . Art. 356
- *See* LIQUIDTIGHT FLEXIBLE NONMETALLIC CONDUIT . . . . Ferm's Finder

Luminaires (Lighting Fixtures) As 410.64
Metal Surface Raceways . . . . Art. 386
- *See* SURFACE METAL RACEWAYS . . . . . Ferm's Finder

No Splices in 300.15
- *See* SPLICES AND TAPS . . . . Ferm's Finder

Number and Size of Conductors in 300.17
- *See* CONDUCTORS, Number of, in . . . . . Ferm's Finder

On Rooftops, Raceways and Cables. . . . . . . 310.15(B)(2)
Outside 225.22
- Services. 230.53

Parallel Runs . . . . 310.10(G)(1)
- Underground. . . . . 300.5(I) Ex. 1-2

Rigid Metal Conduit . . . . Art. 344
- *See* RIGID METAL CONDUIT . . . . Ferm's Finder

Rigid Polyvinyl Chloride Conduit . . . . Art. 352
- *See* RIGID PVC (POLYVINYL CHLORIDE) CONDUIT . . . . Ferm's Finder

Sealing of
- Class I Locations . . . . 501.15(A)
- Class I, Zone 0, 1, and 2 Locations . . . . . . 505.16(A)
- Class II Locations 502.15
- Exposed to Different Temperature . . . . 300.7(A)
- Intrinsically Safe Systems. 504.70
- Service Raceways 230.8
- Spare or Unused . . . . 300.5(G)

Underground Installations (General) . . . . 300.5(G)
Zone 20, 21, and 22 Locations. 506.16
Strut-Type Channel. . . . . Art. 384
- *See* STRUT-TYPE CHANNEL RACEWAYS . . Ferm's Finder

Support for Nonelectrical Equipment. . . . . 300.11(B)
Support of 300.11
Supporting Conductors, Vertical 300.19
Surface Metal Raceways . . . . Art. 386
- *See* SURFACE METAL RACEWAYS . . . . . Ferm's Finder

Surface Nonmetallic Raceways . . . . Art. 388
- *See* SURFACE NONMETALLIC RACEWAYS Ferm's Finder

Temperature Changes of 300.7
Under Roof Decking . . . . 300.4(E)
Underfloor, *See* UNDERFLOOR RACEWAYS. . . Ferm's Finder
Underground
- *See* UNDERGROUND WIRING . . . . . . Ferm's Finder
- Definition of . . . . Art. 100
- Indoors. . . . . 300.6(D)
- Underground. . . . . 300.5(B)
- Use of PVC Conduit . . . . 352.10(G)
- Wet Locations
  - 1000 Volts or Less . . . . 300.5(B)
  - Over 1000 Volts . . . . 300.50(B)

Wireways, Metal . . . . Art. 376
Wireways, Nonmetallic . . . . Art. 378
- *See* WIREWAYS, METAL & WIREWAYS, NONMETALLIC Ferm's Finder

RADIO & TELEVISION DISTRIBUTION (CATV) . Art. 820
See COMMUNITY ANTENNA TELEVISION AND RADIO DISTRIBUTION SYSTEMS . . . . Ferm's Finder

RADIO & TELEVISION EQUIPMENT . . . . . . Art. 810
Amateur Transmitting & Receiving – Antennas Art. 810 Part III
Antenna Discharge Units 810.20
Antenna Lead-In Protectors. 810.6
Clearances 810.18
Grounding Devices to be Listed. 810.7
Grounding Means 810.21
Intersystem Bonding Termination. . . . . 810.21(F)(1)
Material. 810.11
Receiving Equipment – Antenna Systems. . . . . Art. 810 Part II
Supports 810.12
- Grounding 810.15

RADIUS OF BENDS
  Cables
    Armored: Type AC 320.24
    High Voltage Cable 300.34
    Integrated Gas Spacer Cable 326.24
    Metal-Clad Cable: Type MC 330.24
    Mineral-Insulated Metal-Sheathed: Type MI Cable . . 332.24
    Nonmetallic-Sheathed Type NM (Romex) 334.24
    P Cable 337.24
    Service-Entrance Cable: Type SE 338.24
  Conductors
    Auxiliary Gutters, in . . . . 366.58(A)
    Bending Radius, Over 1000 Volts 300.34
    Enclosures for Motor Controllers and Disconnects, in . . 430.10(B)
    Examination of Equipment . . . . 110.3(A)(3)
    Pull and Junction Boxes Not Over 1000 Volts, in . . . 314.28
    Pull and Junction Boxes Over 1000 Volts, in 314.71
    Metal Wireways, in . . . . 376.23(A)
    Nonmetallic Wireways, in . . . . 378.23(A)
    Switch or Circuit-Breaker Enclosures, in . . . . . . 404.3(A)
    Switchboards, Switchgear or Panelboards, in . . . . . 408.3(F)
      Clearance Entering Bus Enclosures 408.5
      Provisions for 408.55
      Terminals of Cabinets, Cutout Boxes, Meter Sockets . 312.6
  Conduit and Tubing
    Electrical Metallic Tubing 358.24
    Electrical Nonmetallic Tubing 362.24
    Flexible Metal Conduit 348.24
    Flexible Metallic Tubing 360.24
    Intermediate Metal Conduit 342.24
    High Density Polyethylene Conduit Type HDPE . . . 353.24
    Nonmetallic Underground Conduit with Conductors . 354.24
    Radius of Conduit and Tubing Bends . . . Chapter 9, Table 2
    Rigid Metal Conduit 344.24
    Rigid PVC Conduit 352.24
    Table 2 Radius of Conduit and Tubing Bends . . . Chapter 9

RAILROAD TRACKS
  Clearance for Overhead Conductors . . . . 225.18(5)

RAILWAY
  Light and Power Not Connected 110.19
  Motors and Controls 555.8
  Not Covered by Code . . . . 90.2(B)
  Outside Branch Circuits and Feeders . . . . 225.18(5)

  Service Conductor Clearance . . . . 230.24(B)(5)

RAINTIGHT, RAINPROOF, OR WEATHERPROOF
  Boxes, Conduit Bodies, and Fittings 314.15
  Cabinets, Cutout Boxes, and Meter Socket Enclosures . . . 312.2
  Definition
    Rainproof . . . . Art. 100
    Raintight . . . . Art. 100
    Watertight . . . . Art. 100
    Weatherproof . . . . Art. 100
  Motor Control Enclosures Selection . . . . Table 110.28
  Switch Enclosures, Weatherproof in Damp or Wet Locations . . 404.4
  See WEATHERPROOF . . . . Ferm's Finder

RANGE HOODS . . . . 422.16(B)(4)
  See CALCULATIONS . . . . Ferm's Finder
  See OVENS AND RANGES . . . . Ferm's Finder

RAPID SHUTDOWN OF SOLAR PHOTOVOLTAIC SYSTEMS
  Rapid Shutdown of PV Systems on Buildings 690.12
    Controlled Conductors . . . . 690.12(A)
    Controlled Limits . . . . 690.12(B)
      Inside the Array Boundary . . . . 690.12(B)(2)
      Outside the Array Boundary . . . . 690.12(B)(1)
    Equipment . . . . 690.12(D)
    Initiation Device . . . . 690.12(C)
  Identification of Power Systems
    Building with Rapid Shutdown Plaques and Directories . . . .
      . . . . 690.56(C)
    Label Notification
      Label for Roof-Mounted PV Systems with Rapid Shutdown
      Figure 690.56(C)
    More Than One Type . . . . 690.56(C)(1)
    Rapid Shutdown Switch . . . . 690.56(C)(2)
    Type . . . . 690.56(C)(1)

RATED LOAD CURRENT
  Definition of 440.2
  Marking on Hermetic Refrigerant Motor-Compressors . . . 440.4

REACTORS . . . . Art. 470

READILY ACCESSIBLE
  See ACCESSIBLE (READILY ACCESSIBLE) . . . Ferm's Finder

REBAR (CONCRETE ENCASED ELECTRODE) 250.52(A)(3)
  Concrete Encased Electrode . . . . 250.52(A)(3)

Accessible (not required). . . . . 250.68(A) Ex. No. 1
Connections. . . . . 250.66(B)
Direct-Current Sole Connection. . . . . 250.166(D)
Existing Building or Structures . . . . 250.50 Ex.
Rebar (Consist of and Size) . . . . 250.52(A)(3)
Rebar Extension . . . . 250.68(C)(3)
Swimming Pool (not allowed for Grounding Electrode) . . . . .
. . . . 250.52(B)(3)

RECEPTACLES

Anesthetizing Locations, Low-Voltage . . . . 517.64(F)
Appliances (GFCI to be Readily Accessible). 422.5
Arc-Fault Circuit-Interrupter
    Branch Circuit Protections . . . . 406.4(D)(4)
    Dormitory Units . . . . 210.12(C)
    Dwelling Units . . . . 210.12(A)
    Extensions and Modifications . . . . 210.12(B)
    Readily Accessible Location . . . . .406.4(D)
    Replacement . . . . .406.4(D)
Attachment Methods (Screws) 406.5
Baseboard Heaters, in. 424.9
    As Required Outlet for Wall Space 210.52
Bathtubs, Within 6 feet, GFCI-Protected . . . . . 210.8(A)(9)
Bathtub Zone- Prohibited. . . . . 406.9(C)
Branch Circuits 210.7
    Required in Dwelling Units 210.52
Calculations . . . . 220.14(A)-(L)
    Demand Factors – Non-Dwelling Units 220.44
    Marinas and Boatyards. 555.6
    See CALCULATIONS. . . . . .Ferm's Finder
    See Notes under Table 220.12 . . . . . . 220.14(J) and (K)
Child Care Facilities, Tamper Resistant . . . . 406.12(3)
    Definition (Child Care Facility) 406.2
CO/ALR Marking Required If Aluminum Wire Is Used 406.3(C)
    See (RTRT). UL Product iQ
Configurations
    Mobile Home Power Supply . . . . 550.10(C)
    Park Trailer Power Supply . . . . 552.44(C)
    Recreational Vehicle Power Supply . . . . 551.46(C)
    Ratings 551.81
Connected to an Equipment Grounding Conductor . . 406.4(B)
Controlled
    Building Automation. . . . . 406.3(E)
    Energy Management . . . . 406.3(E)
    Marking . . . . Figure 406.3(E)

Replacement . . . . 406.4(D)(7)
Countertop Applications, Not Face-Up Position . . . . 406.5(G)
Critical Branch, Health Care Facilities . . . . 517.34(A)
Critical Care (Category 1), Patient Bed Locations. . 517.19(B)(1)
Damp or Wet Locations, Weather Resistant Required. . 406.9(A)
Decks, Dwelling Units . . . . 210.52(E)(3)
Definition of (Selected Receptacles). 517.2
Demand Factors (Non-Dwelling Feeders) 220.44
    Marinas and Boatyards. . . . . Art. 555
    Permitted to Use Lighting Load Demand Factors . .Table 220.42

Disconnecting Means
    Cord- and Plug-Connected Appliances. 422.16
    Cord- and Plug-Connected Motors . . . . 430.109(F)
Dormitories. 210.60
Electrical Service Areas 210.63
Faceplates
    Completely Cover Opening 406.6
    Grounding . . . . 406.6(B)
    Insulated Material . . . . 406.6(C)
    Integral Night Light and/or USB Charger . . . . . . 406.6(D)
    Means of Grounding, Patient Care Spaces . . 517.13(B) Ex. 1
    Metal. . . . . 406.6(A)
    Nonmetallic Plate on Isolated Ground Receptacles Nonmetallic Boxes. . . . . 406.3(D)(2)
    Position of Receptacle Faces . . . . 406.5(G)
    Required on Boxes . . . . 314.28(C)
    Thickness (Metal) . . . . 406.6(A)
    Wet or Damp Locations, Flush with Faceplate . . . . 406.9(E)
Face-up Position (Not Permitted)
    Countertops and Similar Work Surfaces . . . . . . 406.5(G)
    Mobile Home Countertops . . . . 550.13(F)(2)
    Park Trailers Countertops . . . . 552.41(F)(2)
    Recreational Vehicle, Countertops or Similar Surfaces . 551.41(D)
    Under Sinks . . . . 406.5(G)(2)
Floor
    Boxes to Be Listed for the Application . . . . 314.27(B)
    Meeting Room 210.65
    Protection for Floor Receptacle . . . . 406.9(D)
    Seating Areas . . . . 406.5(H)(4)
    Within 18 in. of Wall . . . . 210.52(A)(3)
Foyer, Dwelling Unit . . . . 210.52(I)
General Care (Category 2), Patient Bed Locations . . 517.18(B)
Ground-Fault Circuit-Interrupter, Protection of
    See GROUND-FAULT CIRCUIT-INTERRUPTERS. Ferm's

Finder

Grounding

    Connecting Receptacle Terminal to Box 250.146

    Nongrounding Replacement or Circuit Extension . 250.130(C)

    Receptacles and Cord Connectors . . . . 406.4(A)

    Receptacles, Adapters, Cord Connectors, Plugs 406.4

Guest Rooms and Suites of Hotels and Motels, Tamper-Resistant 406.12(2)

Hazardous (Classified) Locations

    Class I Locations 501.145

        Connections for Process Control Instruments . . 501.105(B)(6)

    Class II Locations 502.145

    Class III Locations 503.145

Health Care Facilities

    Above Hazardous (Classified) Anesthetizing Locations . . . . . 517.61(B)(5)

    Critical Branch

        Receptacle Identification. . . . . 517.31(E)

        Selected Receptacles Supplied by . . . . 517.34(A)

    Ground-Fault Circuit-Interrupter

        Not Required, Certain Critical Care (Category 1) Spaces . 517.21

        Wet Procedure Locations . . . . 517.20(A)

    Grounding

        Critical Care (Category 1) Spaces 517.19

        General Care (Category 2) Spaces 517.18

        Insulated (Isolated) Receptacles . . . . . . 517.160(A)(5)

        Isolated Ground Receptacles. . . . . 250.146(D)

        Patient Care Spaces 517.13

    Hospital Grade Receptacles

        Critical Care (Category 1) Patient Bed Location. . . 517.19(B)

        General Care (Category 2) Patient Bed Location . . 517.18(B)

        Operating Rooms . . . . 517.19(C)(2)

        Note: It is not intended that there be a total, immediate replacement of existing non-hospital-grade receptacles. It is intended, however, that non-hospital-grade receptacles be replaced with hospital-grade receptacles upon modification of use, renovation, or as existing receptacles need replacement . . . . 517.18(B)(Info. Note)

    Hospital Use Type

        Above Hazardous Anesthetizing Locations. . . 517.61(B)(5)

        Other-Than-Hazardous Anesthetizing Locations 517.61(C)(2)

    Insulated Equipment Grounding Conductor . . . 517.13(B)

        Connection Grounding Terminal to Box 250.146

        Critical Care (Category 1) Space Bed Locations, Required. . . . . 517.19(B)

        General Care (Category 2) Space Bed Locations, Required . . . . 517.18(B)

    Insulated Grounding Terminals, Outside Patient Care Vicinity . . . . 517.16(B)

    Permitted, for Reduction of Electrical Noise. . . 250.96(B)

    Laundry Areas

        AFCI Protection. . . . . 210.12(A)

        GFCI Protection. . . . . 210.8(A)(10)

    Low-Voltage Circuits . . . . 517.64(F)

    Number of, in Hospitals

        Certain Rooms and Spaces Exempt from 517.18(B) Ex. 1&2

        Critical Care (Category 1) Space. . . . . 517.19(B)

        General Care (Category 2) Space . . . . 517.18(B)

        Operating Rooms . . . . 517.19(C)

    Other-Than-Hazardous Anesthetizing Locations . . 517.61(C)

    Special Purpose, Grounding . . . . 517.19(H)

    Tamper-Resistant Cover (Listed) Pediatric Locations 517.18(C)

    Tamper-Resistant Receptacles (Listed) Pediatric Locations . . . . 517.18(C)

    Within Hazardous (Classified) Anesthetizing Locations . . . . 517.61(A)(5)

Heating, Air-Conditioning, and Refrigeration Equipment

    Rooftops, Attics and Crawl Spaces 210.63

Hotels & Motels (Non-Dwellings)

    Arc-Fault Circuit-Interrupter . . . . 210.12(C)

    GFCI Bathroom and Rooftop . . . . 210.8(B)

    Guest Rooms, Number of . . . . 210.60(A)

    Placement of . . . . 210.60(B)

    Lighting Loads- General (Unit Load). . . . . . Table 220.12

    Lighting Outlets . . . . 210.70(B)

    Load Calculations for . . . . 220.14(I)

    Permitted Demand Factors. 220.44

    Receptacle Load Permitted to Apply Lighting Demands . . . . Table 220.42

    Tamper-Resistant. . . . . 406.12(2)

    Voltage Not Exceed 120 Volts . . . . 210.6(A)

Hydromassage Bathtubs. . . . . Art. 680 Part VII

    Accessibility of Receptacle 680.73

    Bonding. 680.74

    GFCI Protected- within 6 ft 680.71

    Individual Branch Circuit 680.71

In Damp or Wet Locations . . . . 406.9(A) and (B)

Installation

    Connecting Receptacle Ground Terminal to Box . . 250.146

    Dwelling Units 210.52

    Methods of Equipment Grounding 250.130

Multiple Branch Circuits on Same Yoke 210.7
Required Outlets, General 210.50
Insulated Equipment Grounding Conductor . . . . . 517.13(B)
Connection Grounding Terminal to Box . . . . . 250.146(D)
Critical Care (Category 1) Space Bed Locations, Required . . .
. . . . 517.19(B)
General Care (Category 2) Space Bed Locations, Required. . .
. . . . 517.18(B)
Isolated Grounding Receptacles (Patient Care Vicinity)517.16(A)
Inside Patient Care Vicinity . . . . 517.16(A)
Outside Patient Care Vicinity . . . . 517.16(B)
Isolated Ground Receptacles . . . . .406.3(D)
Permitted, for Reduction of Electrical Noise . . . . 250.96(B)
Kitchens, Dwelling Units . . . . 210.8(A)(6)
Kitchens, Other Than Dwelling Units . . . . 210.8(B)(2)
Loads . . . . 210.21(B)
Maximum Cord- and Plug-Connected . . Table 210.21(B)(2)
Maximum 220.18
Other Loads – All Occupancies 220.14
Permissible 210.23
Ratings for Various Size Circuits . . . . . Table 210.21(B)(3)
Summary 210.24
Locations
General 210.50
Dwelling Units 210.52
Required Outlets in Addition to Those Part of a Luminaire (Fixture) or Appliance or in Cabinets or Cupboards 210.52
Electrical Service Areas 210.63
Guest Rooms 210.60
HACR Equipment Outlet 210.63
Show Windows 210.62
Marinas and Boatyards 555.33
Maximum Cord- and Plug-Connected Load to Table 210.21(B)(2)
Maximum Height [Dwellings 1.7 m (5 1/2 ft)] . . . 210.52(4)
Meeting Rooms 210.65
Mounting 406.5
On Countertops . . . . 406.5(E)
On Covers . . . . 406.5(C)
Orientation .406.5(G)
Work Surfaces . . . . 406.5(F)
Multioutlet Assembly . . . . Art. 380
Calculations . . . . 220.14(H)
Multiwire Branch-Circuit 210.4
See MULTIWIRE BRANCH CIRCUITS . . . . Ferm's Finder
Non-Grounding-Type, Replacement of . . . . . . 406.4(D)(2)

Connection of Equipment Grounding Conductor . . 250.130
Nursing Homes, Identification . . . . 517.42(E)
Operating Rooms . . . . 517.19(C)
Outdoors (Dwellings)
Ground-Fault Circuit-Interrupter Protection Required
. . . . 210.8(A)(3)
Outlet Required . . . . 210.52(E)
Outdoors (Damp or Wet Locations) 406.8
Outlet Assemblies . . . . 210.52(C)(3) and D
Outlet Box Hood, Definition 406.2
Damp or Wet Locations 314.15
Wet Locations (Extra-Duty) . . . . 406.9(B)
Overcurrent Protection 210.20
Maximum Cord- and Plug-Connected Load . . Table 210.21(B)(2)
Summary of Branch-Circuit Requirements . . . . Table 210.24
Ratings (Receptacles) . . . . .210.21(B)(3)
In Motion Picture and TV Studios 530.21
Receptacles for Various Size Circuits . . . Table 210.21(B)(3)
Receptacles, Cord Connectors, and Attachment Plugs . .406.3
Summary Requirements 210.24
Replacements . . . . .406.4(D)
Arc-Fault Circuit-Interrupter Protection . . . . . 406.4(D)(4)
Ground-Fault Circuit-Interrupters . . . . 406.4(D)(3)
Grounding-Type Receptacles . . . . 406.4(D)(1)
Non-Grounding Type Receptacles . . . . 406.4(D)(2)
Tamper-Resistant Receptacles . . . . 406.4(D)(5)
Weather-Resistant Receptacles . . . . 406.4(D)(6)
Connection of Equipment Grounding Conductor   250.130(C)

Note: AFCI, GFCI, Tamper-Resistant and Weather-Resistant protected receptacles must be provided where replacements are made at receptacle outlets that are required be so protected elsewhere in the *NEC*.

Required in Dwellings (General) 210.50
Every 3.6 m (12 ft) or 1.8 m (6 ft) from a Point .210.52(A)(1)
Hydromassage Bathtubs . . . . Art. 680 Part VII
Locations 210.52
Motels & Hotels 210.60
Required in Occupancies . . . . Art. 210 Part III
Rooftop Receptacles 210.63
Show Windows 210.62
Spas and Hot Tubs Indoors . . . . 680.43(A)
Swimming Pools . . . . 680.22(A)
See GROUND-FAULT CIRCUIT-INTERRUPTER PROTECTION . . . . .Ferm's Finder
Seating Areas . . . . .406.5(H)

Separable Attachment Fitting (see Receptacle) . . . . . Art. 100
    Ceiling Fan. . . . . 422.18(2)
    Disconnection Means 422.33
    Outlet Box . . . . 314.27(E)
Separately Derived Systems 60 Volts to Ground 647.1
Show Windows, Required, ... 210.62
    In Elevated Floors of Show Windows . . . . . 314.27(B) Ex.
Shower Stalls, Within 6 feet, GFCI-Protected. . . . 210.8(A)(9)
Shower Zone- Prohibited . . . . 406.9(C)
Spas and Hot Tubs . . . . 680.43(A)
Swimming Pools . . . . 680.22(A)
Tamper Resistant (Pediatric Locations) . . . . 517.18(C)
Tamper Resistant Type
    Assembly Occupancy Areas. . . . . 406.12(6)
    Assisted Living Facilities . . . . 406.12(8)
    Child Care Facilities . . . . 406.12(3)
    Clinics, Medical, and Dental Offices . . . . 406.12(5)
    Dormitories . . . . 406.12(7)
    Dwelling Units . . . . 406.12(1)
    Guest Rooms and Guest Suites (Hotel, Motels) . . 406.12(2)
    Identified Receptacle Types. . . . . 406.12 Info. Note
    Locations- Not Required . . . . . . 406.12 Ex. to (1) thru (4)
    Preschools and Elementary Educational . . . . . . 406.12(4)
Temporary Installations
    Ground-Fault Circuit-Interrupter Requirements 590.6
    Installation of . . . . 590.6(A)
    Wet Locations, Receptacles. . . . . 590.4(D)(2)
Terminals, Identification . . . . 200.10(B)
Theaters
    Control and Overcurrent Protection of Circuits . . 520.21(3)
    For Equipment and Luminaires (Fixtures) on Stage . . 520.45
Torque of Connections . . . . 110.14(D)
USB Charger . . . . 406.3(F)
    Faceplate . . . . .406.6(D)
Voltage Between Adjacent Devices . . . . 406.5(J)
Weather-Resistant Type . . . . 406.9(A) and (B)
Wet or Damp Locations 406.9
Work Surfaces. . . . . 406.5(F)
See (RTDV) . . . . UL Product iQ
See (RTRT) . . . . UL Product iQ

RECEPTACLES & PLUG COMBINATIONS, HAZARDOUS (CLASSIFIED) LOCATIONS
    See (RRAT) . . . . UL Product iQ

RECESSED LUMINAIRES
Clearances, Installation 410.116
Clothes Closets, in 410.16
Electric Radiant Heating Panels (separation) . . . . 424.93(A)(3)
Insulation, Clearance from . . . . 410.116(B)
Raceways, Identified for Through-Wiring . . . . . . 410.64(B)
Swimming Pools- Indoor . . . . 680.43(B)(1)(1)
Thermal Protection of Incandescent-Type . . . . . . 410.115(B)
Used as Raceways 410.64
Wiring 410.117

RECIPROCALS, *See* Formula . *Ferm's Charts and Information*

RECONDITIONED EQUIPMENT
Arc-Fault Circuit-Interrupter 210.15
Attachment Plugs 406.7
Communication Equipment . . . . 800.3(G)
Cord Connectors 406.7
Definition. . . . . Art. 100
Equipment Over 1000 Volts 490.49
Fire Pump and Transfer Switch 695.10
Flanged Surface Devices. 406.7
Ground-Fault Circuit-Interrupter. 210.15
Ground-Fault Protection of Equipment. 210.15
Lighting, Low-Voltage 411.4
Luminaires, Lampholders and Retrofit Kits 410.7
Marking. . . . . 110.21(A)(2)
Overcurrent Protection 240.62, 240.88, 240.102
Panelboards . . . . 408.8(A)
Receptacles 406.3
Switchboards . . . . 408.8(B)
Switchgear . . . . 408.8(B)
Transfer Switches . . . . 702.5(A), 708.24
Transfer Switches, Automatic . . . . 700.5(C), 701.5(C)
X-Ray Equipment, Health Care Facilities 517.75
X-Ray Equipment, Not Health Care Facility Related . . . 660.10

RECREATIONAL VEHICLES . . . . Art. 551
Air Conditioning, Pre-Wiring . . . . 551.47(Q)
Branch Circuit, Pre-Wiring . . . . 551.47(S)
Branch Circuits Required 551.42
Combination Systems. . . . . Art. 551 Part II
Definitions 551.2
Energy Management System . . . . 551.42(C) Ex. 2
Factory Tests . . . . Art. 551 Part V

Generator, Pre-Wiring . . . . 551.47(R)
Labels. . . . . 551.4(C)
   At the Electrical Entrance . . . . 551.46(D)
Nominal 120- or 120/240-Volt Systems . . . . Art. 551 Part IV
Other Power Sources . . . . Art. 551 Part III
Parks (RV) . . . . Art. 551 Part IV
   Calculated Load 551.73
   Clearance for Overhead Conductors 551.79
   Demand Factors . . . . Table 551.73(A)
   Distribution System 551.72
   Grounding . . . . 551.76(A)
   Overcurrent Protection. 551.74
   Protection of Equipment 551.78
   Receptacles 551.81
   Supply Equipment Location 551.77
      Grounding. 551.76
   Type of Receptacles Provided. 551.71
   Underground Conductors 551.80
Receptacle Configuration Types. . . . . . . Figure 551.46(C)(1)
Receptacle Outlets Required 551.41
   Rooftop Decks. 551.41(B)(4)
Reverse Polarity Device . . . . 551.40(D)
Supply Conductors . . . . 551.30(E)
Wiring Methods 551.47
See MOBILE HOMES . . . . Ferm's Finder
See PARK TRAILERS . . . . Ferm's Finder

RECREATIONAL VEHICLE PARKS . . . . Art. 551
   Calculation of Loads 551.73
   Definitions 551.2
   Demand Factors . . . . Table 551.73(A)
   Disconnecting Means . . . . 551.77(B)
   Distribution System. 551.72
      50-Ampere Receptacles from 120/240-Volt, or 208Y/120-Volt System 551.72
      Neutral Conductors . . . . 551.72(D)
      Receptacles. . . . . 551.72(C)
      Systems. . . . . 551.72(A)
      Three-Phase Systems . . . . 551.72(B)
   Grounding (General) 551.76
      Exposed Non-Current-Carrying Metal Parts . . . . 551.76(A)
      Grounding Electrode . . . . 551.76(A)
      Neutral Conductor Not to Be Used As Equipment Ground . . . . 551.76(C)
      No Ground Connection on Load Side of Service . 551.76(D)

Secondary Distribution Systems . . . . 551.76(B)
Site Supply Equipment. 551.76
Marking of Site Equipment. . . . . 551.77(F)
Outdoor Equipment, Protection of 551.78
Overcurrent Protection 551.74
Overhead Conductors, Clearance of 551.79
Pre-Wiring . . . . 551.47(Q), (S)
Receptacle Outlets 551.71
   20-Ampere . . . . 551.71(A)
   30-Ampere . . . . 551.71(B)
   50-Ampere . . . . 551.71(C)
   50-ampere installation requires a 30-ampere . . . . 551.71(C)
   Additional Receptacles . . . . 551.71(E)
   Configuration of Supply Receptacles 551.81
   Ground-Fault Circuit-Interrupter Protection . . . 551.71(F)
   Tent Sites. . . . . 551.71(D)
Site Equipment, Location of . . . . 551.77(A)
Underground Service, Feeder and Branch Circuits 551.80
See UNDERGROUND SERVICE & WIRING . . Ferm's Finder

RECTIFIER
AC Systems 50 to 1000 Volts Not Required to Be Grounded . . . 250.21(A)(2)
Belowground-protected from physical damage 110.55
Class I, Division 1 Locations 501.55
Class II, Division 1 Locations . . . . 502.150(A)(2)
Motion Picture Projection Rooms 540.11
   Listed Enclosures 540.20
Motors- nominal voltage 430.18
   Single Motor 430.22
Two-Wire DC Systems, Derived from . . . . 250.162(A) Ex. 2
Wind Electric Systems
   Components (Interactive System) . . . . . . Figure 694.1(a)
   Components (Stand-Alone System) . . . . . Figure 694.1(b)
   Wind Turbine Output Circuit 694.2

RED BRASS RIGID METAL CONDUIT . . . .
Uses Permitted 344.10

REFRIGERATION & AIR CONDITIONING . . . . Art. 440
See AIR-CONDITIONING & REFRIGERATING EQUIPMENT Ferm's Finder

REINFORCING THERMOSETTING RESIN CONDUIT (FIBERGLASS) (TYPE RTRC). . . . . Art. 355

REINFORCING-BAR GROUNDING ELECTRODES . . . .
. . . . 250.52(A)(3)
See RTRC . . . . Ferm's Finder

RELAYS
Overload, Motor Overcurrent Protection 430.40
Overload, Motor-Compressor and Branch-Circuit Art. 440 Part VI
Reverse-Current, Transformers . . . . 450.11(B)

RELOCATABLE STRUCTURES . . . . Art. 545, PART II
Application Provisions 545.20
Bonding Exposed Metal Parts 545.26
Disconnecting Means 545.24
Ground-Fault Circuit-Interrupters 545.28
Intersystem Bonding 545.27
Power Supplies 545.22
    Feeder . . . . 545.22(A)
    Grounding . . . . 545.22(D)
    Identification . . . . 545.22(C)
    Number of Supplies . . . . 545.22(B)

REMOTE CONTROL CIRCUITS . . . . Art. 725
Access to Electrical Equipment 725.21
Grounding . . . . 250.112(I)
Motors, Control Circuits . . . . Art. 430 Part VI
Safety-Control Equipment 725.31
See CLASS 1, 2, & 3 REMOTE CONTROL CIRCUITS . . . .
. . . . Ferm's Finder
See CONTROL CIRCUITS . . . . Ferm's Finder
See DIFFERENT SYSTEMS IN SAME ENCLOSURE . . . . .
. . . . Ferm's Finder
See Power-Limited Circuit Cable (QPTZ) . . . . UL Product iQ

REPAIR GARAGES (MAJOR AND MINOR) . . . . . Art. 511
See Commercial Garages . . . . Ferm's Finder

REPAIRING NONCOMBUSTIBLE SURFACES 312.4
Cabinets, Cutout Boxes, Meter Enclosures 312.4
Outlet, Device, Pull and Junction Boxes 314.21

RESISTANCE
Appliances . . . . Art. 422
Fixed Electric Heating Equipment for Pipelines and Vessels . . Art. 427
Fixed Electric Space-Heating Equipment . . . . Art. 424
Fixed Outdoor Electric Deicing and Snow-Melting Equipment . .
. . . . Art. 426
Fixed Resistance and Electrode Industrial Process Heating Equipment . . . . Art. 425

Low Contact Resistance, Wet-Niche Luminaires . . 680.23(B)(5)
Of Conductors, Direct-Current . . . . Chapter 9, Table 8
Of Made Electrodes 25 Ohms or Less . . . . 250.53(A)(2) Ex.
Three Single Conductors in Conduit, Alternating-Current . . . .
. . . . Chapter 9, Table 9
Welders . . . . 630 Part III

RESISTIVELY GROUNDED DC SYSTEMS
Caution Signage . . . . 408.3(F)(5)

RESISTORS & REACTORS . . . . Art. 470
Combustible Material within 305 mm (12 in.), Thermal Barrier .
. . . . 470.3
Conductor Insulation 470.4
Location, Not Subject to Physical Damage 470.2
Over 1000 Volts, Nominal . . . . Art. 470 Part II

RETROFIT KIT
Defined . . . . Art. 100
Field-Installed Secondary Wiring, Signs 600.12
Installation Instructions . . . . 600.4(E)
Listing Requirements for Luminaires, Lampholders and Lamps .
. . . . 410.6
Listing Requirements for Signs 600.3
Sign Installations Instructions . . . . 600.4(F)

RETROFIT TRIP UNITS . . . . 490.21(A)(5)

REVERSE FEEDING TRANSFORMERS . . . . . . 450.11(B)

REVERSE POLARITY DEVICE-RECREATION VEHICLE . .
. . . . 551.40(D)

RHEOSTATS
Motion Picture and TV Studios . . . . 530.15(C)
Motor Controllers . . . . 430.82(C)
Theaters and Audience Areas
    Dimmers 520.25
    Installed in Cases or Cabinets 520.7

RIGID METAL CONDUIT: TYPE RMC . . . . Art. 344
Aluminum, Must Be Suitable for the Condition . . . .
. . . . 300.6, 344.10(A)(3), (B)(2)
Shall Not Be Used in Concrete or Earth Burial
    See (DYWV) . . . . UL Product iQ
Bends
    How Made 344.24
    Number of 344.26

Radius of . . . . Chapter 9, Table 2
Couplings and Connectors 344.42
Bushings 344.46
    Conductors 4 AWG and Larger . . . . 300.4(G)
Corrosive Environments . . . . 344.10(B)
Couplings and Connectors, Threadless . . . . 344.42(A)
Definition. 344.2
Dimensions & Percent Area of . . . . Chapter 9 Table 4
Dissimilar Metals 344.14
Field-Cut Threads . . . . 300.6(A) Info. Notes
Listing 344.6
Marking. 344.120
Material of Construction 344.100
Minimum and Maximum Sizes 344.20
Not to Be Used As a Means of Support for Cables or
    Nonelectrical Equipment (General) . . . . 300.11(B)
    CATV and Radio Distribution Systems 800.24
    Class 2 and 3 Circuit Conductors 725.143
    Communications Circuits . . . . 805.133(B)
    Fire Alarm Circuit Conductors 760.143
    Network-Powered Broadband Systems . . . . . 805.133(B)
Number of Conductors in 344.22
    Combinations of Conductors (General) . . Chapter 9 Table 1
        Compact Stranded. . . . . Chapter 9 Table 5A
        Dimensions of Conductors in . . . . . . Chapter 9 Table 5
    Same Size. . . . . . . Informative Annex Tables C8 and C8(A)
Physical Damage (Severe) . . . . 344.10(E)
Reaming & Threading 344.28
Red Brass RMC . . . . 344.10(A)(1)
Running Threads Not Permitted . . . . 344.42(B)
Securing and Supporting of 344.30
Sizes
    Minimum . . . . 344.20(A)
    Maximum . . . . 344.20(B)
Splices and Taps. 344.56
Stainless Steel- only with Stainless Steel fittings 344.14
Supporting Enclosures . . . . 314.23(E) & (F)
Types
    Aluminum . . . . 344.100(2)
    Red Brass. . . . . 344.100(3)
    Stainless Steel. . . . . 344.100(4)
    Steel with Protective Coatings . . . . 344.100(1)
Uses Permitted 344.10
    See (DYIX) . . . . UL Product iQ

RIGID PVC CONDUIT . . . . **Art. 352**
Ambient Temperature Limitations . . . . 352.12(D)
Bends
    How Made 352.24
    Number of 352.26
    Radius of . . . . Chapter 9 Table 2
Boxes and Fittings 352.48
Bushings 352.46
    Conductors 4 AWG and Larger . . . . 300.4(G)
Construction 352.100
Definition. 352.2
Dimensions and Percent Area of . . . . Chapter 9 Table 4
Expansion Fittings 352.44
    Expansion Characteristics of Fiberglass Reinforced . . Table 352.44
    Expansion Characteristics of PVC . . . . Table 352.44
Exposed, Protection from Physical Damage . . . . . 352.10(F)
Grounding (Equipment) 352.60
    See GROUNDING, Fixed Equipment . . . . . . Ferm's Finder
Hazardous (Classified) Location Installation . . . . 352.12(A)
Installation . . . . Art. 352 Part II
Insulation Temperature Limitations . . . . 352.10(I)
Listed 352.6
Marking (Including Material Type) 352.120
Minimum and Maximum Sizes 352.20
Not to Be Used As a Means of Support for Cables or
    CATV and Radio Distribution Systems . . . . . 820.133(B)
    Class 2 and 3 Circuit Conductors 725.143
    Communications Circuits 800.24
    Fire Alarm Circuit Conductors 760.143
    Network-Powered Broadband Systems . . . . . 830.133(B)
    Nonelectrical Equipment (General) . . . . 300.11(B)
Number of Conductors in, Schedule 40 and HDPE Conduit 352.22
    Combinations of Conductors (General) . . Chapter 9 Table 1
        Compact Stranded. . . . . Chapter 9 Table 5A
        Dimension of Conductors in . . . . . . Chapter 9 Table 5
    Same Size. . . . . Informative Annex Tables C10 and C10(A)
Number of Conductors in, Schedule 80 (Same Size)   Informative Annex Table C9
    Combinations of Conductors (General) . . Chapter 9 Table 1
        Compact Stranded. . . . . Chapter 9 Table 5A
        Dimension of Conductors in . . . . . . Chapter 9 Table 5
    Same Size . . . . Informative Annex Tables C9 and C9(A)
Number of Conductors in, ... 352.22
    Combinations of Conductors (General) . . Chapter 9 Table 1

Compact Stranded . . . . .Chapter 9 Table 5A
Dimensions of Conductors in . . . . Chapter 9 Table 5
Same Size. . . . . Informative Annex Tables C11 and C11(A)
Number of Conductors in, Type EB (Encased Burial in Concrete)
. . . . 352.22
   Combinations of Conductors (General) . . Chapter 9 Table 1
      Compact Stranded. . . . . Chapter 9 Table 5A
      Dimensions of Conductors in . . . . . . Chapter 9 Table 5
   Same Size . . . . Informative Annex Tables C12 and C12(A)
Securing and Supporting 352.30
   Other Support Spacings Included in Listing . . . . 352.30(B)
Sizes
   Maximum . . . . 352.20(B)
   Mimimum . . . . 352.20(A)
Splices and Taps. 352.56
Uses Permitted 352.10
   Above Class I Locations
      Commercial Garages. . . . . 511.7(A)(1)
   Essential Electrical System Hospitals, Schedule 40 . . 517.31(C)(2)
   Essential Electrical System Hospitals, Schedule 80   517.31(C)(3)(1)
   Motor Fuel Dispensing Facilities . . . . 514.8, Ex. No. 2
   Places of Assembly
      If Encased in 50 mm (2 in.) of Concrete . . . . . 518.4(A)
      Nonrated Construction . . . . 518.4(B)
      In Certain Spaces with 15-minute Finish Rating. . 518.4(C)
   Protection of Service Cables, Suitable for Location
   Schedule 80 . . . . 230.50(B)(1)(3)
   Protection of Underground Conductors, Schedule 80
      1000 Volts and Below . . . . 300.5(D)(4)
      Over 1000 Volts . . . . 300.50(C)
   Recreational Vehicle, Underground Conductors
      Protection of, Schedule 80. . . . . 551.80(B)
   Support of Conduit Bodies Only. . . . . 352.10(H)
   Theaters and Audience Areas 520.5
   Under Hazardous Areas
      Bulk Storage Plants 515.8
      Class I, Division 1 . . . . 501.10(A)(1)(a) Ex.
      Commercial Garages. . . . . 511.7(A)(1)
      Motor Fuel Dispensing Facilities. 514.8 Ex.2
Uses Not Permitted 352.12
   Ambient Temperatures Over 50°C Unless Listed . . 352.12(D)
   Branch-Circuit Wiring, Patient Care Areas 517.13
   Conductor Insulation Temperatures Exceeding
      Ambient Temperatures . . . . 352.12(D)

Ducts, Plenums, and Other Air-Handling Spaces . . . 300.22
Hazardous (Classified) Locations (Except As Permitted in
Ch. 5 Art.) . . . . 352.12(A)
Support of Luminaires (Fixtures) and Other Equip. 352.12(B)
Where Subject to Physical Damage Unless Identified 352.12(C)
See Reinforced Thermosetting (DZKT) . . . . . . *UL Product iQ*
See Schedule 40 & 80, PVC Type RTRC (DZYR) . . . .
. . . . *UL Product iQ*
See Underground, PVC Type (EAZX). . . . . . . *UL Product iQ*

ROMEX (NONMETALLIC-SHEATHED CABLE) . . Art. 334
See NONMETALLIC-SHEATHED CABLE. . . . Ferm's Finder

ROOF DECKING
Cables, Raceways and Boxes Installed Under, Physical
Protection. . . . . 300.4(E)
Luminaires (Fixtures) Installed Under, Physical Prot . . . 410.10(F)

ROOF ON BUILDING BEFORE INSTALLING EQUIPMENT
. . . . **110.11**
Dry Locations, Definition of . . . . Art. 100 Part I
Nonmetallic-Sheathed Cable, Type NM – Uses Permitted. .334.10(A)

ROOF TOPS
Raceways and Cables Exposed to Sunlight 310.15(B)(2) and Table
310.15(B)(2)

RTRC TYPE CONDUIT . . . . Art. 355
Ambient Temperature Restrictions . . . . 355.12(D)
Bends 355.24
Bushings 355.46
Conductors- Number of 355.22
Construction 355.100
Definition. 355.2
Expansion Fittings 355.44
Grounding 355.60
Joints 355.48
Listing 355.6
Securing and Supporting 355.30
Sizes
   Maximum . . . . 355.20(B)
   Minimum . . . . 355.20(A)
Splices and Taps 355.56
Trimming 355.28
Uses Not Permitted. 355.12
Uses Permitted 355.10

RUNNING THREADS (NOT PERMITTED)
Intermediate Metal Conduit on Connections, Couplings . . . . 342.42(B)

Rigid Metal Conduit on Connections, Couplings. . . 344.42(B)

S

(SCADA) SUPERVISORY CONTROL & DATA ACQUISITION
. . . . Info. Annex G
Definition. 708.2

SAFETY AND EXAMINATION OF EQUIPMENT. . . . 90.7

SCHOOLS
Branch-Circuit, Feeder, & Service Calculations . . . . . Art. 220

General Lighting Load . . . . Table 220.12

Optional Method 220.86

Tamper-Resistant Receptacles

    Preschool and Elementary Education Facilities . . . 406.12(4)

    See CALCULATIONS . . . . Ferm's Finder

SCREW SHELLS
Identification of Terminals, Screw Shells . . . . 200.10(C)

    Screw Shell Devices with Leads. . . . . 200.10(D)

Lampholders, Installation . . . . Art. 410 Part VIII

Polarization of Luminaires (Fixtures) 410.50

SCREWS
Covers and Canopies, Drywall Type Not to be Used . . . 314.25

Receptacles, Drywall Type Not to be Used 406.5

Switches, Drywall Type Not to be Used. . . . . 404.10(B)

Under Roof Decking- Damage by Screws

    Cables, Raceways and Boxes . . . . . 300.4(E) and Info. Note

    Luminaires . . . . 410.10(F)

SEALING, HAZARDOUS (CLASSIFIED) LOCATIONS
Accessible, Class I, Divisions 1 and 2 . . . . 501.15(C)(1)

Accessible, Class II, Divisions 1 and 2. . . . . 502.15

Cable Systems, Class I, Division 1 . . . . 501.15(D)

Cable Systems, Class I, Division 2 . . . . 501.15(E)

Class II, Divisions 1 and 2, Methods of. 502.15

Compound Thickness. . . . . .501.15(C)(3)

Conductor or Optical Fiber Fill Not More Than 25%. . 501.15(C)(6)

Conduit Systems, Class I, Division 1 . . . . 501.15(A)

Conduit Systems, Class I, Division 2 . . . . 501.15(B)

Optical Fiber Cables . . . . 500.8(F)

Splices Not Permitted in Seal Fittings . . . . 501.15(C)(4)

Zones

    Class I, Zone 2 . . . . 505.15(C)(1)(6)

    Optical Fiber Cable . . . . 505.9(F)

    Sealing. 505.16

        Class I, Zones 0, 1, and 2 . . . . 505.16(D)

        Zone 0 . . . . 505.16(A)

        Zone 1 . . . . 505.16(B)

        Zone 2 . . . . 505.16(C)

        Zone 20, 21, and 22 . . . . 506.16

See HAZARDOUS (CLASSIFIED) LOCATIONS . .Ferm's Finder

SEALING RACEWAY FROM WARM TO COLD . . . 300.7(A)
Busway Sections, Barriers and Seals 368.234

SEALING UNDERGROUND RACEWAY
1000 Volts and Less . . . . 300.5(G)

Over 1000 Volts . . . . 300.50(F)

Service Raceways 230.8

SEALTIGHT, LIQUIDTIGHT FLEXIBLE METAL CONDUIT
. . . . .Art. 350

See LIQUIDTIGHT FLEXIBLE METAL CONDUIT . Ferm's Finder

See (DXHR) . . . . UL Product iQ

SEALTIGHT, LIQUIDTIGHT FLEXIBLE NONMETALLIC CONDUIT . . . . Art. 356
See LIQUIDTIGHT FLEXIBLE NONMETALLIC CONDUIT UL Product iQ

See (DXOQ) . . . . UL Product iQ

SEATING AND OTHER SIMILAR AREAS
Receptacle Requirements . . . . .406.5(H)

SECTIONED EQUIPMENT BONDING CONDUCTORS . .
. . . . .310.10(G)(5)

SECURING AND SUPPORTING
Armored Cable, Type AC 320.30

Boxes 314.23

    Nonmetallic 314.43

Busways 368.30

Cabinets, Cutout Boxes, and Meter Socket Enclosures

    Of Cables to . . . . 312.5(C)

Cables & Raceways above Access Panels . . . . 300.4(C)

    CATV Systems 800.21

182 | 2020 Ferm's Fast Finder

Class 1, 2 and 3 Remote Control Circuits 725.21
Communications Circuits 800.21
Fire Alarm Systems. 760.21
Network-Powered Broadband Systems 800.21
Optical Fiber Cables and Raceways. 770.21
Ceiling-Suspended (Paddle) Fans . . . . 314.27(D)
   Boxes. . . . . 314.27(C)
   Canopies for Wire Space 422.19
   Exposed Ceiling Finish, Combustible 422.21
   Outlet Boxes to Be Covered 422.20
   Specific Weight Limitations 422.18
Conductors in Vertical Raceways 300.19
Electrical Metallic Tubing. 358.30
Electrical Nonmetallic Tubing 362.30
Equipment, General . . . . 110.13(A)
Fire-Rated Floor/Ceiling or Roof/Ceiling Assembly . . . 300.11(A)(1)
Flat Cable Assemblies, Type FC. 322.30
Flexible Cords
   Branches from Busways . . . . 368.56(B)
   For Underwater Lighting, Strain Relief. . . . . . . 680.24(E)
   Securing at Terminals. 400.10
Flexible Metal Conduit 348.30
Intermediate Metal Conduit 342.30
Liquidtight Flexible Metal Conduit. 350.30
Liquidtight Flexible Nonmetallic Conduit 356.30
Luminaires . . . . Art. 410 Part IV
   Class I, Division 1 Locations . . . . 501.130(A)(4)
   Class II, Division 1 Locations . . . . 502.130(A)(4)
   Outlet Boxes for . . . . 314.27(A)
   Suspended Ceilings, By. . . . . 410.36(B)
   Swimming Pools, Outdoor Spas and Hot Tubs . . . 680.22(B)
Messenger Supported Wiring . . . . Art. 396
   Messenger Supports . . . . 396.30(A)
Metal-Clad Cable, Type MC 330.30
Non-Fire-Rated Floor/Ceiling or Roof/Ceiling Assembly . . . . .
. . . . 300.11(A)(2)
Nonmetallic-Sheathed Cable (Romex)
   300 mm (12 in.) from Metal Boxes 334.30
   200 mm (8 in.) from Plastic Boxes . . . . 314.17(C) Ex.
   Every 1.4 m (4 1/2 ft) 334.30
   In Unfinished Basements and Crawl Spaces . . . 334.15(C)
Open Wiring on Insulators 398.30
Paddle, Ceiling-Suspended Fans . . . . 314.27(D)
   Canopies for Wire Space 422.19
   Exposed Ceiling Finish, Combustible 422.21
   Outlet Boxes to Be Covered 422.20
   Specific Weight Limitations 422.18
Panelboards, General . . . . 110.13(A)
   Busbars and Conductors in. . . . . 408.3(A)(1)
   Of Cables to . . . . 312.5(C)
Radio and Television, Antennas and Lead-in 810.12
Rigid Metal Conduit 344.30
Rigid PVC Conduit. 352.30
Service Overhead Spans . . . . Art. 230 Part II
   By Service Masts 230.28
   Note: Only power service-drop conductors are permitted to be attached to a service mast. *See* also . . . . 800.44(C)
Strut-Type Channel Raceways 384.30
Wireways
   Metal. 376.30
   Nonmetallic 378.30

SELECTIVE COORDINATION
Critical Operations Power Systems (COPS). 708.54
Definition. . . . . Art. 100
Elevators 620.62
Emergency Systems 700.32
Fire Pumps . . . . 695.3(C)(3)
Information Technology Equipment 645.27
Legally Required Standby Systems 701.32

SELECTIVITY
Critical Operations Power Systems . . . . 708.52(D)
Health Care Facilities 517.17

SELF-CONTAINED DEVICES . . . . **334.40(B)**

SEMICONDUCTOR FUSES
Individual Motor Circuits. . . . . 430.52(C)(5)
Motors with Power Conversion Equipment. . . . 430.130(A)(4)

SENSITIVE ELECTRONIC EQUIPMENT
Sensitive Electronic Equipment . . . . Art. 647
Disconnecting Means, Luminaire, Lockable . . . . . . 647.8(A)
Grounding of 647.6
Lighting Equipment 647.8
Receptacles 647.7
Three-Phase Systems 647.5
Voltage Line-to-Line, Line-to-Ground 647.1
Wiring Methods 647.4

SEPARABLE ATTACHMENT FITTINGS . . . . . . .314.27(E)

SEPARATE BUILDINGS OR OTHER STRUCTURES
Disconnecting Means
    Access to Occupants 225.35
    Construction of 225.38
    Grouping of 225.34
    Identification of 225.37
    Location 225.32
    Maximum Number of Disconnects for Each Supply . . 225.33
    Number of Supplies 225.30
    Rating of 225.39
    Suitable for Use as Service Equipment 225.36
Grounding of 250.32
    Agricultural Buildings 547.9
    Grounding Electrode System . . . . Art. 250 Part III
    Panelboards 408.40
Mobile Home Services 550.32
Overcurrent Protection
    Access to Protective Devices 225.40
    Location in Circuit 240.21
    Panelboards 408.36
Panelboards . . . . Art. 408 Part II

SEPARATELY DERIVED SYSTEMS
Bonding Jumper, System . . . . 250.30(A)(1)
    Size of . . . . 250.28(D)
Definition of . . . . Art. 100 Part I
Disconnecting Means 240.21
    Location of 404.8
For Ungrounded Direct-Current Systems 250.169
Grounded Systems . . . . 250.30(A)
    Bonding . . . . 250.30(A)(8)
    Grounded Conductor . . . . 250.30(A)(3)
    Grounding Electrode . . . . 250.30(A)(4)
    Grounding Electrode Conductor, Multiple System . . 250.30(A)(6)
    Grounding Electrode Conductor, Single System . . . 250.30(A)(5)
    Installation . . . . 250.30(A)(7)
    Supply-Side Bonding Jumper . . . . 250.30(A)(2)
    System Bonding Jumper . . . . 250.30(A)(1)
Not Required to Be Grounded 250.21
Outdoor Source . . . . 250.32(C)
Overcurrent Protection of
    Generators 445.12
    In Supervised Industrial Installations . . . . Art. 240 Part VIII

Location of Overcurrent Protection 240.92
Location in Circuit 240.21
Motor Control Circuit . . . . 430.72(A)
Not Required If Power Loss Hazard Created . . . . . 240.4(A)
Panelboard Supplied Through Transformer . . . . . 408.36(B)
Requirement for . . . . 240.15(A)
Transformer Secondary Conductors by Primary OCD . . 240.4(F)
Transformers 450.3
Required to Be Grounded . . . . 250.20(D)
Systems with 60 Volts to Ground 647.3
Ungrounded Systems . . . . 250.30(B)
    Bonding Path and Conductor . . . . 250.30(B)(3)
    Grounding Electrode . . . . 250.30(B)(2)
    Grounding Electrode Conductor . . . . 250.32(B)(1)

SERIES/PARALLEL CIRCUITS . . . . Ferm's Finder

SERVICE AREA, ELECTRICAL
Receptacle Requirements 210.63

SERVICE CABLES
See SERVICE-ENTRANCE CABLE . . . . Ferm's Finder

SERVICE CONDUCTORS
Over 1000 Volts, Ground-Fault Circuit Conductors Brought to Service Point
    With a Grounded Conductor . . . . 250.186(A)
    Without an Grounded Conductor . . . . 250.186(B)
Overhead
    Clearances 230.24
    Insulation and Covering 230.22
    Means of Attachment 230.27
    Point of Attachment 230.26
    Service Masts as Supports 230.28
    Size and Rating 230.23
    Supports over Buildings 230.29

Underground
Installation . . . . 230.30(A)
Size and Rating 230.31
Protection Against Physical Damage 230.32
Spliced Conductors 230.33
Wiring Methods . . . . 230.30(B)

SERVICE DROPS . . . . Art. 230 Part II
Ampacity of Conductors 230.23

Attachment to Service Mast. 230.28
Clearances
    Above Roofs . . . . 230.24(A)
    From Building Openings. . . . . 230.9(C)
    From Buildings. 230.9
    From Final Grade or Earth . . . . 230.24(B)
    Over 1000 Volts see *Life Safety Code, NFPA 101* . . NFPA 101
    Swimming Pools 680.9
    Vegetation as Support 230.10
Definition of . . . . Art. 100 Part I
Means of Attachment. 230.27
Number of, to a Building 230.2
Overhead Service Locations 230.54
Point of Attachment 230.26
Service Masts
    Attachment. . . . . 230.28(B)
    As Support. 230.28
    Strength . . . . 230.28(A)
    Size and Rating 230.23
        Rated 60°C through 90°C . . . . Table 310.17
        Rated 150°C through 250°C . . . . . . Table 310.18
Support Over Buildings. 230.29

SERVICE ENTRANCE CABLE (TYPES SE AND USE) Art. 338
    Bending Radius 338.24
    Construction 338.100
    Definition. 338.2
    Exterior Installations . . . . 338.10(B)(4)(b)
    Interior Installations . . . . 338.10(B)(4)(a)
    Listing Requirements .338.6
    Marking. 338.120
    Uses Not Permitted 338.12
    Uses Permitted 338.10

SERVICE ENTRANCE CONDUCTORS
    Cable Trays 230.44
    Conductors Entering Building or Structure. 230.52
    Higher Voltage to Ground 230.56
    Insulation 230.41
    Minimum Size and Rating 230.42
    Mounting Supports 230.51
    Number Per Set 230.40
    Overhead Service Locations 230.54
    Protection Against Physical Damage 230.50
    Raceways to Drain 230.53

    Spliced Conductors 230.46
    Wiring Methods for 1000 Volts, Nominal, or Less 230.43

SERVICE EQUIPMENT
Bonding 250.92
Disconnect, Supply Side 230.82
Equipment Connected to Supply Side 230.82
Ground-Fault Protection of Equipment 230.95
    Additional Level Required . . . . 517.17(B)
    Definition of . . . . Art. 100
    Emergency Systems . . . . .700.6(D)
    Fire Pumps . . . . .695.6(G)
    Optional Standby Systems Not Required. 701.31
    Performance Testing . . . . 230.95(C)
Illumination of Service Equipment
    1000 Volts, Nominal or Less . . . . 110.26(D)
    Over 1000 Volts, Nominal . . . . 110.34(D)
        Note: Additional lighting luminaires (fixtures) are not intended where the workspace is illuminated by an adjacent light source.
Location of Service Equipment . . . . Art. 230 Part VI
    Floating Buildings . . . . Art. 555 Part III
    General. 110.26
    Manufactured Homes . . . . 550.32(B)
    Mobile Homes . . . . 550.32(A)
    Nearest the Point of Entrance of the Service Conductors. . . . 230.70(A)(1)
    Over 1000 Volts . . . . Art. 110 Part III
    Overcurrent Device 240.24
Main Disconnects . . . . Art. 230 Part VI
    Combined Rating of 230.80
    Grouped . . . . 230.72(A)
    Hazardous (Classified) Locations (*See* Chap. 5). . . 230.70(C)
    High Voltage Systems . . . . Art. 230 Part VIII
        Location and Type. 230.205
        Overcurrent Device as 230.206
    Location on Premises. 240.24
    Locked or Sealed 230.92
        Specific Circuits 230.93
    Marked Suitable for 230.66
        Suitable for Prevailing Conditions . . . . 230.70(C)
    Marked As Service Disconnect . . . . 230.70(B)
    Minimum Size & Rating. 230.79
    More Than One Building or Other Structure 225.33
    Mounting Height . . . . 240.24(A)
    No Mains Smaller Than 15 Amperes 230.79

Not More Than Six
- For Each Service 230.71
- To Be Grouped 230.72
- For Service-Entrance Conductor Sets 230.40

Provisions for Breaking the Grounded Conductor . . . 230.75
Provisions to Bond the Grounded Conductor 250.26
Rating of Disconnect. 230.79
Readily Accessible Place . . . . 230.70(A)(1)
- Accessible to Occupants . . . . 230.72(C)

Minimum Size & Rating
- Combined Rating of Disconnects 230.80
- Dwelling Unit Service and Feeder Conductors 310.12
- Service Disconnect 230.79
- Service-Drop Conductors 230.23
- Service-Entrance Conductors. . . . . 230.42(A)
- Service-Lateral Conductors, Underground 230.31

Overcurrent Protection of . . . . Art. 230 Part VII
- Location 230.91
- Permitted Rating for Dwelling Services . . . 230.90(A) Ex. 5

Panelboards 408.36
- Service Conductors. . . . . 230.90(A)
- General 240.22
- No Overcurrent Device in Grounded Conductor 230.90(B)

Suitable for Available Short-Circuit Current 110.9
- Cartridge Fuses. . . . . .240.60(C)(3)
- Circuit Breakers . . . . 240.83(C)
- Circuit Impedance and Other Characteristics 110.10

Surge Arresters (Over 1000 Volts). 230.209

Working Space about
- See WORKING SPACE . . . . Ferm's Finder

SERVICE LATERALS . . . . Art. 230 Part III
Ampacity of Conductors . . . . Table 310.15(B)16
Bare Permitted for Grounded Conductor . . . . . 230.30(A) Ex.
Bushing or Terminal Fitting on Open End of Conduit . 300.5(H)
Definition of . . . . Art. 100 Part I
Direct Burial 300.5
- Over 1000 Volts, Nominal 300.50
- Identified for Use. . . . . 310.10(E)
- See OVER 1000 VOLTS, NOMINAL. . . . . . Ferm's Finder

Insulation of 230.30
Protection Against Damage. 230.32
Sealing Underground Raceway 230.8
- General. . . . . 300.5(G)
- Over 1000 Volts . . . . 300.50(F)

Size and Rating 230.31
UF, Type . . . . Art. 340
Under Buildings . . . . 300.5(C)
- Where Considered Outside the Building. 230.6

USE, Type . . . . Art. 338
USE and UF . . . . Table 310.4(A)
See UNDERGROUND WIRING . . . . Ferm's Finder

SERVICE POINT
Definition of . . . . Art. 100 Part I
Distribution Point (Agricultural Buildings) 547.2
Grounded Conductor. 250.186
Premises Wiring, Definition of . . . . Art. 100 Part I
Service Conductors, Definition of . . . . Art. 100 Part I
Shall Not Apply. 230.200

SERVICE STATIONS, GASOLINE (Motor Fuel Dispensing Facilities) . . . . Art. 514
See GASOLINE (MOTOR FUEL) DISPENSING Ferm's Finder

SERVICE-ENTRANCE CABLE . . . . Art. 338
Bends. 338.24
Definition of .338.2
Feeder or Branch-Circuit Wiring . . . . 338.10(B)(4)
- Entering, Exiting, or Attached to Buildings or Structures . 225.11
- In Messenger Supported Wiring 396.10
- Use of Grounded Conductor for Equipment. 250.142
- Wiring on Buildings or Other Structures. 225.10

Grounding Frames of Ranges and Dryers 250.140
Installation Methods
- Exterior . . . . 338.10(4)(b)
- Interior. . . . . 338.10(4)(a)

Physical Protection
- Aboveground. . . . . 230.50(B)
- Underground. . . . . 230.50(A)

Service-Entrance Conductors, As . . . . 338.10(A)
Underground 338.100
Uses Not Permitted, Type SE 338.12
USE Type, Underground or Outside Only . . . . . . 338.12(B)
See (TXKT) . . . . UL Product iQ

SERVICE-ENTRANCE CONDUCTORS . . . Art. 230 Part IV
Bare Permitted for Grounded Conductor . . . . . . 230.41 Ex.
Connections at Weatherhead 230.54

Definition of (Overhead and Underground Systems) Art. 100 Part I
Drip Loops . . . . 230.54(F)
Identification of Higher Voltage to Ground Conductor . . 230.56
Insulation of 230.41
High-Leg Marking 110.15
    Mounting Supports 230.51
Number of Sets 230.40
Overcurrent Protection . . . . 240.21(D)
Protection of Open Conductors
    Aboveground . . . . 230.50(B)
    Underground . . . . 230.50(A)
Raceways to Drain 230.53
Size and Rating 230.42
Splices, Permitted in 230.46
Wiring Methods 230.43

SERVICES . . . . **Art. 230**
Bonding of 250.92
    See BONDING, Service Equipment . . . . . . Ferm's Finder
Calculations for Size of . . . . Art. 220
    Examples . . . . Informative Annex D
    See CALCULATIONS . . . . Ferm's Finder

Conductors
    Considered Outside of Building 230.6
    Example for Sizing . . . . Informative Annex D7
    For Fire Pumps . . . . 695.6(A)
    One Building Not to Be Supplied through Another . . 230.3
    Other Conductors Not in Service Raceway 230.7
    Overcurrent Protection of . . . . Art. 230 Part VII
        Dwelling Services 310.12
        Overhead Service Conductors . . . . Art. 100
        Underground Service Conductors . . . . Art. 100
    Dedicated Space for Equipment
        1000 Volts, Nominal or Less . . . . 110.26(E)
        Over 1000 Volts, Nominal . . . . 110.34(F)
    Grounded Conductor to . . . . 250.24(C)
    Grounding . . . . Art. 250
        Conductors . . . . Art. 250 Part III
        Connections . . . . Art. 250 Part III
        Electrode System . . . . Art. 250 Part III
        High-Voltage Systems (Over 1 kV) . . . . . Art. 250 Part X
        Methods of Equipment Grounding . . . . Art. 250 Part VII
        Raceways and Enclosures . . . . Art. 250 Part IV
        System . . . . Art. 250 Part II

    See BONDING, Service Equipment . . . . Ferm's Finder
    See GROUNDING, Service Equipment . . . . . Ferm's Finder
High-Leg Color Coding
    General Marking Requirements 110.15
    Phase Arrangement and Busbar Arrangement . . . . 408.3(E)
    Service Conductor 230.56
High Voltage Service . . . . Art. 230 Part VIII
    See OVER 1000 VOLTS, NOMINAL . . . . . Ferm's Finder
Identification of Other Services, Feeders and Circuits . 230.2(E)
Manufactured Home . . . . 550.32(B)
Mast Type 230.28
    Note: Only power service-drop conductors are permitted to be attached to a service mast. Antennas are not permitted to be attached to the electric service mast 810.12
    See Drawing SM-1 & SM-2 . . . . *Ferm's Charts and Formulas*
Mobile Home Service . . . . 550.32(A)
    Additional Outside Electrical Equipment . . . . . 550.32(D)
    Additional Receptacles . . . . 550.32(E)
    Marking . . . . 550.32(G)
    Mounting Height . . . . 550.32(F)
    Rating . . . . 550.32(C)
    See MOBILE HOMES . . . . Ferm's Finder
Multiple Occupancy Building
    Access to Occupants . . . . 230.72(C)
    Access to Overcurrent Device . . . . 240.24(B)
    Additional Permitted 230.2
    Number of Disconnects at One Location 230.71
    Number of Service-Entrance Conductor Sets 230.40 Exceptions
    See CALCULATIONS . . . . Ferm's Finder
Multiple Service for Separate Loads . . . . 230.2(A)
    Service-Entrance Conductor Sets . . . . 230.40 Ex. 2
Neutral (Grounded Conductor) to Every Service Disconnect . . 250.24(C)
No Other Conductors in Service Raceway or Cable . . . 230.7
No Solder Lugs in Main Service 230.81
Number of
    Drops or Laterals to a Building 230.2
    Main Disconnects 230.71
        Additional Disconnects . . . . 230.72(B)
        Combined Rating 230.80
    Number of Supplies, More than One Building 225.30
        Number of Disconnects for Each Supply 225.33
    Service-Entrance Conductor Sets 230.40
        Floating Buildings 555.50
    Service-Entrance Conductors Supplied 230.40

Services to a Building 230.2
Over 1000 Volts . . . . Art. 230 Part VIII
    See OVER 1000 VOLTS, NOMINAL . . . . . . Ferm's Finder
Over 35,000 Volts 230.212
Overhead Service Conductors. . . . . Art. 230 Part II
    Defined . . . . Art. 100
Service-Entrance Conductors . . . . Art. 230 Part IV
Raceways for Use at Services 230.43
    Floating Buildings 555.52
Raceways to Drain 230.53
Rating of Disconnect 230.79
    Combined Ratings 230.80
Recreational Vehicles Parks . . . . Art. 551 Part VI
Relative Location of Equipment and Overcurrent Device . . . 230.94
Remote Control of Disconnect, Over 1000 Volts . . 230.205(C)
Remote Disconnecting Means . . . . 230.72(B)
Sealing Raceway from Warm to Cold . . . . 300.7(A)
    Busways 368.234
Sealing Underground Raceway 230.8
    1000 Volts or Less . . . . 300.5(G)
    Over 1000 Volts . . . . 300.50(F)
Separate Building or Structure . . . . Art. 225 Part II
Separate Service Permitted
    Capacity Requirements . . . . 230.2(C)
    Different Characteristics . . . . 230.2(D)
    Special Conditions . . . . 230.2(A)
    Special Occupancies . . . . 230.2(B)
Service-Entrance Cable . . . . Art. 338
    See (TXKT) . . . . UL Product iQ
Service Head (Location)
    Clearance from Building Openings 230.9
    Overhead Service Locations 230.54
    Point of Attachment 230.26
    Physical Protection
        Aboveground . . . . 230.50(B)
        Underground . . . . 230.50(A)
Size & Rating
    Combined Rating of Disconnects 230.80
    Disconnecting Means 230.79
        Mobile and Manufactured Homes . . . . 550.32(C)
    Dwelling Unit Service and Feeder Conductors 310.12
    Example, Service Conductors for Dwellings . . . .
        . . . . Informative Annex D7
    Overhead Service-Drop Conductors 230.24

Service-Entrance Conductors 230.42
Service-Lateral Conductors 230.30
Splices Permitted in Conductors 230.46
Split Bus Panels 408.36
Tap Ahead of Mains
    Equipment Permitted 230.82
    Fire Pumps . . . . 695.3(A)(1)
    Interconnected Power Production Sources 705.12
    Legally Required Standby Systems . . . . 701.12(G)
    Solar Photovoltaic Systems 690.59
Taps Permitted 230.46
Underground Service . . . . Art. 230 Part III
    Conductors Connected at Supply End Only 230.2
    Considered Outside Building 230.6
Underground Service Conductors, Defined . . . . Art. 100
Vertical Clearances of Service Conductors From Final Grade or Earth . . . . 230.24(B)
Wiring Methods for Service Conductors 230.43
    Floating Buildings 555.50
    Over 1000 Volts . . . . 230.202(B)
Working Space about
    See WORKING SPACE . . . . Ferm's Finder

SETTING (of Circuit Breaker)

Adjustable-Trip Circuit Breaker . . . . 240.6(B)
Arc Energy Reduction 240.87
Branch-Circuit Overcurrent Protection 210.20
Branch-Circuit Rating 210.3
Critical Operations Power Systems (COPS) . . . . . . 708.24(E)
Definition of (Circuit Breaker, Setting) . . . . . . Art. 100 Part I
Emergency Systems . . . . 700.5(E)
Feeders and Branch Circuits, Over 1000 Volts, Nominal 240.100
Feeders, Over 1000 Volts 240.101
Ground-Fault Protection of Equipment . . . . 230.95(A)
Inspections and Test
    Feeders 225.56
    Over 1000 Volts 110.41
Legally Required Standby Systems . . . . .701.5(D)
Optional Standby Systems 702.5
Restricted Access Adjustable-Trip Circuit Breakers . . . 240.6(C)
Service Equipment, Over 1000 Volts, Nominal 230.208
Service Equipment Overcurrent Protection
. . . . 230.90, Art. 230 Part VII
Service Ground-Fault Protection of Equipment 230.95
Transformer Overcurrent Protection . . . Table 450.3(A) and (B)

SHEET METAL SCREWS
Covers and Canopies, Not to be Used 314.25
Grounding and Bonding Equipment, Not to be Used. . . 250.8(A)(6)
Receptacles, Not to be Used 406.5
Switches, Not to be Used . . . . 404.10(B)

SHIELDING OF CONDUCTORS **311.44**
Ampacity Values for Shielded Cables . . . . 311.60(D)
Bending Radius, Over 1000 Volts 300.34
Bending Radius, Type MC Cable 330.24
CATV Cable Grounding 820.100
Direct Burial Conductors Over 2001 Volts 311.36 Ex.1
Insulation Shielding Over 1000 Volts. 300.40
Intrinsically Safe Systems . . . . 504.50(B)
Pull and Junction Boxes, Over 1000 Volts, Nominal . . . 314.71
Sealing of Conductors, Class I, Division 1 Locations . . . .
. . . . .501.15(D)(1) Ex.
Sealing of Conductors, Class I, Division 2 Locations . . . .
. . . . .501.15(E)(1) Ex. 2
Splices of Underground Conductors . . . . 300.50(D)
    See OVER 1000 VOLTS, NOMINAL . . . . . . Ferm's Finder
Type FCC Cable . . . . Art. 324
Type PLTC Cable . . . . 725.179(E)

SHORT-CIRCUIT CURRENT AVAILABLE
Circuit Impedance, Short-Circuit Current Ratings and Other Characteristics 110.10
Definition of (Short-Circuit Current Rating) . . . . . . Art. 100
Disconnect
    Feeder . . . . 225.52(B)
    Service (in excess of 1000 V) . . . . .230.205(B)
    Service (not in excess of 1000 V) . . . . 230.82(3)
Equipment to Be Rated for 110.9
Marking of, for Fuses . . . . 240.60(C)
Marking of, for Circuit Breakers . . . . 240.83(C)
Panelboards. 408.6
Series Rated Systems 240.86
    Marking of for End Use Equipment . . . . 240.86(B)
    Motor Contributions. . . . . 240.86(C)
    Selected Under Engineering Supervision . . . . . . 240.86(A)
Switchboards 408.6
Switchgear 408.6

Note: Contact the serving utility company for short-circuit current available at service point, and be able to calculate it up to the service/feeder equipment.

SHOW WINDOWS
Branch Circuits, Load Calculations for . . . . 220.14(G)
Definition of . . . . Art. 100 Part I
Electric Signs and Outline Lighting. . . . . Art. 600
Feeders, Load Calculations for . . . . 220.43(A)
Flexible Cords, Types Permitted. 400.15
Floor Boxes for Use in Elevated Show Windows . 314.27(B) Ex.
Luminaires, Types Permitted 410.14
Receptacle Spacing for 210.62
    In Floor Boxes . . . . 314.27(B)

SHOWCASES (CORD-CONNECTED) **410.59**
Cord Requirements . . . . 410.59(A)
No Other Equipment Permitted . . . . 410.59(D)
Receptacles, Connectors, and Attachment Plugs . . . 410.59(B)
Secondary Circuit(s) of Ballasts . . . . 410.59(E)
Support of Cords . . . . 410.59(C)

SHOWER STALLS
Receptacles Within 6 feet of Bathtub or Shower Stall, GFCI Protection . . . . 210.8(A)(9)

SIGNALING CIRCUITS . . . . **Art. 725**
Class 1, Class2 and Class 3 Remote-Control, Signaling, and Power-Limited Circuits . . . . Art. 725
Energy Management Systems . . . . Art. 750
Fire Alarm Systems . . . . Art. 760
    See FIRE ALARM SYSTEMS . . . . Ferm's Finder
See CLASS 1, 2 & 3 REMOTE CONTROL,

SIGNALING, AND POWER-LIMITED CIRCUITS See **CIRCUITS** . . . . **Ferm's Finder**

SIGNS, ELECTRIC AND OUTLINE LIGHTING . . **Art. 600**
Applications of Power Limited Cable . . . . . Table 600.33(A)(1)
Ballasts 600.22
    Class 2 Power Sources 600.21
    Listing Required . . . . 600.22(A)
    Thermal Protection Required . . . . 600.22(B)
Ballasts, Transformers and Electronic Power Supplies. . . 600.21
    Accessibility . . . . 600.21(A)
    Attic and Soffit Locations . . . . 600.21(E)
    Location . . . . 600.21(B)
    Suspended Ceilings. . . . . 600.21(F)
    Working Space . . . . 600.21(D)
Branch Circuits 600.5
    Computed Load . . . . 220.14(G)

Rating . . . . 600.5(B)
Required . . . . 600.5(A)
Wiring Methods . . . . 600.5(C)
Cable Substitutions . . . . Table 600.33(A)(2)
    Applications of Power Limited Cable. . . . Table 600.33(A)(1)
    Conductors. . . . . 600.33(A)
    Power Sources 600.24
    Secondary Lighting. 600.33
Connected to an Equipment Grounding Conductor . . 600.7(A)
Definitions 600.2
Definition of Outline Lighting . . . . Art. 100 Part I
Disconnecting Means. 600.6
    Marking . . . . 600.5(B)
    Remote Locations . . . . 600.6(A)(4)
Dwelling Occupancies, Voltage Restrictions . . . 600.32(I)
Electric-Discharge Lighting, Definition . . . . Article 100 Part I
Electrode Connections 600.42
Equipment Grounding Conductor
    Connected to . . . . 600.7(A)(1)
    Size. 600.7(A)(2)
Enclosures 600.8
Field-Installed Secondary Wiring 600.12
Field-Installed Skeleton Tubing and Wiring . . Art. 600 Part II
Grounding and Bonding 600.7
In Fountains 680.57
Installation Instructions. . . . . 600.4(E)
Listing Requirements 600.3
    Location of 600.9
    Marking Requirements. 600.4
        Durability. . . . . 600.4(D)
        Portable Signs Exempted . . . . 600.4(E) Ex.
        Visibility. . . . . 600.4(C)
Neon Secondary-Circuit Conductors, 1000 Volts or Less   600.31
Neon Secondary-Circuit Conductors, Over 1000 Volts . 600.32
    Length of High-Voltage Cable . . . . 600.32(J)
Neon Tubing 600.41
Photovoltaic (PV) Powered Sign 600.34
    Definition 600.2
Portable or Mobile Signs 600.10
    GFCI Protection Required . . . . 600.10(C)(2)
Retrofit Kit 600.35
    Defined . . . . Art. 100
    Field-Installed Secondary Wiring, Signs . . . 600.12
    Listing Requirements for Signs. 600.3

Marking . . . . 600.4(B)
Sign Installations Instructions . . . . 600.4(E)
Remote Metal Parts, Bonding . . . . 600.7(B)(1) Ex
Section Signs, Definition of 600.2
Transformers and Electronic Power Supplies . . . 600.23
    Listing Required . . . . 600.23(A)
    Marking Required . . . . 600.23(F)
    Rating of Secondary Current . . . . 600.23(D)
    Secondary Connections . . . . 600.23(E)
    Secondary-Circuit Ground-Fault Protection . . 600.23(B)
    Voltage Limitations. . . . . 600.23(C)
See the IAEI book on *NEON LIGHTING*
See Signs (UXYT). . . . . *UL Product iQ*
See Skeletal Neon Sign and Outline Lighting (UZBL) . *UL Product iQ*

SIGNS AND LABELS
Caution, Meet These Requirements. . . . . 110.21(B)
Danger, Meet These Requirements . . . . 110.21(B)
Field-Applied . . . . 110.21(B)
Warning, Meet These Requirements . . . . 110.21(B)
See ANSI Z535.4-2011, *Product Safety Signs and Labels*
See WARNING SIGNS. . . . . *Ferm's Finder*

SIMPLE APPARATUS
Definition. . . . . Art. 100
Installation . . . . 504.10(D)

SINGLE-POLE SEPERABLE-CONNECTORS 406.13

SINKS
Dwelling Units . . . . 210.8(A)(7)
Other than Dwelling Units . . . . 210.8(B)(5)
Receptacle Outlet Locations . . . . 210.52(C)(3)
Under Sink in Faceup Position (prohibited). . . . 406.5(G)(2)

SKELETON TUBING . . . . Art. 600 Part II
See SIGNS, ELECTRIC AND OUTLINE LIGHTING . . . . .
. . . . *Ferm's Finder*
See (UZBL) . . . . *UL Product iQ*

SKID MOUNTED EQUIPMENT
Equipment Grounding Conductor, Connected to . 250.112(K)

SKIN EFFECT HEATING
Fixed Electric Heating for Pipelines and Vessels . Art. 427 Part VI
Fixed Outdoor Electric Deicing and Snow Melting Equipment .
. . . . Art. 426 Part V

Skin Effect Heating Systems, Defined. . . . . . 426.2 and 427.2

SMOKE ALARM
AFCI Protection of Outlet . . . . 210.12(A)

Installed in Dwelling Unit . . . . 210.12(A) IN 2

See International Residential Code (IRC) Chapter 3- R314

See NFPA 72, 2013, National Fire Alarm and Signaling Code . . .
. . . . NFPA 72

SMOKE DETECTOR . . . . .645.4(2)(3)

SNAP SWITCHES
Accessibility and Grouping 404.8

    Location . . . . 404.8(A)

    Voltage between Adjacent Switches . . . . 404.8(B)

Definition of (General-Use Snap) . . . . Art. 100 Part I

Motors, Stationary 2 HP or Less

    Permitted As Controller . . . . .430.83(C)(2)

    Permitted As Disconnect . . . . 430.109(C)(2)

Mounting of 404.10

Panelboards Containing Switches 30 Amperes or Less 408.36(A)

Provisions for Faceplates . . . . 404.9(A)

    Grounding . . . . 404.9(B)

    Position of . . . . 404.9(A)

Ratings 404.14

Relative Arrangement of Switches and Fuses 408.39

See Snap Switches (WJQR) . . . . *UL Product iQ*

SOLAR PHOTOVOLTAIC SYSTEMS . . . Art. 690, Art. 691
Alternating-Current Modules 690.6

Arc-Fault Circuit Protection, Direct Current 690.11

Back-Fed Circuit Breakers . . . . 710.15(E)

Circuit and Equipment Overcurrent Protection 690.9

Circuit Requirements . . . . Art. 690 Part II

Circuit Routing . . . . 690.31(D)

Circuit Sizing & Current 690.8

    DC-to-DC Converter Output Circuit Current. . . . 690.7(B)

Component Identification . . . . . . Figure 690.1(a) and (b)

Conductors of Different Systems . . . . 690.31(B)

Connection to Other Sources 690.59

    Supply Side of Service Equipment . . . . 230.82(6)

Definition of (Photovoltaic Systems) . . . . Art. 100 Part I

Definitions Applicable to System 690.2

Disconnecting Means . . . . Art. 690 Part III

    Equipment Disconnecting Means . . . . 690.15(C)

    Location . . . . 690.15(A)

    Isolating Device . . . . 690.15(B)

    Types of Disconnects . . . . 690.13(D)

Electronic Power Converters Mounted in Not Readily Accessible Locations . . . . 690.4(F)

Energy Storage Systems

    General 690.71

    Self-Regulated PV Charge Control 690.72

Flexible Cords and Cables . . . . 690.31(C)(4)

    Flexible, Fine Stranded Cables, Listed Connectors Required . .
. . . . 690.31(C)(5)

Functional Grounded PV System 690.2

General Requirements 690.4

Ground-Fault Protection, DC Modules . . . . 690.41(B)

Grounding and Bonding . . . . Art. 690 Part V

    Building or Structures Supporting a PV Array . . . 690.47(A)

    Additional Auxiliary Electrodes for Array Grounding 690.47(B)

Identification of Components in Common Configurations . . . .
. . . . .Figure 690.1(b)

Identification of System Components . . . . Figure 690.1(a)

Inverters (Multiple) . . . . 690.4(D)

Large-Scale Photovoltaic Systems . . . . Art. 691

    Accessible 691.4

    Approval 691.5

    Arc-Fault Mitigation 691.10

    Conformance of Construction of Engineered Design . .691.7

    Definitions 691.2

    Direct Current Operating Voltage 691.8

    Disconnection of Photovoltaic Equipment 691.9

    Engineered Design .691.6

    Fence Grounding 691.11

    Field-Applied Hazard Labels . . . . 691.4(2)

    Not Installed on Buildings . . . . 691.4(5)

    Qualified Personnel . . . . 691.4(1)

Listing and Identification Required — Equipment . . . 690.4(B)

Marking . . . . Art. 690 Part VI

Maximum Voltage 690.7

Multiple Inverters . . . . 690.4(D)

Not Permitted (Bathrooms) . . . . 690.4(E)

Stand-Alone Systems 690.10

Overcurrent Protection 690.9

    Circuits and Equipment . . . . 690.9(A)

    Power Transformers . . . . 690.9(D)

Qualified Persons to Perform Work . . . . 690.4(C)

Rapid Shutdown of PV Systems on Buildings 690.12

    Controlled Conductors . . . . 690.12(A)

Controlled Limits . . . . 690.12(B)
　　Equipment . . . . 690.12(D)
　　Initiation Device . . . . 690.12(C)
　　Plaques and Directories . . . . 690.56(C)
　Scope 690.1
　Separation and Marking Requirements . . . . 690.31(B)
　Wiring Methods . . . . Art. 690 Part IV
　See (QIIO) . . . . UL Product iQ

SOLDERING LUGS & WIRE CONNECTORS
　Electrical Connections 110.14
　Equipment Grounding Conductors to Boxes 250.148
　Fire Pumps- Pump Wiring . . . . 695.6(D)
　Grounding Electrode Conductor- Not Permitted 250.70
　Knob-And-Tube Wiring (Splices and Taps) 394.42
　Low-Voltage Suspended Ceiling Power Distribution Systems . . .
　. . . . . 393.40(A)
　Service Conductors- Not Permitted 230.81

SOUND-RECORDING AND SIMILAR EQUIPMENT Art. 640
　See AUDIO SIGNAL PROCESSING, AMPLIFICATION, AND REPRODUCTION EQUIPMENT . . . . Ferm's Finder

SPACE HEATING EQUIPMENT, FIXED ELECTRIC Art. 424
　See FIXED ELECTRIC HEATING EQUIPMENT FOR PIPELINES AND VESSELS . . . . Ferm's Finder
　See FIXED ELECTRIC SPACE-HEATING EQUIPMENT . . .
　. . . . . Ferm's Finder
　See FIXED OUTDOOR ELECTRIC DEICING AND SNOW-MELTING EQUIPMENT . . . . Ferm's Finder
　See HEAT PUMPS . . . . Ferm's Finder
　See INDUCTION & DIELECTRIC HEATING . Ferm's Finder
　See INFRARED LAMP HEATING APPLIANCES Ferm's Finder
　See (ZMVV) . . . . UL Product iQ

SPAS & HOT TUBS . . . . Art. 680 Part IV
　See HOT TUBS & SPAS . . . . Ferm's Finder
　See SWIMMING POOLS . . . . Ferm's Finder
　See (WBYQ) . . . . UL Product iQ

SPRAY APPLICATION DIPPING, COATING, AND PRINTING PROCESSES
　Classification of Locations
　　Open Containers 516.4
　　Membrane Enclosures 516.18
　　Painting, Dipping, and Coating Processes 516.29
　　Spray Application Processes 516.5

　Definitions . . . . Art. 100, Part III
　Grounding 516.16
　Special Equipment 516.10
　Wiring and Equipment
　　Class I Locations 516.6
　Not Within Classified Locations 516.7

SPECIAL PERMISSION (Definition of) . . . . Art. 100 Part I
　Electric Signs and Outline Lighting 600.3
　Enforcement 90.4
　Fixed Electric Space-Heating Equipment 424.10
　Fixed Industrial Process Heating Equipment 425.10
　Fixed Outdoor Electric Deicing and Snow-Melting Equipment . .
　. . . . . 426.14
　For Certain Utility Installations Exempted from Code Requirements . . . . 90.2(C)
　Knob-And-Tube Use Permitted 394.10
　Number of Services 230.2
　Surge Arrester- Over 1000 Volts . . . . Art. 242 Part III

　Note: Many *Code* sections contain alternate methods that may be used if special permission is granted by the AHJ.

SPLASH PADS
　Definition 680.2
　Requirements 680.50

SPLICES AND TAPS
　Auxiliary Gutters 366.56
　Boxes Not Required . . . . 334.40(B)
　Boxes Required 300.15
　　Temporary Wiring . . . . 590.4(G)
　　See BOXLESS DEVICES . . . . Ferm's Finder
　Cabinets and Cutout Boxes . . . . 312.8(A)
　Cable Trays, Permitted to Project Above Side Rails 392.56
　Carnivals, Circuses, Fairs and Similar Events . . . . . 525.20(D)
　Cellular Concrete Floor Raceways 372.56
　Cellular Metal Floor Raceways 374.56
　Concealed Knob-and-Tube 394.56
　Conduit Bodies (General) 300.15
　　When Permitted . . . . 314.16(C)(2)
　Construction Sites . . . . 590.4(G)
　Deicing and Snow-Melting . . . . 426.24(B)
　Direct Burial Cable . . . . 110.14(B)
　　Box Not Required . . . . 300.5(E)
　　Box Not Required, Over 1000 Volts . . . . 300.50(D)
　Feeder Taps (General) . . . . 240.21(B)

Over 1000 Volts . . . . .240.101(B)
Supervised Industrial Installations . . . . . Art. 240 Part VIII

Flat Cable Assemblies (Splices) . . . . 322.56(A)

Taps . . . . 322.56(B)

Flexible Cords

General 400.13

On Construction Sites . . . . .590.4(G)

General Requirements 110.14

Grounding Electrode Conductor 250.64

Method of Splicing. . . . . 250.64(C)

Splices in Busbars Permitted . . . . .250.64(C)(2)

Taps . . . . .250.64(D)(1)

Heating Cables

For Pipelines and Vessels . . . . 427.23(A)

Outdoor Deicing and Snow-Melting. . . . . 426.24(B)

Space-Heating Cables, Cannot Alter Length 424.40

Space-Heating Cables, Embedded . . . . 424.41(D)

Insulation of . . . . 110.14(B)

Messenger Supported Wiring. 396.56

Motor Feeders 430.28

Panelboards, in . . . . 312.8(A)

Power Distribution Blocks . . . . 376.56(B)

Power Monitoring Equipment . . . . 312.8(B)

Raceways, Not in (Generally) . . . . 300.13(A)

Note: Refer to appropriate raceway article for further information and see also . . . . 300.15(A)

Sealing Fittings, Not in . . . . .501.15(C)(4)

Service-Entrance Conductors 230.46

Swimming Pool Lighting Ground Wire, Splice Not Permitted . . . . 680.23(F)(2)

Transformers . . . . 240.21(C)

Underground, Listed . . . . 110.14(B)

Wireways, in

Metal. 376.56

Nonmetallic 378.56

Power Distribution Blocks (Metallic Wireways) . . 376.56(B)

Note: Use oxide inhibitor on aluminum terminations where required by listing or manufacturer.

See Aluminum Conductor Terminations *Ferm's Charts and Formulas*

See Tightening Torque Information . . *Ferm's Charts and Formulas*

See (DVYW) . . . . *UL Product iQ*

SPLIT-BUS PANELBOARDS **408.36**

SPRAY APPLICATION, DIPPING, COATING & PRINTING PROCESSES. . . . .**Art. 516**

See FINISHING PROCESSES . . . . *Ferm's Finder*

STAGE

Equipment, Defined .520.2

Lighting Hoist, Defined 520.2

Lighting Hoist, Wiring Method 520.40

Switchboard 520.2

See . . . . Article 520

STAND-ALONE SYSTEMS

Back-Fed Circuit Breaker. . . . . 710.15(F)

Conductor Sizing. . . . . 710.15(B)

Equipment Approval. 710.6

Premises Wiring System 710.15

Supply Output. 710.15(A)

Three-Phase Supply. . . . . 710.15(D)

STANDBY SYSTEMS . . . . **Art. 700**

Back-Fed Circuit Breaker. . . . . 710.15(F)

Conductor Sizing. . . . . 710.15(B)

Equipment Approval. 710.6

Identification of Power Sources 710.10

Premises Wiring System 710.15

Stand-Alone Inverter Input Circuit Current 710.12

Supply Output. 710.15(A)

STATIC

Antenna, Radio and Television 810.57

Bulk Storage Plants . . . . 515.16 Info. Note

Electric Discharges (Spray Applications) 516.40

Hazardous (Classified) Locations . . . 500.4 Info. Note 1 and 3

Painting in Aircraft Hangers . . . . . . 513.3(C)(2) Info. Note

Zones 0, 1 and 2 . . . . 505.4 IN 1 and 3

Zones 20, 21 and 22 . . . . 506.4 IN

STEEL TUBE

See ELECTRICAL METALLIC TUBING . . . . *Ferm's Finder*

See FLEXIBLE METALLIC TUBING . . . . . . *Ferm's Finder*

STORABLE SWIMMING POOLS . . . . . . **Art. 680 Part III**

Ground-Fault Circuit-Interrupter Required 680.32

Lighting (Fixtures) Luminaires 680.33

Pumps 680.31

See SWIMMING POOLS . . . . *Ferm's Finder*

See (WCSX). . . . . *UL Product iQ*

See (WBDT) . . . . *UL Product iQ*

STORAGE BATTERIES . . . . Art. 480
Aircraft, Special Equipment 513.10
Audio Signal Processing, Amplification, & Reproduction
    Auxiliary Power Supply Wiring . . . . 640.9(B)
Battery and Cell Terminations 480.4
Battery System, Defined . . . . Art 100 Part I
Charging Equipment, Class III Locations 503.160
Corrosion Prevention . . . . 480.4(A)
DC Disconnect Methods 480.7
Definitions 480.2
Electric Vehicles
    Ventilation Not Required . . . . 625.52(A)
    Ventilation Required . . . . 625.52(B)
Emergency Systems
    As Alternate Power Source, Health Care Facilities . . 517.30(B)(1)
    (*See* 517.2 for definition of *Alternate Power Source*)
    As Power Source . . . . 700.12(A)
    Unit Equipment . . . . 700.12(F)
Equipment 480.3
Garages, Special Equipment 511.10
IEEE Standards . . . . 480.1 Info. Note
Insulation Support 480.8
Listing Requirement 480.3
Location 480.10
Nominal Voltage, (Battery or Cell) Defined 480.2
Racks and Trays (Support Structure) 480.9
Recreational Vehicles . . . . 551.4(B)
Solar Photovoltaic Systems . . . . Art. 690 Part VIII
Support Systems 480.9
Ventilation . . . . 480.10(A)
Vents (Cells) 480.11

STRIKE TERMINATION DEVICES 250.60
Electrode Location and Placement . . . . 250.53(B)
Lightning Protection System 250.106
Radio Receiving Stations . . . . 810.18(A) Info. Note 1
See NFPA 780-2011, Standard for the Installation of Lightning
Protection Systems . . . . NFPA 780

STORAGE BATTERIES . . . . ARTICLE 480
DC Disconnect Methods 480.7
Emergency Disconnect . . . . 480.7(B)

STRUCTURAL JOINT, EXPANSION AND DEFLECTION . . . . 300.4(H)

STRUCTURAL STEEL
As a Grounding Electrode . . . . 250.52(A)(2)
Bonding . . . . 250.52(A)(6)
Bonding Piping Systems and Structural Metal 250.104
Metal In-Ground Support Structure(s) . . . . . . 250.52(A)(2)

STRUCTURES (METAL), OVER 1000 VOLTS . . . 250.194(B)

STRUT-TYPE CHANNEL RACEWAYS . . . . Art. 384
Busways . . . . 368.56(A)(14)
Definition of 384.2
Grounding of 384.60
Listing 384.6
Marking 384.120
Number of Conductors 384.22
Size of . . . . Table 384.22
Size of Conductors 384.21
Support of 384.30
Uses Not Permitted 384.12
Uses Permitted 384.10
See (RIUU) . . . . *UL Product iQ*

SUBMERSIBLE PUMP CABLE
See PUMP HOUSES . . . . Ferm's Finder
See Underground Branch Circuit and Feeder Cable (YDUX) . . . . *UL Product iQ*

SUBSTATION 490.48
Covered by Code, Industrial . . . . 90.2(A)(2)
Definition . . . . Art. 100 Part I
Design, Documentation, and Required Diagrams 490.48
Feeders, Inspection and Test . . . . 225.56(A)(6), 110.41
Fences and Other Metal Structures, Grounding and Bonding 250.194
Grounding for AC Substations 250.191
High-Voltage Fuses . . . . 490.21(B)(7)
Indoor Installations . . . . 110.31(B)(1)
Inspections and Test 110.41
Large-Scale Photovoltaic (PV) Electric Power Production
Facility 691.2
Mobile and Portable Equipment 490.51
Motion Picture Productions, Feeders . . . . 530.18(B)
Motion Picture and Television Studios and Similar Locations . . . . Art. 540 Part VI

Not Covered by Code, Utility . . . . 90.2(A)(4)
Portable or Mobile Equipment, Grounding Exemption. 250.188
Supervised Industrial Installations, Exclusion 240.2
Television Studio Sets, Feeder Sizing 530.19
Tunnels. . . . . 110.51(A)
Warning Signs . . . . 490.48(B)
See ANSI/IEEE 80-2000, IEEE Guide for Safety in AC Substation Grounding

SUBSURFACE ENCLOSURES. . . . . 110.12(B)
Manholes and Enclosures for Personnel Entry . .Art. 110 Part V

SUNLIGHT ON ROOFTOPS
Raceways and Cables Exposed . . . . .Table 310.15(B)(2)

SUPERVISED INDUSTRIAL INSTALLATIONS Art. 240 VIII

SUPERVISORY CONTROL AND DATA ACQUISITION (SCADA) . . . . Informative Annex G
See SCADA . . . . Ferm's Finder

SUPPLEMENTARY OVERCURRENT PROTECTION 240.10
Bathrooms, Permitted in . . . . 240.24(E)
Class I, Division 2, Fuses Internal to (Fixtures) Luminaires . . . . 501.115(B)(4)
Electric Heating Appliance . . . . 422.11(F)
Electric Heating Equipment Disconnecting Means . . 424.19(A)
Electric Heating Equipment Overcurrent Protection   424.22(C)
Motor Control Circuits. . . . . 430.72(A)
Not Required to Be Readily Accessible . . . . . . 240.24(A)(2)
Solar Photovoltaic Source and Output Circuits . . . . .690.9(D)
Supplementary Overcurrent Protective Device (Definition of) . . . . Art. 100, Part I

SUPPLY CONDUCTORS
Ungrounded Conductors . . . . 250.102(C)(2) Info. Note

SUPPLY SIDE BONDING JUMPER, DEFINED. . . .Art. 100
Sizing. Table 250.102(C)(1)

SUPPORTS
See SECURING AND SUPPORTING . . . . . .Ferm's Finder

SURFACE EXTENSIONS 314.22
Grounding
250.86 . . . .

SURFACE METAL RACEWAYS . . . . Art. 386
Combination, Signaling, Lighting and Power Circuits . . 386.70
Conductors
Number of 386.22
Size of 386.21
Extension through Walls and Floors . . . . 386.10(4)
General . . . . Art. 386 Part I
Grounding 386.60
Marking 386.120
Splices and Taps. 386.56
Uses Not Permitted 386.12
Uses Permitted 386.10
See RACEWAYS . . . . .Ferm's Finder
See (RJBT) . . . . UL Product iQ

SURFACE NONMETALLIC RACEWAYS . . . . Art. 388
Combination 388.70
Conductors
Number of 388.22
Size of 388.21
Construction 388.100
Extension through Walls and Floors . . . . 388.10(2)
General . . . . Art. 388 Part I
Marking 388.120
Securing and Supporting 388.30
Splices and Taps. 388.56
Uses Not Permitted 388.12
Uses Permitted 388.10
See (RJTX). UL Product iQ

SURGE PROTECTION (LIGHTNING ARRESTERS)
See LIGHTNING, SURGE ARRESTERS . . . . .Ferm's Finder

SURGE ARRESTORS, OVER 1000 Volts . . . Art. 242 Part III
Connection. 242.50
Definition of . . . . Art. 100 Part I
Emergency Systems. 700.8
Grounding Electrode Conductor 242.56
Installation 242.42
Interconnections 242.54
Location 242.46
Number Required. 242.44
Routing of Grounding Conductors 242.48
Selection 242.42
Uses Not Permitted 242.40
SURGE-PROTECTIVE DEVICES (SPDs), 1000 VOLTS OR LESS . . . . Art. 242 Part II

Definition . . . . Art. 100 Part I
Dwelling Units 230.67
Emergency Systems 700.8
Grounding Electrode Conductor 242.32
Installation . . . . 242.12(A)
Listing 242.8
Location 242.22
Number Required. 242.20
Routing of Connections 242.24
Short Circuit Current Rating 242.10
See (XUPD) . . . . UL Product iQ
Type 1 SPDs 242.12
Type 2 SPDs 242.14
Type 3 SPDs 242.16
Type 4 and Other Component Type SPDs 242.18
Uses Not Permitted 242.6

SWIMMING POOLS . . . . **Art. 680**
Applicability of Article 680 . . . . 680.1
Approval of Equipment 680.3
Branch Circuit Wiring . . . . 680.23(F)(1)
Bonding 680.26
   Equipotential Bonding Grid . . . . 680.26(B)
   Potting Compound Used for Wet-Niche (Fixtures)
   Luminaires . . . . 680.23(B)(2)(b)
   Reinforcing Steel . . . . 680.26(B)(1)(a)
   Water Heaters . . . . 680.26(B)(6)(b)
   See BONDING . . . . Ferm's Finder
   See (WCRY) . . . . UL Product iQ
Conductor Clearances
   Not Covered by Article 680 . . . . 225.18 & 19
   Overhead 680.8
   Underground. 680.11
   See also Life Safety Code, NFPA 101 . . . . . . NFPA 101
Cord- and Plug-Connected Equipment 680.8
   Methods of Grounding. 680.6
   Receptacles for . . . . 680.22(5)
Corrosive Environment 680.14
   Chapter 3 (Noncorrosive Environments) . . . . 680.21(A)(1)
Covers (Electrically Operated)
   Location of Motors and Controllers . . . . . . 680.27(B)(1)
   Wiring Methods . . . . 680.21(A)
      Note: See 680.27(B)(1) Info. Notes 1, 2, & 3 for other applicable sections
Deck Area Heating . . . . 680.27(C)

   Permanently Wired Radiant Heaters . . . . . . 680.27(C)(2)
   Radiant Heating Cables Not Permitted Above . 680.27(C)(2)
   Radiant Heating Cables Not Permitted Below Deck . 680.27(C)(3)
   Unit Heaters . . . . 680.27(C)(1)
Decorative Fireplaces . . . . 680.22(B)(7)
Definitions 680.2
Disconnecting Means (Maintenance) 680.13
Double Insulated Pumps . . . . 680.21(B)
   Bonding Not Required (Other Provisions Required) . . . . . .
   . . . . 680.26(B)(6) Ex.
Emergency (Shut-Off) Switch for Spas and Hot Tubs . . 680.41
Electric Pool Water Heater 680.10
   Bonding & Grounding. . . . . 680.26(B)(6)
Electrically Powered Pool Lifts
   Approval 680.81
   Bonding 680.83
   Definition 680.2
   Nameplate Marking 680.85
   Protection 680.82
   Switching Devices 680.84
   See ELECTRICALLY
   POWERED POOL LIFTS . . . . Ferm's Finder
Enclosures for Transformers, GFCIS or Similar Devices . . 680.24(B)
   Grounding Terminals . . . . 680.24(D)
   Location and Mounting Height . . . . 680.24(B)(2)
   Protection . . . . 680.24(C)
   Sealing of Conduits . . . . 680.24(B)(1)(3)
   Strain Relief . . . . 680.24(E)
Equipment Close to Pool . . . . 680.22(E)
Equipment Room Receptacle (GFCI Protected) . . 680.22(A)(5)
Equipment Rooms & Pits 680.12
Equipotential Bonding
   Copper Grid . . . . 680.26(B)(2)(c)
   Perimeter Surfaces, Pools . . . . 680.26(B)
   Perimeter Surfaces, Spas and Hot Tubs . . . . 680.42(B)
   Pool Water . . . . 680.26(C)
Feeder Wiring Methods . . . . 680.25(A)
Fire Pits (Low Voltage) . . . . 680.22(B)(7)
Fountain Pumps- Non submersible 680.59
Fountains . . . . Art. 680 Part V
   See FOUNTAINS . . . . Ferm's Finder
Forming Shell, Definition of 680.2
   Bonding of . . . . 680.26(B)(4)
Gas-Fired Luminaires . . . . 680.22(B)(7)

Gas-Fired Water Heater 680.28
Grounding and Bonding Terminals 680.7
Ground-Fault Circuit-Interrupter Protection
   Fountains
      For Signs in . . . . 680.57(B)
      See FOUNTAINS . . . . Ferm's Finder
   Hydromassage Bathtubs . . . . Art. 680 Part VII
      Bonding 680.74
      See HYDROMASSAGE BATHTUBS . . . . Ferm's Finder
   Inspection After Installation 680.4
   Motors . . . . 680.21(C)
   Motors (Cord- and Plug-Connected) . . . . . . 680.22(A)(2)
   Motors (Replacement-GFCI Protection) . . . . . 680.21(D)
   No Other Conductors in Boxes or Raceways . . 680.23(F)(3)
      Spas & Hot Tubs . . . . 680.43(A)(2)
      See HOT TUBS & SPAS . . . . Ferm's Finder
   Swimming Pool Area
      Luminaires [Lighting Luminaires (Fixtures) above Water] . .
      . . . . 680.22(B)(4)
      Luminaires [Lighting Luminaires (Fixtures) below Water] .
      . . . . 680.23(A)(3)
      Pool Covers, Electrically Operated . . . . . . 680.27(B)(2)
      Receptacles . . . . 680.22(A)(5)
      Storable Pools 680.32
   Therapeutic Pools, Permanently Installed 680.61
   Therapeutic Tubs (Hydrotherapeutic Tanks) 680.62
   Types of GFCI Protection Permitted 680.5
   See (KCXS) . . . . UL Product iQ
Grounding (General) 680.6
   Methods of Grounding (Feeders) . . . . 680.25(A)
   Use of Potting Compound on Connections . 680.23(B)(2)(b)
   Water Heaters . . . . 680.26(B)(6)
GFCI Protection
   Area Receptacles . . . . 680.22(A)(4)
   Ceiling Fans . . . . 680.22(B)(4)
   Hydromassage Bathtubs 680.71
   Lighting . . . . 680.22(B)(4)
   Outlets, Receptacle or Direct Connection . . . . . 680.21(C)
   Pool Pump Motors . . . . 680.21(A)(B)(C)(D)
   Relamping, Pool Luminaires, Underwater . . . 680.23(A)(3)
   Storable Pools-Equipment 680.32
   Storable Pools - Pump Motors 680.31
Hydromassage Bathtubs . . . . Art. 680 Part VII
   See HYDROMASSAGE BATHTUBS . . . . . . Ferm's Finder

Inspections After Installation 680.4
Junction Boxes, Underwater Lighting 680.24
   Bonding 680.26
   Grounding Terminals
      Continuity Between Raceways . . . . . . 680.24(B)(1)(4)
      Number Required . . . . 680.24(D)
      Methods of Grounding . . . . 680.24(F)
      Location and Mounting Height . . . . 680.24(B)(2)
      Potting Compound, Flush Boxes Low Voltage Contact Limit or
      Less . . . . 680.24(A)(2)(c)
      Protection . . . . 680.24(C)
      Strain Relief for Cord . . . . 680.24(E)
      Threaded Conduit Hubs or Bosses . . . . . . 680.24(B)(1)
      Transformers, GFCI Enclosures, etc. . . . . . . 680.24(B)
   See (WCEZ) . . . . UL Product iQ
Listed 680.3
Low Voltage Contact Limit, Defined 680.2
Low-Voltage Gas-Fired Equipment
   Decorative Fireplace . . . . 680.22(B)(7)
   Fire Pits . . . . 680.22(B)(7)
   Luminaires . . . . 680.22(B)(7)
   Similar Equipment . . . . 680.22(B)(7)
Luminaires (Lighting Fixtures)
   Gas-Fired . . . . 680.22(B)(7)
   General Lighting (above Water Level) . . . . . . 680.22(B)
      Ground-Fault Circuit-Interrupter Protection . 680.22(B)(4)
      Grounding . . . . 680.23(F)(2)
      Methods . . . . 680.23(F)(2) (a) and (b)
   Low-Voltage . . . . 680.22(B)(6) & (7)
   Switching Devices . . . . 680.22(C)
   Underwater (below Water Level) 680.23
      Dry-Niche . . . . 680.23(C)
      GFCI Protection — Relamping . . . . 680.23(A)(3)
      Junction Boxes for . . . . 680.24(A)
      No-Niche . . . . 680.23(D)
      Transformers, Enclosures . . . . 680.24(B)
      Wet-Niche . . . . 680.23(B)
   See (WBDT) . . . . UL Product iQ
Marking Guidelines for Swimming Pool Equipment
   See UL Swimming Pool Marking Guide, Appendix A . . . . .
   . . . . UL Product iQ
Measurements . . . . 680.22(B)(8)
Motors
   Corrosive Environments . . . . 680.21(A)(1)

Electrically Operated Pool Covers . . . . 680.27(B)

(Note: *See* Info. Notes 1, 2, and 3 for other applicable sections)

GFCI Receptacles Providing Power to . . . . 680.22(A)

Methods of Grounding — 12 AWG Conductor 680.21(A)(1)

See MOTORS . . . . Ferm's Finder

Panelboards 680.25

Pool Anchors (Metallic) . . . . 680.26(B)(5)

Pools and Tubs for Therapeutic Use . . . . 680 Part VI

See THERAPEUTIC POOLS AND TUBS . . . Ferm's Finder

Receptacles . . . . 680.22(A)

Circulation and Sanitation Systems . . . . . . 680.22(A)(2)

Ground-Fault Circuit-Protection . . . . 680.22(A)(4)

Motors (Supplied by Outlets) . . . . 680.21(C)

Motors (Supplied by Receptacles) . . . . 680.22(B)

Location of . . . . 680.22(A)(1)-(6)

Other Receptacles . . . . 680.22(A)(3)

Required, General Purpose . . . . 680.22(A)(1)

Maximum Mounting Height . . . . 680.22(A)(1)

Splash Pads

Definition 680.2

Requirements 680.50

Storable Pools

See STORABLE SWIMMING POOLS . . . . Ferm's Finder

Switching Devices [Not within 1.5 m (5 ft)] . . . . 680.22(C)

Transformers (Swimming Pool) . . . . 680.23(A)(2)

Enclosures for . . . . 680.24(B)

See (WDGV) . . . . UL Product iQ

Underground Wiring 680.11

Underwater Audio Equipment . . . . 680.27(A)

Forming Shell and Metal Screen . . . . 680.27(A)(3)

Speakers . . . . 680.27(A)(1)

Wiring Methods . . . . 680.27(A)(2)

Underwater Luminaires 680.23

Dry-Niche (Fixtures) Luminaires . . . . 680.23(C)

Grounding . . . . 680.23(F)(2)

Junction Boxes Directly Connected to Forming Shell . . . . .
. . . . 680.23(F)(2)(b)

Methods of Grounding . . . . 680.23(F)(2)

No-Niche Luminaires . . . . 680.23(D)

Other Enclosures Directly Connected to Forming Shell . . . .
. . . . 680.23(F)(2)

Storable Swimming Pools 680.33

Wet-Niche (Fixtures) Luminaires . . . . 680.23(B)

See (WBDT) . . . . UL Product iQ

Wiring Methods

Dry-Niche (Fixtures) Luminaires . . . . 680.23(F)(1)

Feeders 680.25

Motors . . . . 680.21(A)

No-Niche (Fixtures) Luminaires . . . . 680.23(F)(1)

Other Equipment . . . . 680.27(A)(2)

Panelboards 680.25

Pool Covers . . . . 680.21(A)

Underground Wiring 680.11

Underwater Audio . . . . 680.27(A)(2)

Wet-Niche (Fixtures) Luminaires . . . . 680.23(B)(2)

ENT and EMT Permitted In or On Buildings 680.23(F)(1)

Flexible Connections . . . . 680.23(F)(1) Ex.

For Grounding . . . . 680.23(F)(2)

SWITCHBOARDS, SWITCHGEAR & PANELBOARDS

Arc-Flash Hazard Warning 110.16

Available Fault Current 110.24

Equipment Over 1000 Volts, Nominal . . . . . Art. 490 Part III

Field Identification Required 408.4

Rear or Side Access . . . . 408.18(C)

Signs, Field Applied . . . . 110.21(B)

High-Impedance Grounded Neutral AC System . 408.3(F)(3)

High-Leg Identification . . . . 408.3(F)(1)

Resistively Grounded DC Systems . . . . 408.3(F)(5)

Ungrounded AC Systems . . . . 408.3(F)(2)

Ungrounded DC Systems . . . . 408.3(F)(4)

Working Space . . . . 110.26(A)

See PANELBOARDS & SWITCHBOARDS . . . Ferm's Finder

See SWITCHGEAR . . . . Ferm's Finder

SWITCHES . . . . Art. 404

AFCI Protection- Dwelling Units 210.12

Attachment Methods (Screws) . . . . 404.10(B)

Accessibility and Grouping 404.8

For Busway Installations . . . . 404.8(A) Ex. 1

Agricultural Buildings 547.6

Both as Controller and Disconnecting Means 430.111

Box Not Required . . . . 334.40(B)

Breakers Used to Switch Fluorescent Lights Must Be Marked "SWD" . . . . 240.83(D)

Breakers Used to Switch High-Intensity Discharge Lights Marked "HID" . . . . 240.83(D)

Cascading (Switch to Switch) Enclosure Not Used as Junction Box
. . . . 312.8

CO/ALR Marking Required If Aluminum Wire Is Used 404.14(C)
   See (WJQR) . . . . *UL Product iQ*
Construction – Faceplates . . . . 404.9(C)
Damp or Wet Locations 404.4
Definitions, 1000 Volts and Less . . . . Art. 100
Definitions, Over 1000 Volts . . . . Art. 100 Part II
Disconnecting Means, More Than One Building 225.33
Double Pole (all conductors including grounded) to Fuel Dispensing
   . . . . 514.11(A)
   Tie Bars Not Permitted for Single-Pole Breakers . . 514.11(A)
Electronic Lighting Control Switches 404.22
   Number on Branch Circuit. . . . . 404.2(C) Ex.
   Number on Feeder . . . . 404.2(C) Ex.
Emergency Lighting Circuits (Control) . . . . 700 Part V
Enclosures 404.3, 404.12
Faceplates
   Grounding (General) 250.110
     For Snap Switches . . . . 404.9(B)
     Mobile and Manufactured Homes . . . . 550.15(D)
     Of Enclosures 404.12
     Patient Care Spaces . . . . 517.13(B) Ex. 1
   In Completed Installations 314.25
   Position for Snap Switches . . . . 404.9(A)
For Busways. 368.239
Four-Way Switching . . . . 404.2(A)
Gravity to Open Switches . . . . 404.6(A)
   Indication of Position 404.7
   Up Position to Be On (Breakers Used as Switches) . . . 240.81
Grounded Conductor (not required) . . . . 404.2(C)
Hazardous (Classified) Locations
   Class I, Division 1 . . . . 501.115(A)
   Class I, Division 2 . . . . 501.115(B)
     Part of Luminaires (Fixtures) or Lampholders 501.130(B)(5)
   Class II, Division 1 . . . . 502.115(A)
   Class II, Division 2 . . . . 502.115(B)
     For Utilization Equipment to Be Dusttight 502.135(B)(3)
   Class III 503.115
     For Utilization Equipment . . . . 503.135(C)
Heights Not to Exceed 2.0 m (6 ft 7 in.) (Generally) . . 404.8(A)
Horsepower Rated 430.109
   Marking 404.20
Identification of
   As Disconnecting Means 110.22
   Circuit Directory in Panelboards . . . . 408.4(A)

Source of Supply . . . . 408.4(B)
Knife Switches
   Marking 404.20
   Position and Connection 404.6
   Rated 600 to 1000 Volts 404.26
   Ratings 404.13
   To Be Indicating 404.7
Lighting Load, Grounded Conductor. . . . . 404.2(C)
Location at Tubs and Showers. 404.4
Marking 404.20
Motor Controllers 430.83
Multiple Snap Switches . . . . 404.8(C)
Not Connected to Equipment Grounding Conductor . . . . . .
   . . . . 404.9(B) Ex.2 to (B)
Over 1000 Volts
   Isolating Means 490.22
   Isolating Switches . . . .
   230.204
   Load Interrupters. . . . . 490.21(E)
Rating of 600 to 1000 Volt Knife Switches 404.26
Rating & Use of Snap Switches 404.14
   Mobile and Manufactured Homes . . . . 550.15(G)
   Motor Disconnecting Means 430.109
   See (WJQR) . . . . *UL Product iQ*
Required
   For Lighting Outlets 210.70
   Theater Dressing Rooms (with Pilot Light) 520.73
Switching to Be Done in Ungrounded Conductors . . . 404.2(B)
   Exception . . . . 404.2(B) Ex.
     Dispensing Equipment . . . . 514.11(A)
     Motor Circuit Disconnects (Grounded Conductors) 430.105
Theater Dressing Rooms 520.73
Three-Way Switching . . . . 404.2(A)
Time Switches, Flashers, and Similar Devices 404.5
Transfer Switches
   See TRANSFER SWITCHES . . . . Ferm's Finder
Up Position to Be On 240.81
   Indication of Position 404.7
Voltage Limitations (Snap Switches)
   Between Adjacent Switches . . . . 404.8(B)
   Between Switches and Receptacles . . . . 404.8(B)
   Rating and Use 404.14
Wet Locations 404.4
Wire Bending Space, Deflection 312.6

At Terminals 404.3
See DISCONNECTING MEANS . . . . Ferm's Finder
See SERVICE EQUIPMENT, Main Disconnects . Ferm's Finder

Note: The wiring space and current-carrying capacity of switches are based on the use of 60°C wire where wire sizes 14 AWG through 1 AWG are used and 75°C wire where wire sizes 1/0 and larger are used.

See Switches, Pull Out Type (WFXV) . . . . UL Product iQ

SWITCHGEAR . . . . Art. 408
AC Phase Arrangement . . . . 408.3(E)
Accessible to Unqualified Persons, Over 1000 Volts . 110.31(B)
Arc-Flash Hazard Warning 110.16
Auxiliary Gutter
    Metallic 366.2
    Nonmetallic 366.2
Available Fault Current 110.24
Barriers for Service Use . . . . 408.3(A)(2)
Bonding, Building of Multiple Occupancy . . . . 250.104(A)(2)
Bus Arrangement . . . . 408.3(E)
Construction 408.50
Critical Care (Category 1) Spaces, Equipment Grounding and Bonding . . . . 517.19(E)
DC Bus Arrangement. . . . . 408.3(E)(2)
Damp Locations 408.16
Dedicated Equipment Space . . . . 110.26(E)
Defined. . . . . Art. 100
Disconnects, Branch Circuits and Feeders, No More than Six . . 225.33
Disconnects, Service, No More than Six . . . . 230.71(A)
Emergency Systems, Wiring . . . . 700.10(B)(5)(b)
Energy Management 408.23
Entrances, Over 1000 Volts . . . . 110.33(A)
Enclosure Types. 110.28
Equipment Grounding Conductor, Connected . . 250.112(A)
Equipment Grounding Conductor, Devices Over 1000 Volts 490.37
Equipment Grounding Conductor, Equipment Over 1000 Volts 490.36
Equipment Over 1000 Volts . . . . Art. 490 Part III
Field-Applied Hazard Markings. . . . . 110.21(B)
Field Identification Required 408.4
Fire Pumps, Utility Service Connection . . . . . . 695.3(A)(1)
Gang Operated Switch, High Voltage Fuses. . . . . 490.21(B)(7)
Grounded Conductor Disconnection Means, Feeders 225.38(C)
Grounded Conductor Disconnection Means, Service . . . 230.75
High-Leg Identification . . . . .408.3(F)(1)

Illumination . . . . 110.26(D)
Industrial Installation Secondary Conductors Not over 25 Feet Long . . . . 240.21(B)(3)
Instruments, Meters, and Relays, Grounding . . Art. 250 Part IX
Legally Required Standby Systems, Ahead of Service Disconnect . . . . 701.12(E)
Location to Easily Ignitable Material 408.17
Marking 110.21
Photovoltaic Systems, Bipolar . . . . 690.31(E)
Power Monitoring 408.23
Protection from Foreign Materials . . . . 110.34(F)
Rear Access . . . . 408.18(C)
Reconditioned Equipment 408.8
Side Access . . . . 408.18(C)
Service, Constructed of Substantial Metal, Over 1000 Volt 230.211
Service, Over 35,000 Volts 230.212
Service Equipment, Used as . . . . 408.3(C)
Service Equipment Over 1000 Volts, Used as 490.47
Short-Circuit Current Rating 408.6
Signs
    Field Applied . . . . 110.21(B)
    High-Impedance Grounded Neutral AC System . 408.3(F)(3)
    High-Leg Identification . . . . 408.3(F)(1)
    Resistively Grounded DC Systems . . . . 408.3(F)(5)
    Ungrounded AC Systems . . . . 408.3(F)(2)
    Ungrounded DC Systems . . . . 408.3(F)(4)
Signage for Identification of Systems . . . . 408.3(F)
Substations, Documentation 490.48
Supervised Industrial Installations, Outside Feeder Taps . . 240.92(D)
Supervised Industrial Installations, Overload Protection . . . . .240.92(C)(2)
Taps Not over 10 Feet Long . . . . 240.21(B)(1)(2)
Temporary Installations, Branch Circuits . . . . . . .590.4(C)
Transformer Secondary Conductors Not over 10 Feet Long. . . . . 240.21(C)(2)(2)
Types (Defined) . . . . .Art. 100 Info. Note
Unused Openings. 408.7
Wet Locations 408.16
Wire Bending Space . . . . 408.3(G)
Wind Electric Systems, Number of Disconnects . 694.22(C)(4)
Working Space, 1000 Volts and Below . . . . 110.26(A)
Working Space, above 1000 Volts. . . . . 110.34(A) Ex.
See, Switchgear, Over 1000 Volts . . . . Ferm's Finder
See (WVDA) . . . . UL Product iQ

See SWITCHBOARDS, SWITCHGEAR & PANELBOARDS . . . . Ferm's Finder

SYMBOL
Controlled Receptacle Marking . . . . 406.3(E) Figure
Grounding . . . . 250.126 Info. Note Figure
Grounding-Pole Identification . 406.10(B)(4) Info. Note Figure
Termination Point- Equipment Grounding Conductor . . . . 406.10(B)(4) Info. Note Figure

SYSTEM BONDING JUMPER
Controlled Lighting Loads . . . . 404.2(C) Exception
Critical Operations Powers Systems, No Need To Ground . . . . 708.20(C) Ex.
Definition . . . . Art. 100
Direct Current System 250.168
Grounded System. 250.28
Separately Derived AC Systems . . . . 250.30(A)(1)
Sizing. Table 250.102(C)(1)

SYSTEM ISOLATION EQUIPMENT . . . . . . 430.109(A)(7)
Definition of 430.2

T

TABLES
AC Resistance and Reactance . . . . Chapter 9 Table 9
Air-Conditioning and Refrigeration Equipment, Other Articles . . . . Table 440.3(D)
Ambient Correction Factors . . . . . Table 310.15(B)(1) and (2)
Ambient Temperature Adjustment, Rooftop Conduits Raceways and Cables. . . . . Table 310.15(B)(1) and (2)
Ampacity of
   Conductors 0 through 2000 Volts, 60°C through 90°C
      Bare Conductors. . . . . .310.21 and Table 310.21
      In Free Air. . . . . .310.17 and Table 310.17
      In Raceways, Cables and Earth (Directly Buried) 310.16 and Table 310.16
      Supported on Messenger . . . . .310.20 and Table 310.20
   Conductors 0 through 2000 Volts, 150°C through 250°C
      In Free Air . . . . .310.19 and Table 310.19
      In Raceways or Cables . . . . . .310.18 and Table 310.18
   Conductors 2001 to 35,000 volts   Table 311.60(C)(67) to (86)
   Cords and Cables for Border Lights . . . . 520.44(C)
   Correction Factors, Solar Photovoltaic Systems — Flexible Cords and Cables . . . . 690.31(E)
   Crane & Hoist Conductors . . . . 610.14(A)
   (Fixture) Luminaire Wire . . . . Table 402.5
   Flexible Cord and Cables . . . . . Table 400.5(A)(1) & (2)
      Adjustment Factors 400.5
   High Voltage Conductors, 2001 through 35,000 Volts. . . . . Tables 311.60(C)(67) through (86)
   Integrated Gas Spacer Cable 326.80
   Note: For more than Three Current-Carrying Conductors in a raceway or cable, see . . 310.15(C)(1) and Table 310.15(C)(1)
Application & Insulation of Conductors
   Under 1000 Volts (see Table Note 1) . . . . . Table 310.4(A)
Bare Metal Parts, Spacings between Switchboards and Panelboards 408.56
Bonding Jumper Sizing . . . . Table 250.102(C)(1)
Boxes, Fill (Metal) . . . . Table 314.16(A)
   Volume Allowance Required Per Conductor . Table 314.16(B)
Branch-Circuit Requirements (Summary). 210.24
Branch Circuits, Specific-Purpose. 210.2
Bulk Storage Plants . . . . .Table 515.3
Cabinets & Cutout Boxes
   Width of Wiring Gutters. . . . . Table 312.6(A)
   Wire Bending Space at Terminals. . . . . . . Table 312.6(B)
   Conductor Does Not Enter Opposite Its Terminal . 312.6(B)(1)
   Conductor Enters Opposite Its Terminal . . . 312.6(B)(2)
Cable Markings/Listing Requirements
   Class 1, 2 and 3 Remote Control and Signaling . . . . 725.179(K) and Table
   Communications. 800.113
   Community Antenna Television and Radio 800.113
   Fire Alarm
      NPLFA . . . . 760.176(G)
      PLFA 760.179
   Optical Fiber 770.113
Cable Substitutions
   Class 2 and 3 Remote Control and Signaling . . 725.154(A)
   Communications. 805.154
   Community Antenna Television and Radio . . Table 820.154
   Fire Alarm . . . . 760.154(A)
   Network-Powered Broadband Communications Table 830.154
   Optical Fiber 770.154
Cable Tray Fill
   Multiconductor Cables in Ladder, Ventilated Trough, or Solid Bottom 392.22
   Multiconductor Cables in Solid Channel . . . . . 392.22(3)

Multiconductor Cables in Ventilated Channel . . . 392.22(2)
Single Conductor, Ladder or Ventilated Trough Trays. . 392. 22(B)
Cable Tray for Grounding 392.60
Cable Tray, Wiring Methods Permitted . . . . 392.10(A)
Calculation of Feeder Lighting Load by Occupancies . . 220.42
Chapter 9 (Tables)
   AC Resistance and Reactance for 600-Volt Cables . . Table 9
   Class 2 and 3 AC Power Source Limitations . . . . Table 11(A)
   Class 2 and 3 DC Power Source Limitations. . . . Table 11(B)
   Conductor Properties . . . . Table 8
   Conductor Stranding . . . . Table 10
   Compact Copper and Aluminum Wire Dimensions and Areas . . . . Table 5A
   Dimensions of Insulated Conductors and Fixture Wire Table 5
   Dimensions and Percent Area of Conduit and Tubing Table 4
   PLFA AC Power Source Limitations . . . . .Table 12(A)
   PLFA DC Power Source Limitations . . . . .Table 12(B)
   Percent of Cross-Sectional Area . . . . Table 1
   Radius of Conduit and Tubing . . . . Table 2
   See   Notes to Tables (1) thru (10)
Circuit Breakers (Inverse Time)- Standard Ampere Ratings . Table 240.6(A)
Class 1, Class 2, and Class 3 Remote-Control, Signaling, and Power-Limited Circuits
   Ampacities . . . . Table 725.144
   Applications . . . . Table 725.154
   Cable Marking . . . . Table 725.179(J)
   Cable Substitutions . . . . .Table 725.154(A)
   Clearance of
      Bare Live Parts, Over 1000 Volts above Working Space . . 110.34(E)
      Bare Live Parts Over 1000 Volts, Field-fabricated . . . 490.24
      Bare Metal Parts in Motor Control Centers . . . . 430.97(D)
      Bare Metal Parts in Switchboards and Panelboards . . . 408.56
      Conductors Entering Bus Enclosures 408.5
      Over 1000 Volts
         Over Buildings and Other Structures 225.61
         Over Roadways, Walkways, Rail, Water, and Open Land . 225.60
      Service Conductors and Surfaces — Cables/Open Conductor . . . . 230.51(C)
   Working Space
      Over 1000 Volts, Nominal . . . . 110.34(A)
      Under 1000 Volts . . . . 110.26(A)(1)
Communication Antenna Television and Radio Distribution Systems
   Coaxial Cable in Buildings . . . . Table 820.154
   Coaxial Cable Marking . . . . Table 820.179
   Coaxial Cable Use and Permitted Substitution . Table 820.154
Communication Circuits
   Cable Markings . . . . Table 805.179
   Cable Routing Assemblies . . . . .Table 800.154(c)
   Cable Routing Assembly Marking . . . . . . Table 800.182(a)
   Cable Substitutions . . . . Table 805.154
   Communication Raceways in Buildings . . . Table 800.154(b)
   Communication Raceway Marking . . . . . Table 800.182(b)
   Wires and Cables in Buildings . . . . Table 800.154(a)
Conductors
   Application & Insulation
      (Fixture) Luminaire Wires 402.3
      Overcurrent Protection, Specific Conductor Applications . . . . .240.4(G)
      Under 1000 Volts (see Note 1 below table) . Table 310.4(A)
   Boxes, Number in
      Metal . . . . Table 314.16(A)
      Volume Allowance Required Per Conductor Table 314.16(B)
   Clearances for, Entering Bus Enclosures 408.5
   Combination of, in Raceways
      Areas of Conduit or Tubing . . . . Chapter 9, Table 4
      Compact Aluminum Building Wire Dimensions and Areas . . . . Chapter 9, Table 5A
      Conductor Properties . . . . Chapter 9, Table 8
      Dimensions of Conductors & Luminaire (Fixture) Wires . . . . Chapter 9, Table 5
      Percent of Cross Section for Conductors and Cables (General) . . . . Chapter 9, Table 1
   Conductors based on 30 degrees C (86 degree F) . . . . Table 310.15(B)(1)
   Conduit and Tubing, and Cable Tray Fill for (All the Same Size) Informative Annex C Tables
   Contact Conductor Supports, Cranes and Hoists . 610.14(D)
   Deflection, Minimum Bending Space
      Conductor Does Not Enter Opposite Its Terminal . . 312.6(B)(1)
      Conductor Enters Opposite Its Terminal . . . 312.6(B)(2)
      Terminals of Enclosed Motor Controllers . . . . 430.10(B)
   Dimensions
      Compact or Copper Aluminum Building Wire . . . . Chapter 9, Table 5A
      Conductor Properties . . . . Chapter 9, Table 8
      Insulated, and (Fixture) Luminaire Wires Chapter 9, Table 5
      Rubber and Thermoplastic Covered . . . Chapter 9, Table 5
   Direct Burial (Cover Requirements) 300.5
      Over 1000 Volts, Nominal 300.50
   (Fixture) Luminaire Wires . . . . Chapter 9, Table 5

Flexible Cords and Cables 400.4
    Ampacity . . . . Tables 400.5(A)(1) & (A)(2)
    Adjustment Factors 400.5
Grounded Conductor . . . . Table 250.102(C)(1)
Grounding, Size
    Equipment Grounding Conductor 250.122
    System, Alternating-Current. 250.66
Insulations
    Under 1000 Volts (see note 1 below table) . . Table 310.4(A)
Main Bonding Jumper . . . . Table 250.102(C)(1)
Maximum Number . . . . . . Chapter 9, Table 1 & Notes
Maximum Number in (All the Same Size) Informative Annex C Tables
    Electrical Metallic Tubing . . . . Table C.1
        Compact Conductors . . . . Table C.1(A)
    Electrical Nonmetallic Tubing . . . . Table C.2
        Compact Conductors . . . . Table C.2(A)
    Flexible Metal Conduit . . . . Table C.3
        Compact Conductors . . . . Table C.3(A)
    Metric Designator 12 (3/8") Flexible Metal Conduit 348.22
    Intermediate Metal Conduit. . . . . Table C.4
        Compact Conductors . . . . Table C.4(A)
    Liquidtight Flexible Metal Conduit . . . . . . Table C.7
        Compact Conductors . . . . Table C.7(A)
    Metric Designator 12 (3/8") Liquidtight Flexible Metal Conduit 348.22
    Liquidtight Flexible Nonmetallic Conduit (LFNMC-A). . . . . Table C.6
        Compact Conductors . . . . Table C.6(A)
    Liquidtight Flexible Nonmetallic Conduit (LFNMC-B). . . . . Table C.5
        Compact Conductors . . . . Table C.5(A)
    PVC Conduit, Type EB . . . . Table C.12
        Compact Conductors . . . . Table C.12(A)
    Rigid Metal Conduit . . . . Table C.8
        Compact Conductors . . . . Table C.8(A)
    Rigid PVC Conduit, Schedule 80 . . . . Table C.9
        Compact Conductors . . . . Table C.9(A)
    Rigid PVC Conduit, Schedule 40 & HDPE . . Table C.10
        Compact Conductors . . . . Table C.10(A)
    Rigid PVC Conduit, Type A. . . . . Table C.11
        Compact Conductors . . . . Table C.11(A)
Minimum Size. . . . . 310.3(A)
Over 1000 Volts to 35,000 Volts
    See TABLES, Ampacity of . . . . Ferm's Finder

Overcurrent Protection, Low-Voltage Wiring, Park Trailers . . . . 552.10(E)(1)
Properties of . . . . Chapter 9, Table 8
Size of Amateur Station Outdoor Antenna Conductors . . 810.52
Size of Receiving Station Outdoor Antenna Conductors . 810.16(A)
Conduit and Tubing
    Combination of Conductors . . . . Chapter 9, Table 4
    Compact Conductors (Aluminum). . . . Chapter 9, Table 5A
    Conductor Fill, All the Same Size   Informative Annex C Tables
    Dimensions of Insulated Conductors & (Fixture) Luminaire Wires. . . . . Chapter 9, Table 5
    Fiberglass Reinforced, Expansion Characteristics. . . . 355.44
    Flexible Metal Conduit, Metric Designator 12 (3/8") Size . 348.22
    Integrated Gas Spacer Cable, Dimensions 326.116
        Paper Spacer Thickness 326.112
        Radii of Bends 326.24
    Liquidtight Flexible Metal Conduit, Metric Designator 12 (3/8") Size. 348.22
    Metric Designator and Trade Sizes . . . . 300.1(C)
    Percent of Cross Section . . . . Chapter 9, Table 1
    PVC Rigid, Expansion Characteristics 352.44
    Radius of Bends (General) . . . . Chapter 9, Table 2
        Flexible Metallic Tubing, Fixed . . . . 360.24(B)
        Flexible Metallic Tubing, Infrequent Flexing . . 360.24(A)
        Nonmetallic Underground Conduit with Conductors. . 354.24
    Supports — Rigid Metal Conduit . . . . 344.30(B)(2)
        Rigid PVC Conduit . . . . 352.30(B)
Cooking Appliances, Dwelling Units 220.55
    Other than Dwelling Units 220.56
Cranes & Hoists
    Ampacities of Conductors, Short-Time Rated . . . 610.14(A)
    Demand Factors . . . . 610.14(E)
    Secondary Conductor Rating Factors . . . . 610.14(B)
Demand Factors
    Commercial Cooking Equipment 220.56
    Cranes & Hoists . . . . 610.14(E)
    Elevators 620.14
    Farm Loads, Other Than Dwelling Units 220.102
        Total Farm Loads 220.103
    Household Clothes Dryers 220.54
    Household Ranges 220.55
    Lighting Load Feeders 220.42
    Marinas and Boatyards 555.6
    Mobile Home Park (Feeders & Service Conductors) . . 550.31
    Multifamily Dwellings, Optional (Service) 220.84

Receptacle Loads (Non-Dwelling Feeders) 220.44
Recreational Vehicle Park (Service & Feeders) 551.73
Restaurants, New (Optional) 220.88
Schools (Service Conductors), Optional 220.86
Stage Set Lighting . . . . 530.19(A)
Dwelling Service Conductor Sizing . . . . Example D7
Electric Signs and Outline Lighting
    Class 2 Cable Substitutions. . . . . .Table 600.33(A)(2)
    Power Limited Cable . . . . .Table 600.33(A)(1)
Electric Vehicle Charging Systems
    Minimum Ventilation Required . . . . . 625.52(B)(1) & (2)
Elevator Feeder Demand Factors 620.14
Electrified Truck Parking Spaces
    Demand Factors . . . . Table 626.11(B)
Elevator Feeder Demand Factors 620.14
Enclosure Selection . . . . Table 110.28
Equipment Grounding Conductors. 250.122
Fire Alarm Systems
    Cable Markings . . . . Table 760.179(I)
    Cable Substitutions. . . . . .Table 760.154(A)
    NPLFA Cable Markings . . . . Table 760.176(G)
    PLFA Cables in Buildings . . . . Table 760.154
(Fixture) Luminaire Wires (Same Size) in Conduit and Tubing . .
. . . . Informative Annex C Tables
(Fixture) Luminaire Wire (Types) 402.3
Flexible Cords and Cables 400.4
    Ampacity . . . . Tables 400.5(A)(1) & (A)(2)
    Adjustment Factors. 400.5
Fuses (Standard Ampere Rating) . . . . Table 240.6(A)
General Lighting Unit Load for Non-Dwelling Occupancies 220.12
Grounded Conductor. . . . . Table 250.102(C)(1)
Grounding, Application of Other Articles 250.3
Grounding, General Requirements 250.1
Grounding Conductor Sizing . . . . . . Table 250.102(C)(1)
Grounding Electrode Conductors 250.66
Hazardous (Classified) Locations
    Bulk Storage Plants 515.3
    Class I, Zone 0, 1, and 2
        Gas Classification Groups . . . . .505.9(C)(1)(2)
        Maximum Surface Temperature, Group II Equipment . . .
. . . . 505.9(D)(1)
        Minimum Distance of Obstructions, Flange Openings . . .
. . . . .505.7(D)
        Types of Protection Designation . . . . . .505.9(C)(2)(4)
    Motor Fuel Dispensing Facilities and Commercial Garages . .
. . . . 514.3(B)(1)
        Classified Areas Adjacent to Dispensers . . . . Figure 514.3
        Classified Areas for Dispensing Devices (Gases) 514.3(B)(2)
Safe Operating Temperatures
    Class I, Identification Numbers . . . . 500.8(C)
    Class II . . . . 500.8(D)(2)
High Voltage Conductors (Over 2000 Volts, Nominal)
    Bare Live Parts, Over 1000 Volts above Working Space . . 110.34(E)
    Bare Live Parts Over 1000 Volts, Field-fabricated . . . 490.24
    Thickness of Insulation (2001 Volts and Higher) . . . .
. . . . 311.10(A),(B) and (C)
Household Clothes Dryers 220.54
Household Ranges and Other Cooking Appliances 220.55
Information Technology Equipment
    Cables Under Raised Floors . . . . .Table 645.10(B)(5)
Integrated Electrical Systems, Application of Other Articles 685.3
Lighting Loads by Occupancy 220.12
Lighting Load (Feeder Demand Factor) 220.42
Limitations for Network-Powered Broadband Communication Systems 830.15
Live Parts, Separation
    Air Separation, Indoors 490.24
        Elevation . . . . 110.34(E)
    From Fences, Over 1000 Volts 110.31
    Working Space, Over 1000 Volts, Nominal . . . . 110.34(A)
        Maximum 1000 Volts . . . . 110.26(A)(1)
Main Bonding Jumper Sizing . . . . Table 250.102(C)(1)
Marinas and Boatyards (Demand Factors) 555.6
Maximum Appliance Loads at Receptacles . . . . . 210.21(B)(2)
    Summary 210.24
Messenger Supported Wiring, Permitted Cable Types 396.10(A)
Metric Designators and Trade Sizes . . . . Table 300.1(C)
Minimum Cover Requirements, 0 to 1000 Volts 300.5
Minimum Cover Requirements, Over 1000 Volts 300.50
Minimum Cover Requirements, Network-Powered Broadband Systems . . . . 830.47(C)
Minimum Size of Conductors . . . . 310.3(A)
Minimum Wire Bending Space
    Conductor Does Not Enter Opposite Its Terminal . 312.6(A)
    Conductor Enters Opposite Its Terminal . . . . . . 312.6(B)
    Terminals of Enclosed Motor Controllers . . . . . 430.10(B)
Mobile Home Parks (Demand Factor) 550.31
Motor Fuel Dispensing
    Electrical Equipment . . . . Table 514.3(B)(2)
    Facilities . . . . Table 514.3(B)(1)

Motors
- Conductor Rating Factors, Power Resistors. 430.29
- Conductors for Small Motors . . . . 430.22(G)
  - 16 AWG Copper . . . . 430.22(G)(2)
  - 18 AWG Copper . . . . 430.22(G)(1)
- Control Circuit Overcurrent Protection . . . . . . 430.72(B)
- Control Center Spacing between Bare Metal Parts . . . 430.97
- Controller Enclosure Type . . . . 110.28 and Table
- Direct-Current Motor-Rectified Supplied . . . . . 430.22(A)
- Duty Cycle Service . . . . Table 430.22(E)
  - Secondary Conductor (Wound Rotor). . . Table 430.23(C)
- Full-Load Current in Amperes
  - DC Motors . . . . Table 430.247
  - Single-Phase AC Motors . . . . Table 430.248
  - Two-Phase AC Motors . . . . Table 430.249
  - Three-Phase AC Motors . . . . Table 430.250
- Locked-Rotor Current, Conversions . Tables 430.251(A)&(B)
- Locked-Rotor Indicating Letters . . . . 430.7(B)
- Maximum Rating or Setting of Protective Devices . . . 430.52
- Multispeed Motor . . . . 430.22(B)
- Other Articles 430.5
- Overload Units Required 430.37
- Part-Winding Motors . . . . 430.22(D)
- Terminal, Spacing and Housing
  - Terminal Spacings, Fixed . . . . Table 430.12(C)(1)
  - Usable Volumes, Fixed Terminals . . . Table 430.12(C)(2)
  - Wire-to-Wire Connections . . . . Table 430.12(B)
- Wire Bending Space, Controllers . . . . . Table 430.10(B)
- Wye-Start, Delta-Run Motor . . . . 430.22(C)

Multifamily Dwellings (Demand Factor), Optional . . . 220.84
Multiplying Factors from DC Resistance to AC Resistance . . . . Chapter 9, Table 9
Network-Powered Broadband Communications System
- Cables in Buildings. . . . . Table 800.154(a)
- Cable Substitutions. . . . . Table 830.154
- Cover Requirements (Underground) . . . . . Table 830.47(C)
- Limitations. . . . . Table 830.15

Notes to Tables (Chapter 9). . . . . See Notes
Optical Fiber Cables
- Cable Markings . . . . Table 770.179
- Cable Substitutions. . . . . Table 770.154(b)
- In Buildings . . . . Table 770.154(a)

Optional Load Calculations
- Multifamily Dwelling Units 220.84
- Restaurants, New. 220.88
- Schools. 220.86

Outside Branch Circuits and Feeders, Other Articles 225.3
- Over 1000 Volts
  - Over Buildings and Other Structures 225.61
  - Over Roadways, Walkways, Rail, Water, and Open Land 225.60

Overcurrent Protection, Other Articles 240.3
- Specific Conductor Applications . . . . 240.4(G)

Overcurrent Protective Device, Maximum Rating . . 430.72(B)
Park Trailer Low-Voltage Overcurrent Protection . . 552.10(E)(1)
Percent of Cross Section of Conduit and Tubing for Conductors and Cables. . . . . Chapter 9, Table 1
Power Limitations
- AC Class 2 & 3 Circuits . . . . .Chapter 9, Table 11(A)
- DC Class 2 & 3 Circuits . . . . .Chapter 9, Table 11(B)
- Network-Powered Broadband Systems 830.15
- PLFA AC Fire Protective Circuits . . Chapter 9, Table 12(A)
- PLFA DC Fire Protective Circuits . . .Chapter 9, Table 12(B)

Properties of Conductors . . . . Chapter 9, Table 8
PVC Expansion Rate . . . . Table 352.44
PVC Support . . . . Table 352.30
Radio and Television Equipment
- Outdoor Antenna Conductors . . . . Table 810.52
- Receiving Station Conductors . . . . Table 810.16(A)

Radius of Conduit Bends (General) . . . . . Chapter 9, Table 2
Rating Factors for Power Resistors 430.29
Receptacle Ratings & Load
- Demand Factors, Other Than Dwelling Units 220.44
- Maximum Cord- and Plug-Connected Load . . . 210.21(B)(2)
- Receptacle Ratings . . . . 210.21(B)(3)
- Shore Power, Marinas and Boatyards, for. 555.33
- Summary. 210.24

Recreational Vehicle Park (Demand) 551.73
Reinforced Thermosetting Resin Conduit Expansion Rate . . . . Table 355.44
Reinforced Thermosetting Resin Conduit Support . Table 355.30
Repair Garages
- Major and Minor Fuel Heavier-Than-Air. . . . Table 511.3(C)
- Major Fuel Lighter-Than-Air. . . . . .Table 511.3(D)

Resistance of Conductors . . . . Chapter 9, Tables 8 & 9
Restaurants, Optional Load Calculations for New 220.88
Rigid Metal Conduit Support. . . . . .Table 344.30(B)(2)
Services . . . . Figure 230.1
- Support and Clearance . . . . 230.51(C)

Schools (Service and Feeder Conductors), Optional . . . 220.86

Signs and Outline Lighting
- Class 2 Cable Substitutions. . . . . .Table 600.33(A)(2)
- Power Limited Cable . . . . .Table 600.33(A)(1)

Solar Photovoltaic Systems
- Ambient Temperature Correction . . . . 690.31(A)
- PV Wire Strands- Minimum . . . . 690.31(E)
- Voltage Correction Factors, Silicon Modules 690.7

Standard Ampere Rating for Fuses and Inverse Time Circuit Breakers
. . . . Table 240.6(A)

Strut-Type Channel Raceway, Size and Inside Diameter . 384.22

Supply-Side Bonding Jumper Sizing . . . . Table 250.102(C)(1)

Sunlight on Rooftop, Adjustments . . . . . .Table 310.15(B)(2)

Supports for
- Conductors in Vertical Raceways. . . . . 300.19(A)
- Rigid Metal Conduit . . . . .344.30(B)(2)
- Rigid PVC Conduit 352.44

Swimming Pools
- Overhead Conductor Clearances. 680.8
- Underground Burial Depths . . . . .Table 300.5

System Bonding Jumper Sizing . . . . . . Table 250.102(C)(1)

Theaters, Motion Pictures and TV Studios, and Similar Locations
- Cords and Cables. . . . . .Table 520.44(C)(3)
- More than Three Current-Carrying Conductors . . . .
. . . . Table 520.44(C)(3)(a)

Transformers
- Overcurrent Protection, 1000 Volts and Less. . . . . 450.3(B)
- Overcurrent Protection, Over 1000 Volts . . . . . . 450.3(A)

Ventilation Required, Electric Vehicle Charging Systems . . . . .
. . . . .625.52(B)(1) & (2)

Welders
- Duty Cycle Factors for Arc Welders . . . . 630.11(A)
- Duty Cycle Factors for Resistance Welders . . . . 630.31(A)(2)

Wind Electric Systems
- Working Spaces 694.7

Wire Bending Space
- Conductor Does Not Enter Opposite Its Terminal . 312.6(A)
- Conductor Enters Opposite Its Terminal . . . . 312.6(B)
- Terminals of Enclosed Motor Controllers . . . . . 430.10(B)

Working Clearances
- Over 1000 Volts, Nominal . . . . 110.34(A)
- Under 1000 Volts . . . . .110.26(A)(1)

Working Spaces . . . . .Table 110.26(A)(1)
- Wind Electric Systems . . . . .Table 694.7

TAMPERABILITY
- Cartridge Fuses . . . . 240.60(D)
- Circuit Breakers. 240.82
- Edison-Base Fuses. . . . . 240.51(B)
- Lighting Track Systems . . . . .410.155(A)
- Service Conductors Supplying Specific Load 230.93
- Type S Fuses . . . . 240.54(D)
    - Mobile Homes 550.11

TAMPER-RESISTANT RECEPTACLES
- Assembly Occupancies (such as transportation waiting areas, gymnasiums, skating rinks, and auditoriums) . . . . 406.12(6)
- Assisted Living Centers . . . . 406.12(8)
- Business Offices, Corridors, Waiting Rooms
  (in clinics, medical and dental offices, and outpatient facilities) . .
  . . . . 406.12(5)
- Child Care Facilities . . . . 406.12(3)
- Dormitories (Dormitory Units). . . . . 406.12(7)
- Dwelling Units . . . . 406.12(1)
- Guest Rooms and Guest Suites of Hotels and Motels 406.12(2)
- Preschools and Elementary Education Facilities. . . . 406.12(4)
- Receptacles or Covers, Designated General Care Pediatric Locations . . . . 517.18(C)
- Replacements . . . . 406.4(D)(5)

TAPS
- Battery Conductors . . . . 240.21(H)
- Box Not Required. . . . . 334.40(B)
- Branch-Circuit Taps 210.19
    - Conductor Overcurrent Protection, General . . . . 210.20(B)
    - Motor Branch Circuit Taps. . . . . 240.21(F)
    - Over 1000 Volts . . . . 240.100(A)
    - Overcurrent Protection, Flexible Cords and Cables . 240.5(B)
    - Protection of Conductors . . . . 240.4(E)
    - Single Motor Taps . . . . 430.53(D)
    - Summary Requirements 210.24
    - Tap Conductors . . . . 240.21(A)
- Busway Taps . . . . 240.21(E)
    - Branch Circuits, Overcurrent Protection Rating . . 368.17(D)
    - Branches from Busways 368.56
    - Feeder or Branch Circuits . . . . 368.17(C)
    - Feeder Overcurrent Protection . . . . 368.17(A)
    - Reduction in Ampacity Size of Busway . . . . . . 368.17(B)
- Cascading (Switch to Switch) 312.8
- Feeders . . . . 240.21(B)
    - 3 m (10 ft.) Tap Rule . . . . 240.21(B)(1)

Motor Feeder . . . . 430.28(1)
Transformer Secondary Conductors . . . . . .240.21(C)(2)
7.5 m (25 ft.) Tap Rule . . . . .240.21(B)(2)
Motor Feeder Tap . . . . 430.28(2)
Single Motor Taps . . . . 430.53(D)
Transformer Secondary Conductors . . . . . .240.21(C)(6)
Transformer Secondary Conductors, Industrial Installations
. . . . 240.21(C)(3)
30 m (100 ft.) Tap Rule (High Bay Manufacturing Building) .
. . . . 240.21(B)(4)
Motor Feeder Taps. . . . . 430.28 Ex.
Over 1000 Volts . . . . .240.101(B)
Outside Taps of Unlimited Length . . . . 240.21(B)(5)
Supervised Industrial Installations . . . . . Art. 240 Part VIII
Branch-Circuit and Feeders (General) . . . . . 240.92(A)
Outside Feeder Taps . . . . 240.92(B)
Transformer Secondary Conductors of Separately Derived
Systems . . . . 240.92(C)
Transformer Feeder. . . . . 240.21(B)
Outside Secondary Conductors . . . . . . .240.21(C)(4)
Primary Plus Secondary Not Over 7.5 m (25 ft.) . .240.21(B)(3)
Secondary Conductors from Feeder Tapped . .240.21(C)(5)
Secondary Conductors . . . . 240.21(C)
Secondary Conductors Not Over 7.5 m (25 ft.) . . 240.21(C)(6)
Generator Terminal Taps . . . . 240.21(G)
Ampacity of Conductors 445.13
Grounding Electrode Conductor Taps . . . . 250.64(D)
Separately Derived Systems. . . . . 250.30(A)(6)
Luminaires (Fixtures) . . . . 210.19(A)(4) Ex. 1
(Fixture) Luminaire Wires and Flexible Cords . 210.19(A)(4) Ex. 2
Overcurrent Protection of. . . . . 240.5(B)
Flexible Metal Conduit. . . . . 348.20(A)(2)(c)
Liquidtight Flexible Metal Conduit . . . . . 350.30(A) Ex. 3
Liquidtight Flexible Nonmetallic Conduit . . . 356.20(A)(2)
Manufactured Wiring Systems . . . . . . 604.100(A)(2) Ex. 1
Wiring . . . . 410.117(C)
Motor Circuit Taps . . . . 240.21(F)
Feeders 430.28
Single Motor Tap, Branch Circuit . . . . 430.53(D)
Outside Feeder Taps
Supervised Industrial Installations . . . . 240.92(D)
Transformer Secondary Conductors . . . . . . .240.21(C)(4)
Unlimited Length . . . . 240.21(B)(5)
Service Conductors . . . . 240.21(D)
Spliced or Tapped Permitted 230.46

Equipment Allowed . . . . 230.82(5)
Supervised Industrial Installations, Feeder Taps . . . . 240.92(B)
Ungrounded Conductors from Grounded Systems 215.7
See CONDUCTORS, Splices of or in . . . . Ferm's Finder
See OVERCURRENT PROTECTION . . . . . . Ferm's Finder
See SPLICES AND TAPS. . . . . Ferm's Finder

TASK ILLUMINATION
Critical Branch (control of). . . . . 517.34(B)
Definition of 517.2
Generator Set Locations
Life Safety Branch- Hospitals. . . . . 517.33(E)
Life Safety Branch- Nursing Home and Limited Health Care
Facilities . . . . 517.43(F)
Hospitals . . . . 517.34(A)
Hospital, Life Safety Branch, Generator Set and Transfer Switch
. . . . 517.34(A)
Nursing Homes. . . . . 517.44(A)(1)
Other Health Care Facilities . . . . 517.45(C)

TELEPHONE EQUIPMENT **805.18**
See COMMUNICATIONS CIRCUITS . . . . . .Ferm's Finder

TELEVISION & MOTION PICTURE STUDIOS . . **Art. 530**
See MOTION PICTURE & TELEVISION STUDIOS . . . . .
. . . . Ferm's Finder
See THEATERS, ETC. . . . . Ferm's Finder

TELEVISION & RADIO EQUIPMENT . . . . **Art. 810**
See Community Antenna Television and Radio Distribution Systems
. . . . Art. 820
See COMMUNITY ANTENNA TELEVISION AND RADIO
DISTRIBUTION SYSTEMS . . . . Ferm's Finder
See RADIO & TELEVISION EQUIPMENT . . .Ferm's Finder

TEMPERATURE [AT LUMINAIRES (LIGHTING FIXTURES)]
Conductors in Outlet Boxes 410.21
Flush and Recessed Luminaires (Fixtures). 410.115
Underwater Luminaires (Lighting Fixtures) . . . . 680.23(A)(7)

TEMPERATURE, AMBIENT

TEMPERATURE, AMBIENT See **AMBIENT TEMPERATURE**
Ferm's Finder

TEMPERATURE CLASSIFICATIONS
Hazardous (Classified) Locations . . . . 500.8(C)(4)
Class I, Zone 0, 1, and 2 Locations . . . . 505.9(D)

TEMPERATURE CORRECTION FACTORS Tables 310.15(B)(1) and (2)
   See Informative Annex B for Examples of Formulas . . . .
   . . . . Informative Annex B

TEMPERATURE OF CONDUCTOR INSULATION . . . . .
. . . . 310.14(A)(3)
   0 to 2000 Volts . . . . Table 310.4(A)
   See CONDUCTORS, Temperature Rating of Insulation . . . . .
   . . . . Ferm's Finder

TEMPORARY INSTALLATIONS . . . . Art. 590
   Approval Required . . . . 590.2(B)
   Branch Circuits Protection . . . . 590.4(C)
      Type SE Cable . . . . 590.4(C)(2)
   Carnival, Circuses, Fairs, and Similar Events . . . . . . Art. 525
   Connected to an Equipment Grounding Conductor (Receptacles) . . . . 590.4(D)
   Cords and Cable Assemblies (Not Extension Cords) . . 590.4(J)
   Disconnecting Means . . . . 590.4(E)
   Exhibition Halls . . . . 518.3(B)
   Feeders . . . . 590.4(B)
      Type SE Cable . . . . 590.4(A)(2)
      Without GFCI Protection . . . . . . 215.10, Ex. No. 3
   Ground-Fault Protection for Personnel 590.6
      Assured Equipment Grounding Conductor Program 590.6(B)(2)
      Other Outlets . . . . 590.6(B)
      Receptacles, 15-, 20-, & 30-Ampere, 125-volt, Single-Phase . . . . 590.6(A)
   Grounding (General) . . . . Art. 250
   High Voltage Systems, Guarding 590.7
   Lamp Protection . . . . 590.4(F)
   Overcurrent Protective Devices 590.8
   Protection from Accidental Damage . . . . 590.4(H)
   Receptacles to Be Grounding Type . . . . 590.4(D)
      Not Permitted on Lighting Circuits . . . . 590.4(D)
   Services (per Article 230) . . . . 590.4(A)
   Splices . . . . 590.4(G)
   Support . . . . 590.4(J)
      By Vegetation Such As Trees Prohibited 225.26
   Termination(s) at Devices . . . . 590.4(I)
   Time Constraints 590.3

TERMINAL BAR (GROUNDING)
   For Panelboards 408.40
      Isolated Ground Receptacles . . . . 250.146(D)
      Swimming Pool Equipment . . . . 680.25(B)
      Junction Boxes and Enclosures 680.24
      Underwater Luminaires (Fixtures) 680.23
   Transformer . . . . 450.10(A)

TERMINAL HOUSINGS (MOTORS) 430.12

TERMINALS
   Connections to . . . . 110.14(A)
      Conductors and Equipment 250.8
      Grounding Electrode 250.70
   CO/ALR Marking
      Receptacles . . . . 406.3(C)
      Switches . . . . 404.14(C)
   Electric Discharge Tubing, Neon Electrode Conductors, Signs . .
   . . . . 600.42
   Flat Cable Assemblies . . . . 322.120(C)
   Identification of
      Grounding-Pole of Receptacles . . . . 406.10(B)
      Motors and Controllers . . . . 430.9(A)
      Polarity . . . . 200.9 & 10
      Switches . . . . 404.6(C)
      Wiring Device 250.126
   Receptacles, Grounding Terminal . . . . 406.10(B)
      Connection to Boxes 250.146
      See (RTDV) . . . . UL Product iQ
   Switches . . . . 404.6(C)
      See (WJQR) . . . . UL Product iQ
      Note: Use oxide inhibitor on aluminum conductors where required by manufacturer . . . . 110.3(B)
   Temperature Limitations . . . . 110.14(C)
   See Conductor Termination Compounds (DVYW) UL Product iQ
   See (AALZ) . . . . UL Product iQ

TESTING
   Assured Equipment Grounding Conductor Program 590.6(B)(2)
   Emergency Systems 700.3
   Equipment for Safety 90.7
   Feeders Over 1000 Volts, Pre-Energization and Operating . . 225.56(A)
   Ground-Fault Protection of Equipment . . . . 230.95(C)
      In Health Care Facilities . . . . 517.17(D)
   Inspections and Tests 225.56, 110.41
      Feeders 225.56
      Over 1000 Volts, (complete electrical system design) . . 110.41
   Insulation Resistance, Space-Heating Cable 424.45
   Legally Required Standby Systems 701.3

Mobile Homes 550.17

Park Trailers. 552.60

Performance Tested . . . . 225.56(A)

Pre-energization and Operating Tests 110.41

Recreational Vehicles 551.60

Temporary Installations, Permitted for . . . . 590.3(C)

Transformers Used for Research, Development, etc.   450.1 Ex. 8

Wiring Integrity 110.7

THEATERS, AUDIENCE AREAS OF MOTION PICTURE AND TELEVISION STUDIOS, AND SIMILAR LOCATIONS Art. 520

Construction 520.53

   Interior Conductors . . . . 520.53(E)

   Neutral Terminal . . . . 520.53(B)

   Pilot Light . . . . 510.53(A)

   Single-Pole Separable Connector . . . . 520.53(C)

   Supply Feed-Through . . . . 520.53(D)

Definitions 520.2

Dressing Rooms. . . . . Art. 520 Part VI

   Lamp Guards. 520.72

   Pendant Lampholders Prohibited 520.71

   Pilot Lights Required. 520.74

   Switches Required 520.73

Emergency Systems . . . . Art. 700

Grounding 520.81

Hard Usage Cord- Listed . . . . 520.68(A)(2)

Luminaire Supply Cords . . . . 520.68(A)(3)

Portable Equipment Used Outdoors 520.10

Stage Equipment

   Defined 520.2

   Fixed, Other Than Switchboards . . . . . . Art. 520 Part III

   Portable, Other Than Switchboards . . . . . Art. 520 Part V

Stage Lighting Hoist

   Defined 520.2

   Wiring Information 520.40

Supply Conductors 520.54

Switchboards

   Defined 520.2

   Constant Power . . . . 520.26(D)

   Fixed . . . . Art. 520 Part II

   Intermediate . . . . 520.26(C)

   Manual. . . . . 520.26(A)

   Portable . . . . Art. 520 Part IV

   Remotely Controlled . . . . 520.26(B)

Wiring Methods . . . . 520.5(A)

   Nonrated Construction . . . . 520.5(C)

   Portable Equipment . . . . 520.5(B)

THERAPEUTIC EQUIPMENT . . . . 517.73(B)

THERAPEUTIC POOLS & TUBS
Art. 680 Part VI

Application of Article 680 Part VI 680.60

Permanently Installed Pools. 680.61

Note: Permanent Installations to comply with 680 Parts I & II

See Exception for Luminaires . . . . 680.61 Ex.

Portable Therapeutic Appliances . . . . Art. 422

Therapeutic Tubs (Hydrotherapeutic Tanks) 680.62

   Bonding . . . . 680.62(B)

      Methods. 680.62(C)

   Ground-Fault Circuit-Interrupter (GFCI) Protection

      Listed Units . . . . 680.62(A)(1)

      Other Units . . . . 680.62(A)(2)

      Outlets . . . . 680.62(A)

      Receptacles . . . . 680.62(E)

   Grounding . . . . 680.62(D)

   Luminaires . . . . 680.62(F)

      Exception from Limitations . . . . 680.61 Ex.

      In Area to Be Totally Enclosed Type . . . . . . 680.62(F)

See THERAPEUTIC POOLS & TUBS . . . . . Ferm's Finder

THERMAL DEVICES 240.9

Fluorescent Luminaires . . . . 410.130(E)

High-Intensity Discharge Luminaires . . . . 410.130(F)

Motors

   Continuous-Duty . . . . 430.32(A)(2) and (B)(2)

Recessed Incandescent Luminaires . . . . 410.115(B)

Thermal Protector(s) (Motors), Definition . . . . Art. 100 Part I

THERMALLY PROTECTED

Definition of, (as applied to motors) . . . . Art. 100

Fluorescent Lamp Ballasts . . . . 410.130(E)

High-Intensity Discharge Luminaires (Fixtures). . . . 410.130(F)

Luminaires (Fixtures), Recessed Incandescent . . . . 410.115(B)

THERMOSTATS

Accessible . . . . 424.20(A)

Appliances . . . . Art. 422

Cables Green Color, Okay – Not as Equipment Ground . . . . .

. . . . 250.119 Ex. 1

Installation (Class 2 Wiring) . . . . Art. 725 Part III
Fixed Electric Space-Heating Equipment
    As Switching Devices. 424.20
    Loads Permitted . . . . 424.4(B)
  See ELECTRIC HEAT (SPACE) . . . . Ferm's Finder

THREADS (CONDUITS)
Explosionproof Equipment
    Divisions Metric- (5 fully engaged). . . . . . . .500.8(E)(2)
    Divisions NPT- (5 fully engaged). . . . . .500.8(E)(1)
    Zones Metric- (5 fully engaged) . . . . .505.9(E)(2)
    Zones NPT- (5 fully engaged) . . . . .505.9(E)(1)
FIELD-CUT (other than at the factory) . . 300.6(A) Info. Note
Running
    Intermediate Metal Conduit . . . . 342.42(B)
    Rigid Metal Conduit . . . . 344.42(B)

THREE OVERLOAD UNITS, MOTORS . . . . . Table 430.37

TIE BARS (Circuit-Breaker Handle Ties)
Circuit Breaker as Overcurrent Device . . . . 240.15(B)
Disconnecting Means, Temporary Wiring . . . . . . 590.4(E)
Feeders . . . . 225.33(B)
Mobile Homes . . . . 550.11(C)
Multiple Branch Circuits 210.7
Multiwire Branch Circuits . . . . .240.15(B)(1)
Service Disconnects. . . . . 230.71(B)
    Simultaneous Opening of Poles 230.74
Ungrounded Conductors Tapped from Grounded System 210.10
Screw Tightness. . . . . Informative Annex I

TIRE INFLATION MACHINES (For Public Use)   422.5(A)(4)

TOOLS
Cord- and Plug-Connected, Double Insulated . . . 250.114 Ex.
Industrial Machinery . . . . Art. 670
Motor-Operated, Hand-Held, Grounding . . . . 250.114(4)(c)

TORQUING INFORMATION
Electrical Connections . . . . 110.14(D)
Motor Controller Terminals . . . . 430.9(C)
Recommended Tightening Torque Tables . Informative Annex I
Tightness of Screws. . . . . Informative Annex I

TRACK, LIGHTING . . . . Art. 410 Part XIV
Defined (Lighting Track) . . . . Article 100

See FIXTURES (LUMINAIRES) LIGHTING, Lighting Track. .
. . . . Ferm's Finder

TRAFFIC SIGNALS
Ungrounded Signal Conductors with Green Insulation
Permitted . . . . 250.119 Ex. 3

TRAILERS, TYPES OF
Park Trailers. . . . . Art. 552
Recreational Vehicles . . . . Art. 551
    Travel Trailers, Definition 551.2

TRANSFER SWITCHES
Definition of . . . . Art. 100 Part I
Emergency Systems
    Documentation . . . . 700.5(E)
    Electrically Operated and Mechanically Held . . . . 700.5(C)
    Fire Protection of. . . . . 700.10(D)
    Identified as Component of Emergency System . . 700.10(A)
    Inadvertent Parallel Operation Shall be Avoided . . . 700.5(B)
    Maintenance and Repair . . . . 700.3(F)
    Normal Power Wiring Permitted in . . . . 700.10(B)
    Provided for Generator Set. . . . . .700.12(D)(1)
    Shall Supply Only Emergency Loads. . . . . 700.5(D)
    To Be Listed for Emergency System Use . . . . . . 700.5(A)
Essential Electrical Systems
Ambulatory Health Care Centers 517.45
Clinics, Medical and Dental Offices, etc. 517.25
Hospitals
    Equipment Branch Connection to Alternate Source . . 517.35
    Illumination of Means of Egress (Life Safety Branch) 517.33(A)
    Location of Components 517.30
    Maximum Time for Transfer (10 Seconds) 517.31
    Note: For typical systems, See Info. Notes Figures 517.31(a) & (b)
    Number of Transfer Switches. . . . . 517.31(B)
        Optional Loads from Separate Transfer Switches 517.31(B)(1)
        Separation of Wiring . . . . .517.31(C)(1)
        Subdivision of Critical Branch . . . 517.34(C) Info. Note
Nursing Homes and Limited Care Facilities
    Connection to Equipment Branch 517.44
    Illumination of Means of Egress (Life Safety Branch) 517.43(A)
    Location of Components . . . . 517.41(C)
    Note: For typical systems, See . . . . Figure 517.42(a) & (b)
    Number of Transfer Switches . . . . 517.42(B)
    Separation of Wiring . . . . 517.42(D)(1)

Relative Location of Switches and Ground-fault Protection . . . . 517.17(B)

Fire Pumps
    Continuity of Power 695.4
    Equipment Location 695.12
    Equipment to Be Listed for Fire Pump Service 695.10
    No Other Loads Permitted to be Supplied from . . . 695.6(E)
    Remote Devices, not to prevent operation of . . . 695.14(C)
    Fuel Cell Systems 692.59

Health Care Facilities (General)
    Critical Care (Category 1) Spaces
        Patient Bed Location Branch Circuits . . . 517.19(A) Ex. 2
        Patient Bed Location Receptacles . . . . . 517.19(B)(1)(2)
    General Care (Category 2) Areas Patient Bed Location Circuits . . . . 517.18(A) Ex. 3

Legally Required Standby
    Electrically Operated and Mechanically Held . . . . 701.5(C)
    Inadvertent Parallel Operation Shall be Avoided . . . 701.5(B)
    Provided for Generator Set . . . . 701.12(D)
    To Be Automatic and Identified for the Use . . . . 701. 5(A)

Meter Mounted (Not Permitted for Emergency Use) . . 700.5(A)

Optional Standby Power Systems
    Transfer Equipment 702.5
    See (WPTZ) . . . . UL Product iQ

TRANSFORMER VAULTS
    Doorways. 450.43
        Fire Ratings . . . . 450.43(A)
        Locks and Panic Hardware . . . . 450.43(C)
        Sill or Curb Required. . . . . 450.43(B)
    Drainage 450.46
    Enclosures and Access 110.31
    Fire Resistivity, Over 1000 Volts . . . . 110.31(A)
    Illumination . . . . 110.34(D)
    Location of 450.41
    Storage in, Prohibited. 450.48
    Ventilation Openings 450.45
    Walls, Roofs and Floors, Construction and Fire Rating . . 450.42
    Water Pipes and Accessories. 450.47
    Working Space, Entrance and Access 110.33
    Working Space and Guarding. 110.34

TRANSFORMERS AND TRANSFORMERS VAULT . Art. 450
    Access (Limited Working Space) . . . . 110.26(A)(4)
    Alternating Current Systems to Be Grounded 250.20

Alternating Current Systems Not Required to Be Grounded 250.21
(Arc) Electric Welders. . . . . Art. 630
Askarel-Insulated, Indoors 450.25
Autotransformers, 1000 Volts or Less 450.4
    Audio Signal Processing . . . . 640.9(D)
    Ballasts 410.138
    Branch-Circuits 210.9
    Feeders 215.11
    For Grounding 3-Phase, 3-Wire Systems 450.5
    Motor Starting . . . . 430.82(B)
    Overcurrent Protection. . . . . 450.4(A)
    Prohibited in Park Trailers . . . . 552.20(E)
    Prohibited in Recreational Vehicles. . . . . 551.20(E)
    Transformer Field-Connected as Autotransformer . . 450.4(B)

Buck & Boost . . . . 210.9 Ex. 1 & 2
    Common Use for Autotransformers 450.4

Calculations for, See CALCULATIONS . . . . . . Ferm's Finder
Carnivals, Circuses, Fairs, and Similar Events . . . . . 525.10(B)
Definitions 450.2
Disconnecting Means. 450.14

Dry-Type
    Indoors
        Not Over 112 1/2 kVA . . . . 450.21(A)
        Over 112 1/2 kVA . . . . 450.21(B)
        Over 35,000 Volts to Be in Vault . . . . 450.21(C)
    Outdoors. 450.22

Electrically Actuated Fuse . . . . Table 450.3(A) Note 4
Fire Pumps, for 695.5
    Feeder Source . . . . .695.5(C)
    Overcurrent Protection of . . . . 695.5(B)
    Sizing. . . . . .695.5(A)

Fuse, Electrically Actuated . . . . Table 450.3(A) Note 4

Grounding of
    Dry-Type Transformer Enclosures . . . . 450.10(A)
    Exceptions from, for Certain Systems . . 250.30(A)(4) Ex 2.
    Metal Parts, Fences, Guards, Etc. . . . . 450.10(B)
    Other Metal Parts . . . . 450.10(B)
    Outside Transformer Supplying Service . . . . 250.24(A)(2)

Guarding 450.8

Hazardous (Classified) Locations
    Class I Locations 501.100
        Control Transformers 501.120
    Class II Locations 502.100
        Control Transformers 502.120

Signaling, Alarm, Etc. Systems. 502.150

Class III Locations 503.100

Control Transformers 503.120

High Leg

Feeders 110.15

See HIGH LEG . . . . Ferm's Finder

Installation of 450.13

Specific Provisions Applicable to Different Types of
Transformers . . . . Art. 450 Part II

Isolating Transformers

Class III Locations . . . . 503.155(A)

Electrolytic Cells . . . . 668.20(B)

Power Supply Circuits, Portable Equipment . . . 668.21(A)

Induction and Dielectric Heating 665.5

Integrated Electrical System, Control Circuits 685.14

Isolated Power Systems, Health Care Facilities 517.160

Lighting Systems 30 Volts or Less 411.7

Optional Protective Technique, Critical Care (Category 1) Spaces
. . . . 517.19(F)

Runway Tracks As Circuit Conductor . . . . . . 610.21(F)(2)

Systems Not Required to Be Grounded 250.21

Underwater Luminaires (Lighting Fixtures) . . 680.23(A)(2)

Wet Procedure Locations, Health Care Facilities 517.20

Less-Flammable, Liquid-Insulated

Indoor Installations . . . . 450.23(A)

Outdoor Installations . . . . 450.23(B)

Location of 450.13

Marking of

General. . . . . 450.11(A)

Source Marking (Reverse Feed). . . . . 450.11(B)

Modification of Existing Transformers 450.28

Motor Control Circuit 430.74

Nonflammable Fluid-Insulated 450.24

Oil-Insulated

Indoor Installations 450.26

Outdoor Installations 450.27

Overcurrent Protection of. 450.3

1000 Volts or Less . . . . 450.3(B)

Autotransformers, 1000 Volts or Less. . . . . . . 450.4(A)

Control Circuit Secondary Conductors 430.74

Electrically Actuated Fuse . . . . . . Table 450.3(A) Note 4

Ground Reference for Fault Protection Devices . 450.5(B)(2)

Grounding Autotransformers Three-Phase 4-Wire 450.5(A)(2)

Over 1000 Volts . . . . 450.3(A)

Current Transformers . . . . 240.100(A)(1)

Panelboard Supplied from . . . . 408.36(B)

Secondary Conductors (General). . . . . 240.4(F)

Secondary Conductors (Specific) . . . . 240.21(C)

Secondary Ties . . . . 450.6(B)

Solar Photovoltaic Systems . . . . 690.9(D)

Supervised Industrial Installations . . . . . Art. 240 Part VIII

Supplying Fire Pumps . . . . 695.5(B)

Supplying Transformer, Primary Plus Secondary
Not Over 7.5 m (25 ft.) . . . . 240.21(B)(3)

Voltage (Potential) Transformers . . . . 450.3(C)

Wind Electric Systems . . . . 694.15(B)

Parallel Operation 450.7

Park Trailer Combination Systems . . . . 552.20(B)

Potential (Voltage) . . . . 450.3(C)

Power-Limited and Signaling Circuits

Class 1 Circuits . . . . 725.41(A)(1)

Class 2 and 3 Circuits . . . . 725.121(A)(1)

Recreational Vehicle Combination Systems . . . . . 551.20(B)

Autotransformers. . . . . 551.20(E)

Remote Control, Circuits for

Control Transformer in Controller Enclosure . . . 430.75(B)

Class 1 Circuits. . . . . 725.41(A)(1)

Class 2 and 3 Circuits . . . . 725.121(A)(1)

Research and Development Use, Exempt . . . . . . 450.1 Ex. 8

Reverse Feeding. . . . . 450.11(B)

Secondary Ties 450.6

Separately Derived Systems 250.30

See SEPARATELY DERIVED SYSTEMS . . . . Ferm's Finder

Short-Circuit Current Available

Circuit Impedance 110.10

Signs and Outline Lighting

Ballasts, Transformers and Electronic Power Supplies and Class 2
Power Sources 600.21

Enclosures 600.8

Transformers and Electronic Power Supplies 600.23

Specific Provisions for Different Types . . . . . . Art. 450 Part II

Storage, Top Surfaces (Prohibited) 450.9

Supplied from Feeders, Ampacity of Conductors . . 215.2(B)(1)

Taps, See TAPS . . . . Ferm's Finder

Terminal Bar Not Installed over Vented Area 450.10

Terminal Wiring Space 450.12

Three-Wire to Three-Wire . . . . 240.4(F)

Two-Winding, Underwater Lighting . . . . 680.23(A)(2)

Two-Wire to Two-Wire . . . . 240.4(F)

Vaults . . . . Art. 450 Part III

Vented Portion, Terminal Bar not Installed 450.10

Ventilation of 450.9

Ventilation, Top Storage (Prohibited) 450.9

Voltage (Potential) . . . . 450.3(C)

Working Space . . . . 110.26(A)

    Limited Access . . . . .110.26(A)(4)

See Transformers (includes many types of) (XNWX) . . . .
. . . . *UL Product iQ*

TRAVEL TRAILERS, RECREATIONAL VEHICLES, AND VEHICLE PARKS . . . . .Art. 551

    Definition 551.2

    See RECREATIONAL VEHICLE . . . . Ferm's Finder

TRAY CABLE, POWER AND CONTROL, TYPE TC .Art. 336

    Ampacity of 336.80

    Bends of 336.24

    Conductors

        Class 1 Circuits. . . . . .336.104(C)

        Fire Alarm Systems. . . . . .336.104(A)

        Thermocouple Circuits. . . . . .336.104(B)

    Construction of . . . . 336 Part III

    Definition of 336.2

    Hazardous (Classified) Location Cable 336.130

    Installation and Support 336.10

    Jacket (flame retardant) 336.116

    Listing 336.6

    Marking. 336.120

    Uses Not Permitted 336.12

    Uses Permitted 336.10

    See Instrumentation Tray Cable, Type ITC . . . . Art. 727

TREES

    Not for Support of Overhead Conductor Spans. 225.26

    Permitted for Support of Outdoor Lighting Luminaires (Fixtures)
. . . . 410.36(G)

    Support, Holiday Lighting Branch Circuits . . . . 590.4(J) Ex.

    Vegetation as Support. 230.10

TUBING

    Electric Discharge, Signs, etc. (Neon). 600.41

    Electrical Metallic. . . . . Art. 358

    See ELECTRICAL METALLIC TUBING. . . . Ferm's Finder

    Electrical Nonmetallic . . . . Art. 362

    See ELECTRICAL NONMETALLIC TUBING Ferm's Finder

    Field-Installed Skeleton . . . . .Art. 600 Part II

    Flexible Metallic . . . . Art. 360

    See FLEXIBLE METALLIC TUBING. . . . . . Ferm's Finder

    Neon, Definition of. 600.2

    Skeleton, Definition 600.2

TURBINE *See* **WIND ELECTRIC SYSTEMS** . . . Ferm's Finder

TV AND MOTION PICTURE STUDIOS AND SIMILAR LOCATIONS . . . . Art. 530

See THEATERS, AUDIENCE AREAS OF MOTION PICTURE AND TELEVISION STUDIOS, AND SIMILAR LOCATIONS Ferm's Finder

TV AND RADIO DISTRIBUTION . . . . Art. 820

See COMMUNITY ANTENNA TELEVISION AND RADIO DISTRIBUTION . . . . Ferm's Finder

TV AND RADIO EQUIPMENT . . . . Art. 810

See RADIO AND TELEVISION EQUIPMENT . Ferm's Finder

TYPE P CABLE (ARMORED AND UNARMORED) .Art. 337

    Ampacity of. 337.80

    Armor. 337.116

    Bending Radius 337.24

    Conductors 337.104

    Definition. 337.2

    Equipment Grounding Conductor 337.108

    Installation . . . . .Art. 337 Part II

    Insulation 337.112

    Jacket 337.115

    Listing Requirements 337.6

    Marketing. 337.120

    Scope 337.1

    Securing and Supporting 337.30

    Shield 337.114

    Single Conductors 337.31

    Uses Not Permitted 337.12

    Uses Permitted 337.10

TYPES OF CONSTRUCTION . . . . . Informative Annex E

U

UF WIRE (Direct Burial) . . . . Art. 340
See UNDERGROUND FEEDER AND BRANCH-CIRCUIT
CABLE . . . . Ferm's Finder

UFER GROUND (Concrete-Encased Electrode) . 250.52(A)(3)
See Rebar . . . . Ferm's Finder

UNBALANCED INTERCONNECTIONS
Energy Storage Systems . . . . 706.16(D)
Fuel Cell Systems 692.64
Interconnected Electric Production Sources
   Single Phase . . . . 705.45(A)
   Three Phase . . . . 705.45(B)

UNDER BUILDINGS
Cables to Be in Raceway . . . . 300.5(C)
Service Conductors 230.6

UNDERFLOOR RACEWAYS . . . . Art. 390
Ampacity of Conductors 390.23
Connections to Cabinets, Wall Outlets 390.76
Covering 390.15
Dead Ends 390.73
Definition. 390.2
Discontinued Outlets 390.57
Inserts. 390.75
Junction Boxes 390.74
Laid in Straight Lines 390.70
Markers at Ends. 390.71
Maximum Number of Conductors 390.22
Size of Conductors 390.20
Splices & Taps 390.56
Uses Not Permitted 390.12
Uses Permitted 390.10

UNDERGROUND CABLES
Optical Fiber Cables 770.47
Type MC . . . . 300.5(C) Ex. 2
Type MI . . . . 300.5(C) Ex. 1

UNDERGROUND ENCLOSURES
Manholes . . . . Art. 110 Part V
Handhole Enclosures 314.30
   Over 1000 Volts . . . . 314.70(C)
Subsurface Enclosures, Accessibility 314.29

UNDERGROUND FEEDER AND BRANCH-CIRCUIT CABLE, TYPE UF . . . . Art. 340
Ampacity of 340.80
Bending Radius 340.24
Conductor Construction and Applications . . . . Table 310.4(A)
Construction . . . . Art. 340 Part III
Definition. 340.2
Ground Movement . . . . 300.5(J)
Installed as Nonmetallic-Sheathed Cable . . . . . . 340.10(4)
Listing Requirements 340.6
Protection of 300.5
Uses Not Permitted 340.12
Uses Permitted 340.10
See (YDUX) . . . . UL Product iQ

UNDERGROUND, RACEWAYS, WET LOCATIONS
1000 Volts or Less . . . . 300.5(B)
Over 1000 Volts . . . . 300.50(B)

UNDERGROUND SERVICE
Entrance Cable Type USE . . . . Art. 338

UNDERGROUND SERVICE CONDUCTORS Art. 230 Part III
See Service Conductors, Underground . . . . Ferm's Finder

UNDERGROUND WIRING
Aluminum Conduit (Not Permitted without Supplementary
   Corrosion Protection) . . . . 344.10(A)(3) and (B)(2)
   See (DYWV) . . . . UL Product iQ
Ampacity of Conductors Underground
   0 to 2000 Volts . . . . Table 310.16
      Under Engineering Supervision — Formula . . 310.14(B)
   2001 to 35,000 Volts   Tables 311.60(C)(77) to 310.60(C)(86)
      Application of Tables 311.60
   Note: For examples of formulas, See . . Informative Annex B
Buried Conductors (Direct Burial)
   1000 Volts or Less 300.5
   Conductor Application and Insulation . . . . . Table 310.4(A)
   Corrosive Conditions . . . . 310.10(F)
   Identified for Use. . . . . 310.10(E)
   Over 1000 Volts 300.50
   Type UF . . . . 340.10(1)
   Type USE 338.2
Concrete Encased (Raceways) . . . . Table 300.5 Note 2
Conductor Types in Wet Locations . . . . 310.10(C)
Cover Requirements

1000 Volts or Less . . . . Table 300.5
Over 1000 Volts . . . . Table 300.50
Earth Movement . . . . 300.5(J)
Over 1000 Volts . . . . 300.50(C)
Electrical Metallic Tubing . . . . 358.10(B)
Supplementary Corrosion Protection Required
See (FJMX) . . . . *UL Product iQ*
High Density Polyethylene Conduit (HDPE) . . . . 353.10(4)
Installation, 1000 Volts and Less 300.5
Backfill . . . . 300.5(F)
Definition of Cover . . . . .Table 300.5 Note 1
Minimum Cover Requirements . . . . Table 300.5
Installation, Over 1000 Volts 300.50
Backfill . . . . 300.50(E)
Definition of Cover . . . . Table 300.50 Note [a]
Minimum Cover Requirements . . . . Table 300.50
Intermediate Metal Conduit . . . . 342.10(B)
Lighting System- Low Voltage . . . . Table 300.5 Note[a]
Liquidtight Flexible Metal Conduit . . . . 350.10(3)
Liquidtight Flexible Nonmetallic Conduit . . . . . . 356.10(4)
Listed Low-Voltage Lighting System . . . . . Table 300.5 Note[b]
Nonmetallic Underground Conduit with Conductors . . 354.10
Over 1000 Volts, Nominal 300.50
See OVER 1000 VOLTS, NOMINAL . . . . . .*Ferm's Finder*
Pool, Spa, and Fountain Lighting . . . . . . Table 300.5 Note[b]
Protection of . . . . 300.5(D)
Over 1000 Volts . . . . 300.50(C)
Recreational Vehicle Parks 551.80
Reinforced Thermosetting Resin Conduit . . . . . . 355.10(G)
Rigid Metal Conduit . . . . 344.10(B)
Aluminum (Not Permitted without Supplementary Corrosion Protection)
See (DYWV) . . . . *UL Product iQ*
Rigid PVC Conduit . . . . 352.10(G)
See RIGID PVC CONDUIT . . . . *Ferm's Finder*
Rock Encountered (reduction of depth) . . . .Table 300.5 Note 5
"S" Loops . . . . 300.5(J) IN
Services . . . . Art. 230 Part III
Protection against Physical Damage . . . . 230.50(A)
Splices and Taps . . . . 110.14(B)
1000 Volts or Less . . . . 300.5(E)
Over 1000 Volts . . . . 300.50(D)
Swimming Pool Area 680.11
Under Hazardous (Classified) Areas

Aircraft Hangars 513.8
Sealing 513.9
Bulk Storage Plant 515.8
Sealing 515.9
Class I, Division 1 . . . . 501.10(A) Ex.
Commercial Garages, Major 511.8
Commercial Garages, Minor 511.8
Motor Fuel Dispensing Facilities 514.8
Sealing 514.9
Wet Locations . . . . 310.10(C)
Definition of . . . . Art. 100
Listing for . . . . 300.5(B)
Over 1000 Volts . . . . 300.50(B)

Note: Installations underground, in concrete slabs, or in direct contact with the earth are considered to be "Locations, Wet" Art. 100

UNGROUNDED
Definition of . . . . Art. 100

UNGROUNDED SYSTEMS
3-Phase and 2-Phase Systems . . . . .240.15(B)(3)
3-Wire Direct Current Circuits . . . . .240.15(B)(4)
AC Systems Not Required to Be Grounded 250.21
AC Systems, Signage . . . . .408.3(F)(2)
Circuits Not to Be Grounded 250.22
Cranes and Hoists . . . . 610.21(F)(2)
Critical Care (Category 1) Spaces (Optional) . . . . . 517.19(F)
Isolated Power System Grounding . . . . 517.19(G)
DC Systems . . . . 250.162(A) Ex.
DC System, Signage . . . . 408.3(F)(4)
Direct-Current Ground-Fault Detection . . . . . . 250.167(A)
Electric Cranes, Hoists, Class III Locations . . . . . 503.155(A)
Electrolytic Cells . . . . Art. 668
Equipment Grounding Conductor Connections . . .250.130(B)
Feeder Identification
Alternating-Current Systems, Ungrounded Conductors 215.12(C)(1)
Direct-Current Systems, Ungrounded Conductors 215.12(C)(2)
General Requirements . . . . 250.4(B)
Ground Detectors
AC Systems 50 to 1000 Volts . . . . 250.21(B)
Critical Care (Category 1) Spaces (Optional) . . . . 517.19(F)
Electric Cranes and Hoists, Class III Locations . 503.155(A)
Isolated Power Systems . . . . 517.160(B)

Wet Procedure Locations, Hospitals, etc. . . . . . 517.20(B)

Ground-Fault Detections . . . . 250.167(A)

Induction and Dielectric Heating, Generator Output. 665.5

Integrated Electrical Systems, DC. 685.12

    Control Circuits 685.14

Isolated Power Systems, Hospitals . . . . Art. 517 Part VII

    Circuits in Anesthetizing Locations . . . . 517.63(F)

    Inhalation Anesthetizing Locations. . . . 517.61(A)(1) & (2)

    Location of Isolated Power Systems . . . . 517.63(E)

Marking of . . . . 250.21(C)

Lighting Systems Operating at 30 Volts or Less . . . . 411.6(A)

Separately Derived Systems, AC . . . . 250.30(B)

Separately Derived Systems, DC 250.169

Swimming Pool Luminaire (Lighting Fixture) Transformers . . .
. . . . .680.23(A)(2)

Switchboards and Panelboards . . . . 408.3(F)

Voltage to Ground (Definition of) . . . . Art. 100 Part I

Wet Procedure Locations, Hospitals, etc. . . . . 517.20(B)

UNINTERRUPTIBLE POWER SUPPLY EQUIPMENT

Critical Operations Power Systems . . . . 708.20(G)

Definition. 708.2

Emergency System Power Supply . . . . 700.12(C)

Information Technology Equipment 645.11

Legally Required Standby Systems . . . . 701.12(C)

See (YEDU) . . . . UL Product iQ

UNIT EQUIPMENT

Battery-Powered Lighting Units. 517.2

Emergency Systems . . . . 700.12(F)

    Components . . . . 700.12(F)(1)

    Installation . . . . 700.12(F)(2)

Legally Required Standby Systems . . . . 701.12(G)

Modular Data Systems 646.17

UNIT SUBSTATIONS

See SERVICE EQUIPMENT. . . . . Ferm's Finder

See (YEFR) . . . . UL Product iQ

See (Over 1000 Volts, Nominal) (YEFV) . . . . . UL Product iQ

UNUSED OPENINGS TO BE CLOSED . . . . . . 110.12(A)

Cabinets, Cutout Boxes, and Meter Socket Enclosures   312.5(A)

Outlet, Device, Pull and Junction Boxes, Conduit Bodies .   314.17(A)

USB CHARGER

Faceplate . . . . .406.6(D)

Receptacle. . . . . 406.3(F)

USE OF EQUIPMENT **110.3**

USE WIRE (Direct Burial) **338.2**

See SERVICE ENTRANCE CABLE . . . . Ferm's Finder

See (TYLZ) . . . . UL Product iQ

UTILITY, ELECTRIC

Connections, Meter Enclosures 230.82

Fire Pump . . . . 695.3(A)(1)

Ground-Fault Protection for Personnel 590.6

Installations Covered by NEC. . . . . .90.2(A)

Installation Not Covered by NEC. . . . . .90.2(B)

Meter Sockets (not required to be listed) . . . . . . 230.66 Ex.

Point of Demarcation (see Service Point) . . . Art. 100 Info Note

Special Permission . . . . .90.2(C)

Transfer Equipment. 702.5

UTILITY-INTERACTIVE INVERTER

Definition of . . . . Art. 100

Energy Storage Systems

    Charge Control . . . . 706.33(B)(3)

    Inverter Utilization Output Circuit (Defined) 706.2

    Loss of Power. . . . . 706.16(C)

Fuel Cell Systems, Point of Connection. 692.65

    Output Circuit (Defined) 692.2

Interconnected Electric Power Production Sources 705.12

    Mounted in Not-Readily-Accessible Locations 705.70

Output Circuit, Interactive Inverter

    Definition of 705.2

    Maximum Current Rating. . . . . 705.30(B)

Overcurrent Protection. 705.65

Permitted Without Grounded Conductor . . . . . . .200.3 Ex.

Solar Photovoltaic Systems

    Mounted in Not-Readily-Accessible Locations . . . 690.15(A)

    Output Circuit. . . . . 690.6(B)

Wind Electric Systems . . . . Art. 694

    Disconnection 694.24

    Inverter System Identification . . . . Figure 694.1(a)

    Listed 694.60

    Marking Requirements. 694.50

UTILIZATION EQUIPMENT

Aircraft Hangars, Portable Utilization Equipment . . . .
. . . . 513.10(E)(2)

Boxes

    Depth of 314.24

    For Support of . . . . 314.27(D)

Branch Circuits, Permissible Loads 210.23

Bulk Storage Plants, Portable Utilization Equipment . . 515.7(C)

Class I Locations 501.135

Class II Locations 502.135

Class III Locations 503.135

Definition of . . . . Art. 100

Spray Areas, Class I Locations. . . . . 516.6(D)

Utilization Equipment, Device Fill in Boxes . . . . 314.16(B)(4)

UL's STANDARDS FOR SAFETY CATALOG OF STANDARDS

All published and proposed STANDARDS FOR SAFETY are listed both alphabetically and by UL subject number in UL's CATALOG OF STANDARDS FOR SAFETY. Issued January and July of each year, the catalog includes the latest edition and revision dates of all standards, description of UL's Revision Subscription Service, and a listing of UL standards covering specific product areas that are available as Sets.

UL's Product iQ tells what products are covered by each UL standard, and which UL standard covers what products. Over 500 generic product categories are listed, and the index gives more than 3,000 examples from the 12,000 specific product types covered in the scope of UL's standards. There is also an alphabetical listing of keywords from the titles of all UL standards. It is available online at https://iq.ulprospector.com/info/

STANDARDS GLOSSARY

UL's STANDARDS GLOSSARY is a compilation of terms and their definitions taken from the glossary sections of UL standards.

V

VACUUM MACHINES

Central Units Within Single Family Dwellings 422.15

GFCI Protection . . . . 422.5(A)(1)

VAPORS, FLAMMABLE

See HAZARDOUS (CLASSIFIED) LOCATIONS Ferm's Finder

VAULTS

Access to 110.76

Capacitors . . . . 460.3(A)

Equipment Work Space. 110.73

Film Storage .Art. 530 Part V

General . . . . .Art. 110 Part V

Over 1000 Volts, Nominal, Enclosures for Installations. . 110.31

    Electrical Vaults . . . . 110.31(A)

Service Conductors Over 35,000 Volts 230.212

Services Considered Outside the Building . . . . 230.6(3)

Strength, of Structure. 110.71

Transformer . . . . Art. 450 Part III

    Specific Provisions for Different Types of Transformers. . . . . .Art. 450 Part II

    Ventilation 110.77

        Guarding of Ventilating Grating 110.78

VEHICLES

Commercial Garages, Repair and Storage . . . . Art. 511

Covered by Code . . . . 90.2(A)(1)

Garage (for storage of)

    Definition . . . . Art. 100

    GFCI Protection

        Dwelling Units . . . . 210.8(A)

        Other Than Dwelling Units . . . . 210.8(B)(8)

    Receptacle Outlet . . . . .210.52(G)(1)

Not Covered by Code. . . . . 90.2(B)(1)

See ELECTRIC VEHICLES . . . . Ferm's Finder

See Electrified Truck Parking Spaces. . . . . Ferm's Finder

See RECREATIONAL VEHICLES. . . . . Ferm's Finder

VENDING MACHINES

Ground-Fault Circuit-Interrupter to be Readily Accessible . . . . 422.5(A)(5)

    See ANSI/UL 541-2016, Standard for Refrigerated Vending Machines,

    or ANSI/UL 751-2016, Standard for Vending Machines

VENTILATED (Definition of) . . . . Art. 100 Part I

VENTILATING DUCTS, WIRING IN Ducts, Plenums, etc. 300.22

Spread of Fire or Products of Combustion 300.21

VENTILATING PIPING FOR MOTORS

Class II Locations 502.128

Class III Locations 503.128

VENTILATION

Aircraft Hangars . . . . .513.3(B) and (D)

Ammonia (Refrigerant Machinery Rooms)

    Class and Division . . . . 500.5(A)

Zones . . . . 505.5(A)
Battery Locations 480.10
Bulk Storage Plants . . . . Table 515.3
Commercial Garages, Repair & Storage. . . . 511.3(C) and (D)
Electric Vehicles. 625.52
   Minimum Required Cubic Feet Per Minute (CFM) . . . . . .
   . . . . Table 625.52(B)(1)(b)
   Minimum Required Cubic Meters Per Minute (m³/min) . . .
   . . . . Table 625.52(B)(1)
Equipment, General . . . . 110.13(B)
Information Technology, under Raised Floors. . . . . 645.5(E)
Motors . . . . 430.14(A)
   Exposed to Dust Accumulations 430.16
Motor Fuel Dispensing Facilities . . . Table 514.3(B)(1) and (2)
Spray Application, Dipping & Coating 516.5
   Spray Booth (Defined) . . . . 516.2 Info. Note
Spread of Fire 300.21
System Controls, Tunnels, Over 1000 Volts 110.57
Transformers 450.9
   Vaults 450.45
      Ventilation to Open Air 110.77
      Guarding of Ventilation Grating. 110.78
   Zone 0, 1, and 2 Locations. . . . . Art. 505
   Zone 20, 21, and 22 Locations. . . . . Art. 506

VERTICAL SURFACES (BOXES) . . . . 314.27(A)(1)

VESSELS & PIPELINES, HEATING . . . . Art. 427
See FIXED ELECTRIC HEATING EQUIPMENT FOR PIPE-
LINES AND VESSELS . . . . *Ferm's Finder*

VISIBLE (IN SIGHT FROM, WITHIN SIGHT) . . . .Art. 100
VOLTAGE
Branch Circuit, Limitations 210.6
Considered 110.4
Definition of
   High Voltage 490.2
   Low Voltage 551.2
   Low Voltage Contact Limit 680.2
   Medium Voltage
   Photovoltaic System Voltage 690.2
   Voltage (Nominal) . . . . Art. 100 Part I
   Voltage (of a Circuit). . . . . Art. 100 Part I
   Voltage for Calculations . . . . 220.5(A)
   Voltage to Ground . . . . Art. 100 Part I
Drop

See VOLTAGE DROP . . . .*Ferm's Finder*
Electric Vehicle Charging System, Nominal 625.4
Elevator Equipment Working Clearances . . . . . . .620.5(D)
Less than 50 Volts . . . . Art. 720
   Class 1, 2, & 3 Remote Control, Signaling Circuits . Art. 725
   Grounding of . . . . 250.20(A)
Lighting Equipment Installed Outdoors 225.7
Limitations . . . . 300.2(A)
   Between Adjacent Snap Switches. . . . . 404.8(B)
   Between Adjacent Receptacles/Devices . . . . 406.5(J)
   Electric-Discharge Lighting Systems (Over 1000 Volts) . . . .
   . . . . Art. 410 Part XIII
   Electric-Discharge Lighting Systems (Under 1000 Volts) . . .
   . . . . Art. 410 Part XII
   Elevators, Etc. (Article 620) 620.3
   Rating and Use, Snap Switches. 404.14
   See (WJQR) . . . . *UL Product iQ*
   Swimming Pool Luminaires (Lighting Fixtures) . 680.23(A)(4)
Outdoor Overhead Conductors over 1000 Volts . . . . Art. 399
Over 1000 Volts, Nominal
   See OVER 1000 VOLTS, NOMINAL . . . . . .*Ferm's Finder*
   Photovoltaic System Voltage, Maximum. . . . . . . 690.7(A)

VOLTAGE DROP
AC Resistance and Reactance . . . . Chapter 9 Table 9
Branch Circuit 3% Recommended . . . .210.19(A) Info. Note 4
Conductors 0-2000 Volts, General . . 310.14(A)(1) Info. Note 1
DC Resistance of Conductors . . . . Chapter 9 Table 8
Feeder 3% (Feeder Plus Branch Circuit 5%) 215.2(A)(1)(b) Info.
Note 2
Fire Pumps 695.7
Sensitive Electronic Equipment. . . . . 647.4(D)
See Voltage Drop Tables . . . . *Ferm's Charts and Formulas*
Note: When conductors are adjusted in size, such as to compensate for voltage drop, equipment ground wires, where required, shall be adjusted proportionately in size according to circular mil areas.
See . . . . 250.122(B)
See Voltage Drop Section
   See Ampere-feet Method . . . . . . *Ferm's Charts and Formulas*
   See Equipment Grounding Conductor Adjustment*Ferm's Charts and Formulas*
   See Voltage Drop Formulas. . . . . *Ferm's Charts and Formulas*

VOLTAGE (POTENTIAL) TRANSFORMER . . . . . 450.3(C)

VOLTAGE VARIATIONS
See Multiplier Formula . . . . *Ferm's Charts and Formulas*

See Multiplier Table . . . . *Ferm's Charts and Formulas*

VOLT-AMPERES
Class 1 Power-Limited Circuits . . . . 725.41(A)(2)
Class 2 & 3 AC Power-Limited Circuits   Chapter 9 Table 11(A)
Class 2 & 3 DC Power-Limited Circuits   Chapter 9 Table 11(B)
Lighting Track, 150 Volt-Amperes Per 600 mm (2 ft)   220.43(B)
Power-Limited Fire Alarms, AC, PLFA . . Chapter 9 Table 12(A)
Power-Limited Fire Alarms, DC, PLFA . . Chapter 9 Table 12(B)
    See CLASS 1, 2 & 3 REMOTE CONTROL CIRCUITS. . .
    . . . . *Ferm's Finder*
Receptacle Load, Nondwelling Use 180 Volt-amperes   220.14(I)
Used for Load Calculations . . . . Art. 220

W

WADING POOLS (Definition of) 680.2
Applicability of Article 680 to. 680.1

WARNING RIBBON
Underground Conductors
    Over 1000 Volts (Direct-Buried Cables)
        . . . . Table 300.50 Note b and d
    Under 1000 Volts (Service Conductors) . . . . . 300.5(D)(3)

WARNING SIGNS
Aircraft Hangers
    Aircraft Battery Charging and Equipment . . . . . 513.10(B)
    External Power Sources for Energizing Aircraft . . 513.10(C)(2)
    Mobile Servicing Equipment with Electrical Components . . .
        . . . . 513.10(D)
    Mobile Stanchions . . . . 513.7(F)
ANSI Standard
    See . . . . ANSI Z535.4-2011, Product Safety Signs and Labels
Arc-Flash Hazard Warning 110.16
Assembly Occupancies (Temporary Wiring in Cable Trays) . . . .
    . . . . 518.3(B) Ex.
Available Fault Current (Non-Dwelling Units) 110.24
Boilers
    Fixed Electric Space Heating Equipment . . . . . . 424.86(5)
    Fixed Resistance and Electrode Industrial Process Heating Equipment . . . . 425.86(5)
Cable Trays
    Containing Conductors Over 600 Volts . . . . . . 392.18(H)
    Containing Service Conductors 230.44
Capacitors, Isolating or Disconnecting Switches with No Interrupting Rating. . . . . 460.24(B)(2)
Disconnecting Means, Series Rated Combination Systems 110.22
Electrically Heated Pipelines and Vessels 427.13
Electroplating 669.7
Electrostatic Spraying Equipment . 516.10(A)(8), 516.10(B)(4)
Elevators, Escalators, etc.
    More Than One Driving Machine in a Machine Room . . . .
        . . . . 620.51(D)
    Power Circuits (Exceeding 1000 Volts) . . . . 620.3(A)
    Power from More Than One Source . . . . 620.52(B)
Emergency Sources . . . . 700.7(A)
    Grounding Connection . . . . 700.7(B)
        Enclosure for Electrical Installations In Places Accessible to Unqualified Persons (Indoors) . . . . 110.31(B)(1)
Entrances to Rooms and Other Guarded Locations . . 110.27(C)
Feed-Through Conductors in Switch and Overcurrent Device Enclosures . . . . 312.8(A)
Field-Applied Hazard Markings . . . . 110.21(B)
Fire Pumps
    Disconnect . . . . 695.4(B)(3)
    Electric Utility Service Connection. . . . . 695.3(A)(1)
Fuel Cell Systems
    Disconnecting Means 692.17
    Point of Connection . . . . 692.65, 705.12(D)(2)
    Stored Energy 692.56
Generator, Power Inlet Use . . . . 702.7(C)
Hazardous (Classified) Locations
    Class 1 Locations
        Meters, Instruments, and Relays (Cord-and-Plug Connection)
            . . . . 501.105(B)(6)
Induction and Dielectric Heating Equipment 665.23
Interconnected Electric Power Production Sources
    Inverter Output Connection . . . . 705.12(B)
Intrinsically Safe Wiring . . . . 504.80(B)
Knife Switches (Energized in the Open Position) . . 404.6(C) Ex.
Legally Required Standby Systems . . . . 701.7(A)
    Remote Grounding Connection . . . . 701.7(B)
Marinas and Boatyards
    Electric Shock Hazard- Swimming. 555.10
Mobile Home
    Outside Heating/Air-Conditioning Equipment . . 550.20(B)
    Mobile Home Service Equipment (Grounding Electrode) . . .
        . . . . 550.32(B)(7)

Mobile Home Service Equipment (125/250-Volt Receptacle) . . . . 550.32(G)
Motion Picture Projection Rooms [Cellulose Acetate (Safety) Film] . . . . 540.11(B) Ex. No. 1
Motors
   Control Circuits (Disconnection) . . . . 430.75(A) Ex. No. 1
   Controller Disconnecting Means (Over 1000 Volts) 430.102(A) Ex. 1
   Energy from More Than One Source. 430.113
Optional Standby Systems . . . . 702.7(A)
   Remote Grounding Connection . . . . 702.7(B)
Outdoor Electric Deicing and Snow-Melting Equipment 426.13
Outside Heating/Air-Conditioning Equipment. . . . 552.59(B)
   Prewiring for Air-Conditioning . . . . 552.48(P)
Over 1000 Volts, Nominal
   Conductor Access in Conduit and Cable Systems . . . 300.45
   Distribution Cutouts and Fuse Links, Expulsion Type . . . . 490.21(C)(2)
   Fused Interrupter Switches (Backfeed) . . . . . . 490.44(B)
   High-Voltage Fuses (More Than One Source). . . 490.21(B)(7) Ex.
   Load Interrupters (More Than One Switch) . . . . 490.21(E)
   Mobile and Portable Equipment Enclosures 490.53
   Outside Branch Circuit or Feeder Disconnecting
      Means (Fused Cutouts). . . . . 225.52(B) Ex.
   Power and Cable Connections to Mobile Machines . . 490.55
   Pull and Junction Boxes . . . . 314.72(E)
   Rooms and Enclosures . . . . 110.34(C)
   Substations 490.48
   Transformer with Exposed Live Parts . . . . 450.8(D)
Park Trailer
   Electrical Entrance (Correct Ampere Rating) . . . 552.44(D)
   Outdoor Heating and Air Conditioning Outlets . . 552.59(B)
   Prewiring for Air-Conditioning . . . . 552.48(P)
   Prewiring for Other Circuits . . . . 552.48(Q)
Power Inlet, Portable Generator Use . . . . 702.7(C)
Recreational Vehicles
   Electrical Entrance (Correct Ampere Rating) . . . 551.46(D)
   Prewiring for Air-Conditioning . . . . 551.47(Q)
   Prewiring for Branch Circuits . . . . 551.47(S)
   Prewiring for Generators . . . . 551.47(R)
   Site Supply Equipment . . . . 551.77(F)
Signage Requirements . . . . 110.21(B)
Signs
   Disconnect . . . . 600.6(A)(1) Ex. 2
   Retrofit Illumination Systems . . . . 600.4(B)
   Servicing Purposes . . . . 600.6(A)(2)
Spas and Hot Tubs (Emergency Switch) 680.41
Sensitive Electronic Equipment (Technical Power) . 647.7(A)(2)
Solar Photovoltaic Systems
   DC PV Source and Output Circuits Inside a Building
      (Wiring Methods and Enclosures Containing PV Circuits) . . . . 690.31(D)(2)
   Disconnecting Means . . . . 690.13(B)
   Equipment Disconnecting Means . . . . 690.15(C)
   Stand-Alone Systems 690.56
Spray Application, Dipping, and Coating Processes 516.10(A)(8)
   Handle (Grounding of Persons) . . . . 516.10(B)(3)
Stand-Alone Systems- 120-Volt Supply . . . . 710.15(C)
Switches, Position and Connection (Energized) . . . 404.6(C) Ex.
Temporary Wiring, Cable Trays in Exhibition Halls (Cable Trays) . . . . 518.3(B) Ex.
Theaters, Motion Picture and Television Studios
   Single-Pole Separable Connectors 530.22
   Single Primary Stage Switchboard (Dimmer Bank) . . . . 520.27(A)(3)
Transformers
   Guarding of Transformers . . . . 450.8(D)
   Remote Disconnecting Means 450.14
   Voltage Warning (Exposed Live Parts) . . . . 450.8(D)
Welding Cable in Cable Trays. . . . . 630.42(C)
Wind Electric Systems
   Disconnecting Means . . . . 694.22(A)
See ANSI Z535.4-2017 . . . . ANSI Z535.4

Note: It is important to pay careful attention to these references, as they provide a form of visual written instructions that are designed to keep people from getting injured or electrocuted.

See PERMANENT WARNING SIGNS (REQUIRED) . . . . Ferm's Finder

WASHERS, HIGH-PRESSURE SPRAY . . . . . . 422.5(A)(3)

WATER HEATERS, STORAGE OR INSTANTANEOUS-TYPE
Branch-Circuit Rating 125% . . . . 422.10(A)
   Continuous Load 422.13
   Swimming Pool Water Heaters 680.10
Controls 422.47
Disconnecting Means . . . . Art. 422 Part III
   Cord- and Plug-Connected 422.33
   Permanently Connected 422.31
   Unit Switch(es) 422.34

Grounding 250.110
    See GROUNDING, Fixed Equipment. . . . . .Ferm's Finder
Overcurrent Protection 422.11
    Rated More Than 48 Amperes, ASME Rated Vessel 422.11(F)
    Swimming Pool Water Heaters. 680.10

WATER METERS, FILTERING DEVICES, AND SIMILAR EQUIPMENT
Bond around to Maintain Continuity. . . . . . . .250.53(D)(1)

WATER PIPE USED AS GROUNDING ELECTRODE
. . . . .250.52(A)(1)
Interior and Connected Within 5 feet. . . . . 250.68(C)
Must be Supplemented . . . . 250.53(D)(2)
See GROUNDING, Electrodes. . . . . .Ferm's Finder

WATER PUMPS
Motor-Operated . . . . 250.112(L)
Swimming Pools
    Double Insulated. . . . . .680.26(B)(6)
    Receptacle Location . . . . .680.22(A)(2)

WEATHERHEAD, ABOVE POWER COMPANY
Point of Attachment . . . . 230.54(C)

WEATHERPROOF
Definition of . . . . Art. 100 Part I
Agricultural Buildings
    Equipment Enclosures, Boxes, Conduit Bodies, and Fittings
        Installed in Damp or Wet Locations . . . . 547.5(C)(2)
Cabinets, Cutout Boxes, and Meter Socket Enclosures . . 312.2
Carnivals, Circuses, Fairs and Similar Events
    Portable Distribution or Termination Boxes . . . . 525.22(A)
    Services . . . . 525.10(B)
Circuit Breaker (Surface-Mounted). . . . . 404.4(A)
Enclosure Types, Motor Controllers . . . . Table 110.28
Enclosures for Overcurrent Devices 240.32
Enclosures for Switches or Circuit Breakers . . .404.4(A) and (B)
Fixed Electric Heating Equipment for Pipelines and Vessels
    Impedance Heating 427.25
Fixed Outdoor Electric Deicing and Snow-Melting Equipment
    Impedance Heating 426.30
Marinas and Boatyards
    Receptacles, Shore Power. . . . . 555.33(A)
Panelboards 408.37
Park Trailers, Direct Wired . . . . .552.48(O)(2)

Receptacles in Damp or Wet Locations 406.9
Recreational Vehicles, Direct Wired. . . . . 551.47(P)(2)
Roof on Building before Installing the Equipment 110.11
Switch (Surface-Mounted) . . . . 404.4(A)
Switchboards 408.16
Transformers, Dry-Type Installed Outdoors 450.22
Tunnel Installations over 1000 Volts (Enclosures). 110.59

WELDERS . . . . Art. 630
Arc Welders . . . . Art. 630 Part II
    Ampacity of Supply Conductors
        Group of Welders . . . . 630.11(B)
        Individual . . . . 630.11(A)
    Disconnecting Means 630.13
    Duty Cycle . . . . Table 630.11(A)
    Overcurrent Protection
        For Conductors . . . . 630.12(B)
        For Welders . . . . 630.12(A)
    Listing 630.6
Resistance Welders . . . . Art. 630 Part III
    Ampacity of Supply Conductors
        Groups of Welders. . . . . 630.31(B)
        Individual Welders, General . . . . 630.31(A)(1)
        Individual Welder, Specific Operation . . . . 630.31(A)(2)
    Disconnecting Means 630.33
    Duty Cycle. . . . . Table 630.31(A)(2)
    Overcurrent Protection
        For Conductors . . . . 630.32(B)
        For Welders . . . . 630.32(A)
Welding Cable . . . . Art. 630 Part IV
Cable Tray Installation 630.42

WELDING
Of or to Metal Raceways Not Permitted . . . . 300.18(B)
Splices . . . . 110.14(B)

WELDING CABLE
Cable Support . . . . 630.42(A)
Conductors 630.41
Fire Spread and Products of Combustion . . . . 630.42(B)
Installation in Cable Tray 630.42
Signage- Cable Tray. . . . . 630.42(C)
See (ZMAY) . . . . UL Product iQ

WELDING, EXOTHERMIC

Concrete-Encased Electrode . . . . 250.52(3)
Connection of Grounding and Bonding Equipment . . . 250.8
Equipotential Bonding Grid Connection . . . . 680.26(B)
    Swimming Pools . . . . 680.26(B)
    Natural and Artificial Made Bodies of Water . . . . 682.33(C)
Fireproofed Structural Metal . . . . 250.68(A) Ex. 2
Grounding Conductor Connection to Electrodes 250.70
Hold-Down Bolts (to Concrete-Encased Electrode) 250.68(C)(2)
Multiple Separately Derived Systems . . . . . 250.30(A)(6)(c)(3)
Natural and Artificial Made Bodies of Water . . . . . 682.33(C)
Splice in Grounding Electrode Conductor Permitted . . . .
. . . . 250.64(C)(1)
Swimming Pools . . . . 680.26(B)

WELLS
See GROUNDING, Fixed Equipment . . . . Ferm's Finder
See MOTORS . . . . Ferm's Finder
See PUMP HOUSES . . . . Ferm's Finder
    Thermoplastic-Insulated (ZLGR) . . . . *UL Product iQ*
    Thermoset-Insulated (ZKST) . . . . *UL Product iQ*
    Underground Feeder and Branch-Circuit Cable (YDUX) . . .
. . . . *UL Product iQ*
See UNDERGROUND WIRING . . . . Ferm's Finder

WET PROCEDURE LOCATIONS, HEALTH CARE FACILITIES
. . . . **517.20**
Definition 517.2
Isolated Power Systems . . . . 517.20(B)
Receptacles and Fixed Equipment . . . . 517.20(A)

WET LOCATIONS
Audio Signal Processing, Flexible Cords and Cables. . 640.42(E)
Above Grade — Raceways, Interior of
    1000 Volts and Less 300.9
    Over 1000 Volts 300.38
Auxiliary Gutters (Nonmetallic) . . . . 366.6(A)(2)
Boxes in 314.15
Cabinets, Cutout Boxes & Meter Socket Enclosures in . . .312.2
Cables and Conductors in Raceways or Enclosures Underground
. . . . 300.5(B)
Class 1, Class 2, and Class 3 Circuits . . . . 725.3(L)
Class 1, 2, 3 Remote Control . . . . 725.179(E)
Conductors, Types . . . . 310.10(C)
    Applications . . . . Table 310.4(A)
    See Thermoplastic-Insulated (ZLGR) . . . . . . *UL Product iQ*
    See Thermoset-Insulated (ZKST) . . . . *UL Product iQ*
    See Underground Feeder and Branch-Circuit Cable (YDUX)
. . . . *UL Product iQ*
Critical Operations Power Systems, Floodplain Protection . . . .
. . . . .708.10(C)(3)
Definition of . . . . Art. 100 Part I
    For Health Care Facilities, Patient Care Spaces 517.2
    Deteriorating Agents 110.11
    Electric Welders, Cord Assemblies . . . . 626.32(B)
Drainage Openings, Field Installed (Boxes, etc.) 314.15
Electrical Metallic Tubing, in . . . . 358.10(C)
Enameled Equipment. . . . . 300.6(A)(1)
Fire Alarm . . . . .760.3(D)
(Fixtures) Luminaires in . . . . 410.10(A)
Flexible Metal Conduit, in (not permitted) . . . . . . 348.12(1)
Health Care Facilities, Wet Procedure Locations 517.20
Indoors. . . . . .300.6(D)
Intermediate Metal Conduit, in . . . . 342.10(D)
Irrigation Cable . . . . 675.4(A)
Liquidtight Flexible Metal Conduit, in . . . . 350.10(1)
Liquidtight Flexible Nonmetallic Conduit, in. . . 356.10(2), (3)
Metal Clad Cable . . . . 330.10(A)(11)
Metal Wireways, in . . . . 376.10(3)
Mineral-Insulated, Metal-Sheathed Cable. . . . . . . 332.10(3)
Mounting of Equipment in . . . . 300.6(A) through (D)
Natural and Artificial Made Bodies of Water, Wiring Methods
. . . . 682.13
Nonmetallic Wireways . . . . 378.10(3)
NPLFA Cables 760.176
Neon, Electrode Enclosures . . . . 600.42(H)(2)
Neon Secondary Conductors in Raceways . . . . . . 600.42(G)
PLFA Cables 760.179
Other Than Dwelling Units . . . . 210.8(B)(6)
Outdoors, Dwelling Units . . . . 210.8(A)(3)
Outside Branch Circuits and Feeders 225.4
Panelboards, in 408.37
Raceways, Exterior Surfaces 225.22
Raceways, Interior Surfaces
    Abovegrade (1000 Volts or Less) 300.9
    Abovegrade (Over 1000 Volts) 300.38
    Underground (1000 Volts or Less) . . . . 300.5(B)
    Underground (Over 1000 Volts) . . . . 300.50(B)
Receptacles in . . . . 406.9(B)
Rigid Metal Conduit, in . . . . 344.10(D)
Rigid PVC Conduit, in . . . . 352.10(D)

Service Heads . . . . 230.54(A)&(B)
Signs and Outline Lighting . . . . 600.9(D)
Switchboards, in 408.16
Switches, in 404.4
See WEATHERPROOF . . . . Ferm's Finder

Note: (Location, Wet) Installations underground or in concrete slabs in direct contact with the earth shall be considered a "wet location." Art. 100 Part I

Underground - Raceways, Interior of
   1000 Volts and Less . . . . 300.5(B)
   Over 1000 Volts . . . . 300.50(B)

WIND ELECTRIC SYSTEMS . . . . Art. 694
Circuit Requirements. 694.10
Circuit Sizing & Current 694.12
Components
   Interactive . . . . Figure 694.1(a)
   Stand-Alone . . . . .Figure 694.1(b)
Connection to Other Sources . . . . Art. 694 Part VII
Definitions Applicable to System 694.2
Disconnecting Means. . . . . Art. 694 Part III
Diversion Load Controllers. . . . . 694.7(C)
Energy Storage or Backup Power System Requirements. Art. 480
Energy Storage Systems . . . . Art. 706
Flexible Cords and Cables . . . . 694.30(B)
Grounding and Bonding . . . . .Art. 694 Part V
Equipment in General . . . . 694.40(A)
   Tower . . . . 694.40(B)
Guy Wires . . . . .694.40(B)(4)
Identification of Power Sources 694.54
Installation 694.7
Listing and Identification Required — Equipment . . . 694.7(B)
Marking . . . . .694.22(A), 694.22(C)(2)
Maximum Voltage 694.10
Overcurrent Protection 694.15
   Circuits and Equipment 694.12
Raceways, Tower Supports used as . . . . 694.7(F)
Receptacle (for Maintenance) . . . . 694.7(E)
Stand-Alone Systems . . . . 694.12(C)
Storage Batteries . . . . Art. 480
Supply Side of Service Equipment . . . . 230.82(6)
Surge Protective Device . . . . 694.7(D)
Tower Grounding. . . . . 694.40(B)
Turbine Disabling. 694.56
Turbine Shutdown 694.23
Wind Turbine Output Circuits . . . . 694.10(A)
Wiring Methods . . . . Art. 694 Part IV
Working Clearances. . . . . 694.7(G)
   Working Spaces . . . . .Table 694.7
See (ZGEN). . . . . UL Product iQ

WIRE
See CONDUCTORS . . . . Ferm's Finder
See (ZGZX) . . . . UL Product iQ

WIRELESS POWER TRANSFER EQUIPMENT (ELECTRIC VEHICLE POWER TRANSFER SYSTEM)
. . . . Art. 625, Part IV
AC Receptacles within Electric Vehicles. 625.60
Charger Power Converter 625.2
Cords and Cables 625.17
Connection to Premises Wiring System 625.44
Definitions 625.2
Disconnecting Means 625.43
Ground-Fault Circuit-Interrupter Protection for Personnel (connection of vehicle) 625.54
Ground-Fault Circuit-Interrupter Protection for Personnel (within vehicle) . . . . 625.60(D)
Grounding 625.101
Individual Branch Circuit. 625.40
Installation . . . . Art. 625, Part III
Installation- General (Charger Power Converter) . . .625.102(B)
Interactive Systems 625.48
Listed 625.5
Loss of Primary Sourse 625.46
Overcurrent Protection 625.41
Personal Protection System 625.22
Primary Pad. . . . . .625.102(C)
Protection of Pad Output Cable . . . . 625.102(D)
Rated for Load Served 625.42
Ventilation 625.52
Voltages. 625.4

WIRE BENDING SPACE
Auxiliary Gutters . . . . 366.58(A) and (B)
Back Wire-Bending Space . . . . 408.55(C)
Bare Metal Part, Minimum Spacing. 408.56
Bending Radius Over 1000 Volts 300.34
Enclosures for Motor Controllers and Disconnects . . 430.10(B)
Examination of Equipment . . . . 110.3(A)(3)
Manholes 110.74

Metal Wireways	376.23(A)
Motor Control Centers	430.97(C)
Motor Terminal Housings	430.12(B) & (C)
Nonmetallic Wireways	378.23(A) and (B)
Pull and Junction Boxes Not Over 1000 Volts	314.28
Pull and Junction Boxes Over 1000 Volts	314.71
Side Wire-Bending Space	408.55(B)
Switchboards and Panelboards	408.3(G)
Clearance Entering Bus Enclosures	408.5
Provisions for	408.55
Switch or Circuit Breaker Enclosures	404.3(A)
How Measured	404.28
Terminals of Cabinets, Cutout Boxes, Meter Sockets	312.6
Top and Bottom Wire-Bending Space	408.55(A)
Transformers	450.12

WIRE CONNECTORS

Direct Burial Use	110.14(B)
Electrical Connections	110.14
Terminal	110.14(A)
Splices	110.14(B)
Examination of Equipment	110.3
Fire Pump (shall not be permitted)	695.6(D)
See TERMINALS	Ferm's Finder
See (ZMKQ)	UL Product iQ
WIRE, Luminaire (FIXTURE)	Art. 402
See FIXTURE WIRES, Fixture	Ferm's Finder

WIRED LUMINAIRE (FIXTURE) SECTIONS . 410.137(C)

WIREWAYS, METAL . . . . .Art. 376

Conductors, Size	376.21
Number of	376.22
Conductors Connected in Parallel	376.20
Dead Ends	376.58
Definition of	376.2
Deflected Insulated Conductors	376.23
Electrical and Mechanical Continuity	376.100(A)
Extensions from	376.70
Extensions through Walls	376.10(4)
Live Parts Required to be Covered	376.56(B)(4)
Marking	376.120
Power Distribution Blocks	376.56(B)
Listed For Line Side of Service Equipment	376.56(B)(1)
Smooth Rounded Edges	376.100(C)
Splices and Taps	376.56
Substantial Construction	376.100(B)
Supports	
Horizontal	376.30(A)
Vertical	376.30(B)
Use	
Not Permitted	376.12
Permitted	376.10

WIREWAYS, NONMETALLIC . . . . Art. 378

Conductors, Size	378.21
Number of	378.22
Conductors Connected in Parallel	378.20
Dead Ends	378.58
Definition of	378.2
Deflected Insulated Conductors	378.23
Expansion Fittings	378.44
Extensions from	378.70
Extensions through Walls	378.10(4)
Grounding	378.60
Listed Required	378.6
Marking	378.120
Splices and Taps	378.56
Supports	
Horizontal	378.30(A)
Vertical	378.30(B)
Use	
Not Permitted	378.12
Permitted	378.10

WIRING METHODS

Agricultural Buildings	547.5
Audio Systems	Art. 640
Boxes (Required)	300.15
See BOXLESS DEVICES	Ferm's Finder
Cables	
See CABLES	Ferm's Finder
Carnivals, Circuses, Fairs, and Similar Events	Art. 525 Part III
Corrosion Protection	300.6
Cranes and Hoists	610.11
Ducts, Plenums, and Other Air-Handling Spaces	300.22
Electric Vehicle Charging Systems	Art. 625 Part II
Electroplating	669.6
Elevators, Dumbwaiters, etc.	620.21
Emergency Systems	Art. 700 Part II

Exit Enclosures (Stair Towers)................ 300.25
Fire Alarm Systems
    Non-Power Limited Fire Alarm (NPLFA) Circuits .. 760.46
    Power-Limited Fire Alarm (PLFA) Circuits
        Load Side of Power Source................ 760.130
        Supply Side of Power Source............. 760.127
Fire Pumps
    Control Wiring .... 695.14(E)
    Generator Control .... 695.14(F)
    Power Wiring 695.6
Floating Buildings, Feeders 555.51
Floating Buildings, Services.................. 555.50
Fuel Cell Systems..................... Art. 692 Part IV
General.. Art. 300
Hazardous (Classified) Locations
    See .... HAZARDOUS (CLASSIFIED) LOCATIONS for
    Specific Items or Locations............ Ferm's Finder
Health Care Facilities................. Art. 517 Part II
Information Technology Equipment 645.5
Intrinsically Safe Systems 504.20
Limitations 300.2
Manufactured Buildings 545.4
Marinas, Boatyards and Docking Facilities 555.34
Mobile and Manufactured Homes 550.15
Motion Picture and Television Studios .... Art. 530
Over 1000 Volts
    See OVER 1000 VOLTS .... Ferm's Finder
Park Trailers
    Low-Voltage .... 552.10(C)
    120 or 120/240-Volt Systems 552.48
Places of Assembly 518.4
Planning 90.8
Raceways
    See RACEWAYS .... Ferm's Finder
Recognized As Suitable 110.8
Recreational Vehicles
    Low-Voltage 551.40
    120 or 120/240-Volt Systems 551.47
Remote-Control, Signaling, and Power-Limited Circuits
    Class 1 .... 725.46
    Class 2 or 3
        Load Side of Power Source............. 725.130
        Supply Side of Power Source........... 725.127
    Sensitive Electronic Equipment 647.4

Signs and Outline Lighting................ 600.5(C)
Solar Photovoltaic Systems .... Art. 690 Part IV
Stair Towers (Exit Enclosures) 300.25
Swimming Pools, Fountains, and Similar Installations . Art. 680
Temporary 590.2
Theaters, Audience Areas 520.5
Underground Installations .... 300.5(B)
    See UNDERGROUND WIRING .... Ferm's Finder
X-Ray Equipment....................... 660.4

WOODEN PLUGS (Not Permitted) .... 110.13(A)

WOODWORKING PLANTS .... 500.5(D)(1) Info. Note 1
Class III Locations .... Art. 503
See HAZARDOUS (CLASSIFIED) LOCATIONS Ferm's Finder

WORKING SPACE
1000 Volts and Less....................... 110.26
Clear Spaces.......................... 110.26(B)
Depth of Working Space .... 110.26(A)(1)
Entrance to and Egress from Working Space ..... 110.26(C)
Headroom .... 110.26(E)
Height of Working Space .... 110.26(A)(3)
Illumination .... 110.26(D)
Illumination-Outdoor Assembly Occupancies 518.6
Limited Access .... 110.26(A)(4)
Locked Electrical Equipment Rooms or Enclosures....... 110.26(F)
Outdoors.......................... 110.26(E)(2)
Personnel Doors .... 110.26(C)(3)
Separation from High-Voltage Equipment ..... 110.26(A)(5)
    Width of Working Space............. 110.26(A)(2)
Over 1000 Volts .... Art. 110 Part III
    Depth of Working Space .... 110.34(A)
    Entrance to Enclosures and Access to Work Space ... 110.33
    Enclosure for Electrical Installations 110.31
    Personnel Doors................... 110.33(A)(3)
    Work Space about Equipment 110.34
    See Life Safety Code, NFPA 101........... NFPA 101

WORKMANLIKE INSTALLATION
CATV and Radio Distribution Systems........... 800.24
Circuits and Equipment Operating at Less Than 50 Volts 720.11
Class 1, 2, and 3 Remote Control, Signaling Circuits .. 725.24
Communications Circuits 800.24
Fire Alarm Systems................... 760.24

General 110.12
Less Than 50 Volts 720.11
Low-Voltage Suspended Ceiling Power Distribution Systems . . . . 393.14(A)
Network-Powered Broadband Systems 800.24
Optical Fiber Cables and Raceways 770.24
Sign Illumination Systems- Class 2 . . . . 600.33(B)(1)

X

X-RAY EQUIPMENT
  For Health Care Facilities . . . . Art. 517 Part V
    Applicability 517.70
    Connection to Supply Circuit 517.71
    Control Circuit Conductors 517.74
    Disconnecting Means 517.72
    Equipment, Approved Type 517.75
    Guarding & Grounding 517.78
    High-Tension Cables 517.77
    Operating at more than 1000 Volts . . . . 517.71(C)
    Overcurrent Protection 517.73
    Rating of Supply Conductors 517.73
    Transformers and Capacitors 517.76
  For Industrial, Nonmedical and Nondental Use . . . . Art. 660
    Capacitors 660.36
    Connection to Supply Circuit 660.4
    Control . . . . Art. 660 Part II
    Definitions 660.2
    Disconnecting Means 660.5
    Equipment, Approved Type 660.10
    Guarding & Grounding . . . . Art. 660 Part IV
    Hazardous Locations 660.3
    Minimum Size of Conductors 660.9
    Operating at more than 1000 Volts . . . . 660.4(C)
    Overcurrent Protection 660.6
    Rating of Supply Conductors 660.6
    Transformers 660.35
    Wiring Terminals 660.7

Note: Radiation safety and performance requirements of several classes of X-ray equipment are regulated under Public Law 90-602 and are enforced by the Department of Health and Human Services. 660.1 Info. Note 1

Note: information on radiation protection by the National Council on Radiation Protection and Measurements is published as Reports of the National Council on Radiation Protection and Measurement. These reports can be obtained from NCRP Publications, 7910 Woodmont Ave., Suite 1016, Bethesda, MD 20814. 660.1 Info. Note 2

Y

YOKE
  Device or Equipment Fill, Boxes . . . . 314.16(B)(4)
  Grounding by Contact Devices or Yokes . . . . . . 250.146(B)
    Provisions for . . . . 404.9(B) Ex. 1 & Ex. 2
    Surface Mounted Box . . . . 250.146(A)
  Load Calculations for Multiple Receptacles on Same . 220.14(I)
  Multiple Branch Circuits, All Occupancies, Disconnect for
    Devices or Equipment on Same 210.7
  Nonmetallic Type, No Equipment Grounding Conductor Required . . . . 404.9(B) Ex. 2
  Receptacle, Definition of . . . . Art. 100 Part I
  Receptacle Mounting 406.5
  Snap Switches, Mounting of, in Box . . . . 404.10(B)
  Utilization Equipment . . . . 314.27(D) Ex.
  Vertical Surface Outlets . . . . 314.27(A)(1) Ex.

Z

ZONE
  Ammonia, Refrigerant Machinery Rooms
    Zone 0, 1, and 2 Locations . . . . 505.5(A)
  Bathtubs and Showers . . . . 410.10(D)
    Receptacles, Shower Zone, Dwellings . . . . . . 406.9(B)(2)
    Receptacles, Shower Zone, Mobile Homes . . . . 550.13(F)
    Receptacles, Shower Zone, Recreational Vehicles 552.41(F)(1)
  Capacitor Overcurrent Protection . . . . 460.25(D)
    See *NSI/IEEE 18 Shunt Power Capacitors* for Definition of Safe Zone
  Dedicated Equipment Space . . . . 110.26(E)
  Definitions
    Zone 0, 1, and 2 Location Guidance   505.1 and Informational Notes
    Zone 20, 21, and 22 Location Guidance 506.1 and Informational Notes
  Electrolytic Cell Line Working 668.10

Definition of 668.2
  Grounding Not Required in . . . . 668.3(C)(3)
Fire Ladders, for, Outside Branch Circuits and Feeders 225.19(E)
Information Technology Equipment Rooms 645.10
Process, Fixed Electrostatic Equipment, Signs. . 516.10(A)(8)(1)

ZONE 0, CLASS I HAZARDOUS (CLASSIFIED) LOCATIONS
Art. 505

Note for All Class I, Zone 0, 1, and 2 Applications: All persons involved in the design, construction and inspection of installations using the zone classifications are advised to pay careful attention to the information and referenced standards contained in fine print notes throughout Article 505.

Classification of Area . . . . 505.5(B)(1)
Classification of Locations, General. . . . . 505.5(A)
Documentation Required 505.4
Drainage . . . . 505.16(E)
Dual Classification of Areas . . . . 505.7(B)
Equipment, Group Marking . . . . 505.20(D)
Equipment Listed and Marked for Use in . . . . . . 505.20(A)
Grounding and Bonding 505.25
Implementation of Zone Classification System . . . . . 505.7(A)
Listing, Marking, and Docu., Equipment Suitability 505.9
Manufacturer's Instructions. . . . . 505.20(E)
Marking of Equipment for . . . . 505.9(C)
Material Groups (Atmospheres) 505.6
Protection Techniques 505.8
  Designation . . . . Table 505.9(C)(2)(4)
  Note: Intrinsic Safety Only Protection Technique Permitted in Zone 0
Qualified Person Required . . . . 505.7(A)
Reclassification Permitted . . . . 505.7(C)
Sealing . . . . 505.16(A)
Seals, General . . . . 505.16(D)
Spray Application, Dipping, Coating, and Printing Processes Using Flammable
Combustible Materials 516.5
Threading . . . . 505.9(E)
Wiring Methods for. . . . . 505.15(A)

ZONE 1, CLASS I HAZARDOUS (CLASSIFIED) LOCATIONS
Art. 505

Available Short-Circuit Current for Type of Prot. "e" . . 505.7(F)
Classification of Area . . . . 505.5(B)(2)
Classification of Locations, General. . . . . 505.5(A)
Documentation Required 505.4
Drainage . . . . 505.16(E)
Dual Classification of Areas . . . . 505.7(B)
Equipment, Group Marking . . . . 505.20(D)
Equipment Listed for Use in . . . . 505.20(B)
Flexible Cords. 505.17
Grounding and Bonding 505.25
Implementation of Zone Classification System . . . . . 505.7(A)
Increased Safety "e" Motors and Generators. 505.22
  Copper Conductors Required . . . . 505.18(A)
Listing, Marking, and Docu., Equipment Suitability 505.9
Manufacturer's Instructions. . . . . 505.20(E)
Marking of Equipment for . . . . 505.9(C)
Material Groups (Atmospheres) 505.6
Optical Fiber Cable . . . . 505.15(B)(1)(h)
Protection Techniques 505.8
  Designation . . . . Table 505.9(C)(2)(4)
Qualified Person Required . . . . 505.7(A)
Reclassification Permitted . . . . 505.7(C)
Sealing . . . . 505.16(B)
Seals, General . . . . 505.16(D)
Spray Application, Dipping, Coating, and Printing Processes Using Flammable
Combustible Materials . . . . 516.5(A)(2)
Threading. . . . . 505.9(E)
Wiring Methods for. . . . . 505.15(B)

ZONE 2, CLASS I HAZARDOUS (CLASSIFIED) LOCATIONS
. . . . .Art. 505

Classification of Area . . . . 505.5(B)(3)
Classifications of Locations, General . . . . 505.5(A)
Documentation Required. 505.4
Drainage . . . . 505.16(E)
Dual Classification of Areas . . . . 505.7(B)
Equipment, Group Marking . . . . 505.20(D)
Equipment Listed for Use in . . . . 505.20(C)
Flexible Connections . . . . .505.15(C)(2)
Flexible Cords and Connections 505.17
Grounding and Bonding 505.25
Implementation of Zone Classification System . . . . . 505.7(A)
Instrumentation Connections . . . . 505.17(B)
Listing, Marking, and Docu., Equipment Suitability 505.9
Manufacturer's Instructions. . . . . 505.20(E)
Marking of Equipment for . . . . 505.9(C)
Material Groups (Atmospheres). 505.6
Optical Fiber Cable . . . . 505.15(C)(1)(h)
Protection Techniques. 505.8

Designation	Table 505.9 (C)(2)(4)
Qualified Person Required	505.7(A)
Reclassification Permitted	505.7(C)
Sealing	505.16(C)
Seals, General	505.16(D)
Spray Application, Dipping, Coating, and Printing Processes Using Flammable	
Combustible Materials	516.5
Threading	505.9(E)
Wiring Methods for	505.15(C)

ZONE 20, 21, and 22 LOCATIONS FOR COMBUSTIBLE DUSTS, FIBERS, AND FLYINGS . . . . Art. 506

ZONE 20, CLASS I HAZARDOUS (CLASSIFIED) LOCATIONS

Classification of Area	506.5(A)
Classifications of Locations, General	506.5(B)
Specific	506.5(B)(1)
Dual Classification of Areas	506.7(B)
Equipment, Group Marking	506.6
Equipment Listed for Use in	506.20(A)
Equipment Temperature Classification	506.9(D)
Grounding and Bonding	506.25
Implementation of Zone Classification System	506.7(A)
Listing, Marking, and Docu., Equipment Suitability	506.9
Manufacturer's Instructions	506.20(E)
Marking of Equipment for	506.9(C)
Protection Techniques	506.8
Designation	Table 506.9(C)(2)(3)
Qualified Persons	506.7(A)
Reclassification Permitted	506.7(C)
Sealing	506.16
Simultaneous Presence of Flammable Gases and Combustible	
Dusts or Fibers/Flyings	506.7(D)
Spray Application, Dipping, Coating, and Printing Processes Using Flammable	
Combustible Materials	516.5
Threading	506.9(E)
Wiring Methods for	506.15(A)

ZONE 21, CLASS I HAZARDOUS (CLASSIFIED) LOCATIONS

Classification of Area	506.5(A)
Classifications of Locations, General	506.5(B)
Specific	506.5(B)(2)
Dual Classification of Areas	506.7(B)
Equipment, Group Marking	506.6
Equipment Listed for Use in	506.20(B)
Equipment Temperature Classification	506.9(D)
Grounding and Bonding	506.25
Implementation of Zone Classification System	506.7(A)
Listing, Marking, and Docu., Equipment Suitability	506.9
Manufacturer's Instructions	506.20(E)
Marking of Equipment for	506.9(C)
Protection Techniques	506.8
Designation	Table 506.9(C)(2)(3)
Qualified Persons	506.7(A)
Reclassification Permitted	506.7(C)
Sealing	506.16
Simultaneous Presence of Flammable Gases and Combustible	
Dusts or Fibers/Flyings	506.7(D)
Spray Application, Dipping, Coating, and Printing Processes Using Flammable	
Combustible Materials	516.5
Threading	506.9(E)
Wiring Methods for	506.15(B)

ZONE 22, CLASS I HAZARDOUS (CLASSIFIED) LOCATIONS

Classification of Area	506.5(A)
Classifications of Locations, General	506.5(B)
Specific	506.5(B)(3)
Dual Classification of Areas	506.7(B)
Equipment, Group Marking	506.6
Equipment Listed for Use in	506.20(C)
Equipment Temperature Classification	506.9(D)
Grounding and Bonding	506.25
Implementation of Zone Classification System	506.7(A)
Listing, Marking, and Docu., Equipment Suitability	506.9
Manufacturer's Instructions	506.20(E)
Marking of Equipment for	506.9(C)
Protection Techniques	506.8
Designation	Table 506.9(C)(2)(3)
Qualified Persons	506.7(A)
Reclassification Permitted	506.7(C)
Sealing	506.16
Simultaneous Presence of Flammable Gases and Combustible	
Dusts or Fibers/Flyings	506.7(D)
Spray Application, Dipping, Coating, and Printing Processes Using Flammable	
Combustible Materials	516.5
Threading	506.9(E)
Wiring Methods for	506.15(C)

Energize your CEU success

Enroll in

IAEI
2021
DIGITAL EDUCATION CLASSES

- Comprehensive Training
- Expert Instructors
- TDLR & ICC Approved

Register Today at...

iaeicourses.org

Ferm's Fast Finder Index

Fifteenth Edition

President/CEO: Rudy Garza

Director of Technical Services: L. Keith Lofland

Director of Digital Education: Joseph Wages, Jr.

Cover Design: Fred Nash

Project Manager: Laura L. Hildreth

Technical Edit and Review:

Laura L. Hildreth

L. Keith Lofland

Joseph Wages, Jr.